Reinforced
Concrete
Design

Part-I

Reinforced Concrete Design

Part-I

S. K. Solomon

B.E. (Civil), M.E. (Structures), Ph.D., MICE
Associate Professor in Structural Engineering
Faculty of Engineering
University of Jodhpur
Jodhpur

CBS

CBS Publishers & Distributors Pvt. Ltd.

New Delhi • Bengaluru • Chennai • Kochi • Kolkata • Mumbai
Hyderabad • Uttarakhand • Nagpur • Patna • Pune • Jharkhand

ISBN: 978-81-239-1596-8

First Edition: 1996
Reprint: 2008, 2009, 2010, 2012, 2014, 2019

Published by **Satish Kumar Jain** and produced by **Varun Jain** for
CBS Publishers & Distributors Pvt. Ltd.,
4819/XI Prahlad Street, 24 Ansari Road, Daryaganj, New Delhi - 110002
delhi@cbspd.com, cbspubs@airtelmail.in • www.cbspd.com
Ph.: 23289259, 23266861, 23266867 • Fax: 011-23243014

Corporate Office: 204 FIE, Industrial Area, Patparganj, Delhi - 110 092
Ph: 49344934 • Fax: 011-49344935
E-mail: publishing@cbspd.com • publicity@cbspd.com

Branches:
• *Bengaluru:* 2975, 17th Cross, K.R. Road, Bansankari 2nd Stage,
  Bengaluru - 70 • Ph: +91-80-26771678/79 • Fax: +91-80-26771680
  E-mail: cbsbng@gmail.com, bangalore@cbspd.com
• *Chennai:* No. 7, Subbaraya Street, Shenoy Nagar, Chennai - 600030
  Ph: +91-44-26681266, 26680620 • Fax: +91-44-42032115
  E-mail: chennai@cbspd.com
• *Kochi:* Ashana House, 39/1904, A.M. Thomas Road, Valanjambalam,
  Ernakulum, Kochi • Ph: +91-484-4059061-65
  Fax: +91-484-4059065 • E-mail: cochin@cbspd.com
• *Kolkata:* 6-B, Ground Floor, Rameshwar Shaw Road, Kolkata - 700014
  Ph: +91-33-22891126/7/8 • E-mail: kolkata@cbspd.com
• *Mumbai:* 83-C, Dr. E. Moses Road, Worli, Mumbai - 400018
  Ph: +91-9833017933, 022-24902340/41 • E-mail: mumbai@cbspd.com

Representatives:

• Hyderabad: 0-9885175004	• Nagpur: 0-9021734563
• Patna: 0-9334159340	• Pune: 0-9623451994
• Jharkhand: 0-9811541605	• Uttarakhand: 0-9716462459

Printed at:
J.S. Offset Printers, Delhi (India)

To
My Wife Mukul
and
Children Benison, Pratibha and Sameer

*"Little things make perfection
and perfection is no trifle"*

Michelangelo

PREFACE

Advances have taken place in the field of Structural Engineering Design like in any other field of scientific study. New concepts and techniques of analysis as well as understanding of design for strength and serviceabil:ty are reflected in the recent IS Codes. This book presents in a simple and lucid manner the basic principles and the theories involved in the design of Reinforced Concrete Structures as well as an understanding and application of the current IS Code to the same. It is intended to serve as the main text for the University and College courses in the Design of Reinforced Concrete Structures. Practicing Structural Engineers will also find this volume as a useful reference for the design problems.

This part of the book is based on the Working Stress Design method. The first two chapters deal with the strength and deformation parameters of concrete and steel, and the types and properties of cements and the design of concrete mix. Subsequent chapters deal with the design principles for structural components like beams, columns and slabs. Chapters on design of staircases, footings and retaining walls follow. In the later chapters the analysis and design of beams curved in plan, spherical and conical domes are presented. A large number of examples have been included to illustrate the analytical principles and the design procedures.

This book has come out as a result of long experience of university teaching gained by the author in this subject. It is an effort towards sharing knowledge and experience with pupils and professionals alike in this field of study. It is hoped that the book will be of service to the profession and practice of Structural Engineering Design.

Jodhpur   **S.K. SOLOMON**
January 1991

ACKNOWLEDGEMENTS

Clauses, tables and figures on pages 1.5, 1.8, 1.9, 1.11, 1.12 - 1.22, 2.5 - 2.24, 2.28 - 2.44, 3.2, 3.3, 3.11 - 3.15, 3.21 - 3.24, 3.33 - 3.39, 3.45 - 3.51, 4.1, 5.2, 5.3, 6.1, 6.7, 7.8, 7.9, 8.2 - 8.14, 8.22, 9.2, 9.4, 10.2 - 10.9, 11.2 - 11.15, 12.3 - 12.8, 13.5 - 13.7, 14.15 - 14.8 of this publication have been reproduced with the permission of Bureau of Indian Standards from Indian Standards 456, 875, 269, 8041, 1489, 455, 8112, 8043, 4031, 383, 2386, Part I, II and III, and 516 to which references are invited for further details. These standards are available for sale from Bureau of Indian Standards, New Delhi and its Regional Branches and Inspection Offices at Ahmedabad, Bangalore, Bhopal, Bhubaneshwar, Bombay, Calcutta, Chandigarh, Hyderabad, Jaipur, Kanpur, Madras, Patna, Pune and Trivandrum.

S.K. SOLOMON

SYMBOLS

Meaning of the letter symbols generally used in the text are explained below. Other letter symbols used in restricted sense are explained at the appropriate place in the text.

A_g = gross area of section

A_c = net sectional area of concrete in a section

A_{st} = sectional area of tensile reinforcement

A_{sv} = Sectional area of shear reinforcement

a = arm of the internal couple, lever arm

B = width in plan

b = width of a member

b_f = width of flange of a section

b_w = width of web of a section

C = compressive force in concrete in a section

c = coefficient of variation

D = overall depth of beam or slab

D_c = diameter of the core of a column

D_o = outer diameter of a column

D_f = thickness of flange

DL = dead load

d = effective depth i.e. the distance from the highly compressed edge of concrete to the centre of tension reinforcement

E_c = elastic modulus of concrete (short term)

E_s = elastic modulus of steel

EL = earthquake load

e = eccentricity

F = a force

f_c = extreme fibre compressive stress in concrete

f_{cc} = direct compressive stress in concrete

f_{cbc} = bending compressive stress in concrete

f_{ck} = characteristic compressive strength of concrete

f_- = flexural tensile strength of concrete (modulus of rupture)

f_{ct} = direct tensile strength of concrete

f_y = characteristic strength of reinforcing steel; yield point stress or 0.2% proof stress as the case may be

f_{st} = tensile stress in steel reinforcement

I = second moment of area of a section

I_e = second moment of area of effective section

I_g = second moment of area of gross section

I_{cr} = second moment of area of cracked section

jd = lever arm for a singly reinforced section

j = lever arm factor

k = stiffness of a member $(= \dfrac{I}{L})$, a constant

k = coefficient or factor, a number

L = length of a member

LL = live load

l = effective span of beam or slab, unsupported length of column

l_c = clear span of beam or slab

l_d = development length of reinforement bar

l_o = distance between points of zero moments in a beam

l_x = shorter effective span of slab

l_y = longer effective span of slab

l_{ex} = effective length of column considering buckling about X-axis

l_{ey} = effective length of column considering buckling about Y-axis

M = bending moment at a section

M_r = moment of resistance of a section

M_x = bending moment in a strip of unit width, along span lx

M_y = bending moment in a strip of unit width, along span ly

m = modular ratio

N = number of samples, any number

n = depth of neutral axis factor, natural frequency of vibrtion

P = axial force

s = standard deviation, spacing c/c

s_v = spacing of stirrups in a beam

T = tensile force in reinforcement

V = shear fore

V_c = strength of concrete in shear

V_s = strength of shear reinforcement

W = concentrated load

WL = wind load

w = intensity of distributed load, width of support

w_d = distributed dead load intensity

w_l = distributed live load intensity

X = depth of neutral axis (=nd)

x = an estimate of true mean value

μ = true mean value

ρ = unit weight

γ_f = partial safety factor for load

γ_m = partial safety factor for material strength

ε_c = extreme fibres compressive strain in concrete

ε_{st} = strain in tension reinforcement

σ_{cbc} = permissible stress in concrete in bending compression

σ_{cc} = permissible stress in concrete in direct compression

σ_{sc} = permissible compressive stress in reinforcement

σ_{st} = permissible tensile stress in reinforcement

σ_{sv} = permissible tensile stress in shear reinforcement

σ_s = design axial stress in a reinforcement bar

τ_{bd} = permissible bond stress in concrete without shear reinforcement

τ_c = permissible shear stress in concrete without shear reinforcement

τ_{cmax} = maximum permitted shear stress in concrete with shear reinforcement

τ_v = nominal shear stress in concrete

τ_{vc} = nominal equivalent shear stress in concrete

ϕ = nominal diameter of reinforcement bar

ERRATA

ADDENDUM

$$\beta_1 = \frac{\Sigma\, k_c}{\Sigma\, k_c + \Sigma\, k_b}$$

for the top joint

$$\beta_2 = \frac{\Sigma\, k_c}{\Sigma\, k_c + \Sigma\, k_b}$$

for the bottom joint

k_c = flexural stiffness for column

k_b = flexural stiffness for beam

CONTENTS

Index

1

Concrete and Steel Reinforcement

1.1 INTRODUCTION

Concrete has a high compressive strength. It can provide adequate sectional dimensions not only to bear compressive stresses safely, but also to ensure the lateral stability of a member. However, concrete has a very low tensile strength. It cracks under the tensile stresses generally encountered in practice. On the other hand steel has a high tensile strength and bars of small sectional dimensions can take up tensile forces of high magnitude. However, under compressive forces the bars tend to buckle off at low stresses on account of their slender proportions.

In reinforced concrete construction the concrete is used to bear compressive stresses while the steel is used to take up the tensile stresses, thus, providing an ideal combination of the two materials.

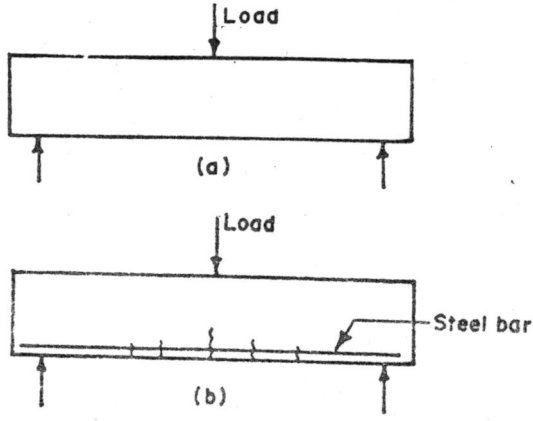

Fig. 1.1

Consider a simple beam of rectangular section, made of plain concrete, loaded as shown in Fig. 1.1 (a). The extreme fibres of concrete in the beam section at midspan will have the maximum bending stresses; compressive at top and tensile at bottom. As the tensile strength of concrete is very low compared with the compressive strength, failure would occur as soon as the flexural stress in the lower fibres equals the tensile strength of concrete, f_{cr}. Thus, the tensile strength of concrete is the critical factor. Bending strength of the section, assuming concrete to be homogeneous and elastic, is given by

$$M_r = f_{cr}z \tag{1.1}$$

Bending strength as given above works out to be very low on account of small magnitude of f_{cr}. When the beam fails, the flexural compressive stress in the upper fibres of concrete remains very much below the flexural compressive strength of concrete. Thus, the full strength of concrete in flexural compression cannot be utilized. However, if steel bars are introduced in the lower region of the concrete as shown in Fig 1.1 (b), they will bear the tension in that region even after the concrete is cracked and thus, inhibit the failure of the beam on account of flexural tension in concrete. The steel so provided is said to have reinforced the concrete in tension. On account of the reinforcement of concrete in tension in the lower region of the beam, it is now possible to attain high compressive stresses in the concrete in the upper region, and utilize the full compressive strength of concrete, which was not otherwise possible. Hence, a much higher bending strength can be obtained from a reinforced concrete beam than from a plain concrete beam.

The materials required and the skill needed for the reinforced concrete construction are readily available. As a material of construction, reinforced concrete is strong, durable and economical. It has the adaptability to take up almost any architectural shape. It can be designed to take up safely both static and dynamic loading and to withstand the various environmental conditions without damage.

1.2 CONCRETE

1.2.1 General. The main ingredients required for making concrete are coarse aggregate, fine aggregate, cement and water.

Coarse aggregate provide the main bulk while fine aggregate fill the voids in them, and the hydrated cement acts as the binder.

Generally, crushed rock or the material derived from the natural sources, such as gravel and sand, provide the aggregates for concrete. Fine aggregate is the aggregate most of which passes the 4.75 mm I.S. sieve. Coarse aggregate is the aggregate most of which is retained on 4.75 mm I.S. sieve. For reinforced concrete work the coarre aggregate of nominal maximum size of 20 mm is generally considered to be satisfactory. Ordinary Portland Cement is the most commonly used cement for reinforced concrete construction, though, special situations call for the use of other types namely, Rapid Hardening, Blast Furnace, Low Heat, Sulphate Resisting and White Portland cements. There are also cements which have considerably different composition than that of the Portland cement. Thus, High Alumina cement has a dark colour and it is characterised by high early strength, high heat of hydration and resistance to chemical attack. Supersulphate cement has a remarkably low heat of hydration and a high resistance to chemical attack.

By suitabty adopting the aggragate/cement ratio and the water/cement ratio in a mix design, concretes of variable strengths and qualities can be produced. Additives and admixtures are sometimes added in order to impart certain desirable properties to the freshly mixed as well as to the hardened concrete. An admixure may produce pure physical effect or may have a chemical reaction with the cement which may produce the desired effect. These desired effects may be intended to accelerate or retard the setting time of cement, improve workability and reduce bleeding of concrete, reduce the rate of heat evolution, enhance impermeability, reduce shrinkage and improve the durability of concrete.

The strength of hardened concrete increases as the water/cement ratio decreases, with increasing fineness of cement, curing time and the age of concrete. Concrete mix design seeks to suitably proportion the aggregates, cement and water in order to attain the required degree of workability, strength, durability and economy. For the detailed study of the properties of cements, aggregates, additives, freshly mixed and the hardened concretes and for the methods of concrete mix design, a text book specialised in Concrete Technology should be consulted.

1.2.2   Compressive strength.   The compressive strength of hardened concrete is found by testing to failure 150 mm cubical specimens after 28 days of standard curing.   At least three specimens are made for testing at a specified age.   The average of three test results may be taken as the compressive strength provided an individual result does not differ from the average value by more than ± 15%.

The compressive strength of concrete can also be found by testing to failure cylindrical specimens of diameter = 150 mm and height = 300 mm.   This strength is denoted by the symbol $f_c{}'$.   The ratio of the cylinder strength to the cube strength depends upon the concrete strength level.   An empirical formula for the ratio is

$$0.76 + 0.2 \ \log_{10} \left(\frac{f_{ck}}{19.6} \right) \tag{1.2}$$

An average value of the ratio may be taken to be 0.8.

1.2.3   Characteristic strength of concrete.   Laboratory tests on similar specimens show variation in the observed compressive strength.   This may be due to the variation in the basic factors during the manufacture of concrete and in the quality of the ingredients.   This variation can be allowed for by fixing up a Characteristic Strength which is calculated by using statistical principle.

The Gaussian or Normal distribution curve showing the plot of number of specimens (i.e. the frequency) against the observed strength is shown in Fig. 1.2.

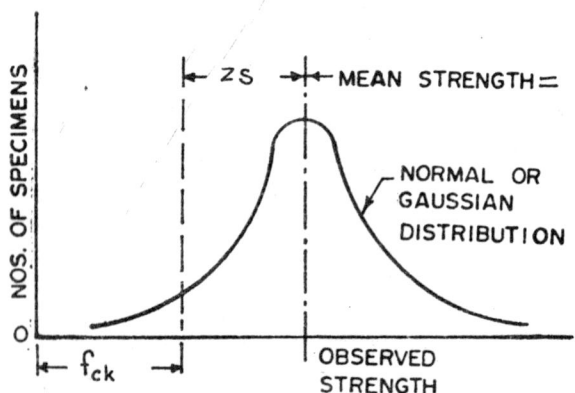

Fig. 1.2 Plot of numbers of specimen against observed strength for normal probability distribution

Ideally, the individual values are equally distributed about the mean and the extreme values are situated at $\pm \infty$. The normal distribution is believed to be sufficiently close to the actual frequency distribution curve for the strength of concrete. The dispersion of strength about the mean is a function of the standard deviation 's' which is given by

$$s = \sqrt{\frac{\Sigma (x - \mu)^2}{N}} \qquad (1.3)$$

where N = total number of test specimens
x = the observed strength of a specimen
μ = true mea . of the observed strength
$= \dfrac{\Sigma x}{N}$

In a laboratory investigation the number of specimens is limited and as such, their mean $\bar{x}$ is only an estimate of the true mean μ. If the variation is now counted from $\bar{x}$ instead of μ, a correction is applied and the standard deviation is written as

$$s = \sqrt{\frac{\Sigma (x - \bar{x})^2}{N - 1}} \qquad (1.4)$$

The coefficient of variation is the ratio expressed as

$$c = \frac{s}{\bar{x}} \times 100\%$$

The equation for the Normal curve is given by

$$y = \frac{1}{s\sqrt{2\pi}} \cdot e^{-(x - \mu)/s^2} \qquad (1.5)$$

The Characteristic Strength takes into account the statistical variation of the strength of concrete. It is defined by the equation,

$$f_{ck} = \mu - zs \qquad (1.6)$$

where f_{ck} = the Charactreistic Strength
z = a coefficient which depends upon the acceptable probability of test results falling below the value f_{ck}.

For a value of $z = 1.65$, the probability of the test results falling below the characteristic value is 5%. Hence, the Characteristic Strength of concrete is defined to be that value below which not more than 5% of the test results are expected to fall.

Example 1.1

The compressive strengths observed from 30 cubical test specimens made from a batch of concrete are as follows:

Specimen No.	Observed strength, N/mm²	Specimen No.	Observed strength, N/mm²
1.	29.1	16.	34.4
2.	35.3	17.	27.1
3.	27.4	18.	27.3
4.	32.9	19.	25.3
5.	30.4	20.	33.0
6.	30.8	21.	33.9
7.	27.4	22.	27.6
8.	31.6	23.	37.0
9.	26.3	24.	27.6
10.	31.1	25.	29.0
11.	32.9	26.	27.3
12.	30.2	27.	31.3
13.	29.7	28.	33.2
14.	28.7	29.	25.4
15.	28.0	30.	34.8

Calculate mean strength, standard deviation and the characteristic strength.

Solution

The sum of all the values is

$$\Sigma x = 906$$

Number of samples is

$$N = 30$$

Hence, the mean value is

$$x = \frac{906}{30}$$

$$= 30.2 \text{ N/mm}^2$$

Calculated deviations and squares of deviations are as follows:

Sample No.	$(x-\bar{x})$	$(x-\bar{x})^2$	Sample No.	$(x-\bar{x})$	$(x-\bar{x})^2$
1.	-1.1	1.21	16.	4.2	17.64
2.	5.1	26.01	17.	-3.1	9.61
3.	-2.8	7.84	18.	-2.9	8.41
4.	2.7	7.29	19.	-4.9	24.01
5.	0.2	0.04	20.	2.8	7.84
6.	0.6	0.36	21.	3.7	13.69
7.	-2.8	7.84	22.	-2.6	6.76
8.	1.4	1.96	23.	6.8	46.24
9.	-3.9	15.21	24.	-2.6	6.76
10.	0.9	0.81	25.	-1.2	1.44
11.	2.7	7.29	26.	-2.9	8.41
12.	0	0	27.	1.1	1.21
13.	-0.5	0.25	28.	3.0	9.00
14.	-1.5	2.25	29.	-4.8	23.04
15.	-2.2	4.84	30.	4.6	21.16

Now $\sum (x - \bar{x})^2 = 288.42$

$\therefore$ Standard deviation $S = \sqrt{\dfrac{288.42}{30 - 1}}$

$$= 3.15 \text{ N/mm}^2$$

∴ Characteristic strength

$$f_{ck} = 30.2 - 1.65 \times 31.5$$
$$= 25 \ N/mm^2$$

1.2.4 Grades of concrete. Grade designation of concrete is according to its characteristic strength. The letter **M** refers to the mix and the succeeding number to the specified characteristic compressive strength of 150 mm cube specimens at 28 days, expressed in N/mm^2. See Table 1.1.

Grades of concrete lower than M15 should not be used in reinforced concrete. Concretes of grades M5 and M7.5 may be used for plain concrete foundations and bases only.

Table 1.1

GRADES OF CONCRETE

Grade designation	Characteristic compressive strength at 28 days, N/mm^2
M 10	10
M 15	15
M 20	20
M 25	25
M 30	30
M 35	35
M 40	40

Note: The above Table is based on Table 2, IS Code 456—1978.

1.2.5 Increase in strength with age. The age of concrete is counted from the date of its casting. It has been observed that the strength of concrete increases with its age. Hence, in order to estimate the increased strength at the time when full design loading is expected on the concrete, its characteristic strength should be multiplied by an age factor. Table 1.2 below gives the values of the age factors as recommended by the IS Code 456

The permissible stress and the design strength for concrete should be based on the increased compressive strength so obtained.

Table 1.2
AGE FACTORS FOR CONCRETE

Age when full design load is expected, (months)	Age factor
1	1.0
3	1.1
6	1.15
12	1.2

1.2.6 Stress-strain curve for concrete. For the experimental determination of the Stress-Strain curve for concrete, cylindrical specimens of diameter = 150 mm and height = 300 mm are used. However, cylindrical specimens of other dimensions as well as rectangular prisms may also be used provided the height to width ratio is $\geqslant 2.0$. The specimens are normally tested in compression after 28 days of curing. The stress-strain curve for concrete as obtained from the short term tests on cylindrical specimens is non-linear. See Fig. 1.3. However, the lower portion of the curve is relatively straight for a stress up to 40% of the ultimate stress.

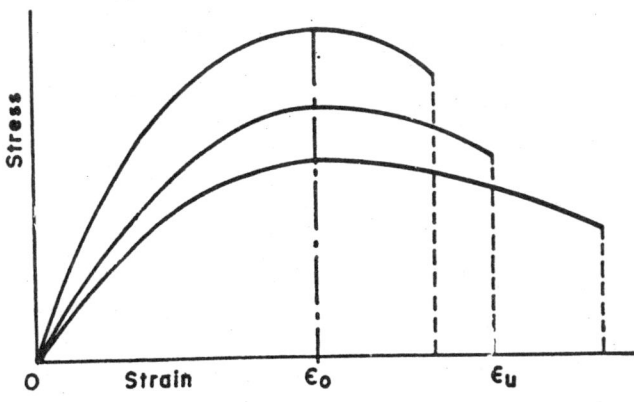

Fig. 1.3 Stress-strain curve for concrete

The curve has an ascending part and a descending part separated by a peak which represents the ultimate strength of concrete in compression. For higher strength concretes the stress-strain curve has a sharp peak whereas for lower strength concretes the curve is comparatively flat. The maximum stress in different strength concretes has been found to occur at nearly the same strain, ϵ_0. It has been observed experimentally that $\epsilon_0 \approx 0.002$. The descending part of the curve terminates at a point which corresponds to the crushing of concrete at an ultimate strain $= \epsilon_u$. It has been observed that concretes of lower strength fracture at a higher ultimate strain than concretes of higher strength.

Attempts have been made in past to standardise the stress-strain curve for concrete in compression. However, of more interest has been the case for concrete in flexural compression than in direct compression. One well known standardization for stress-strain curve for concrete in flexural compression, as given by Hognestad in 1951, is shown in Fig. 1.4.

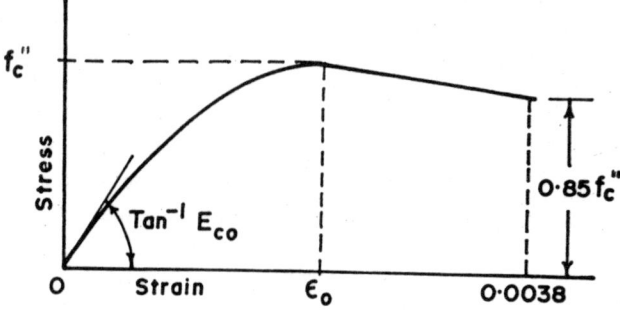

Fig. 1.4 Standard stress-strain curve for concrete in flexural compression. by Hognestad, E., University of Illinois, U.S.A., 1951.

The salient values in the stress-strain curve are:

$$f_c'' = 0.85 f_c'$$

= ultimate compressive stress in concrete in flexural compression.

f_c' = cylinder compressive strength of concrete

E_{co} = slope of the tangent drawn to the curve at the origin.

ϵ_0 = strain corresponding to maximum stress

$$= \frac{2f_c''}{E_{co}}$$

ϵ_u = 0.0038

The ascending part of the curve is taken to be a parabola defined by the equation:

$$f_c = f_0'' \left[\frac{2\epsilon}{\epsilon_0} - \left(\frac{\epsilon}{\epsilon_0} \right)^2 \right] \tag{1.7}$$

The descending part is taken to be a straight line having negative slope. The coefficient 0.85 accounts for the difference in the strength of concrete in an actual structure and that found from a cylindrical specimen in a well controlled laboratory test, the effect of size and shape and that of repetitive loading on the strength of concrete. This curve has been adopted by the ACI code with a modification that $\epsilon_u = 0.003$.

The international recommendations of the FIP Sixth Congress, June, 1970 adopted a design stress-strain curve for concrete in the form of a second degree parabola extended by a straight horizontal line. The factor γ_m represents the safety factor for material strength i.e. for concrete. This curve is shown in Fig. 1.5.

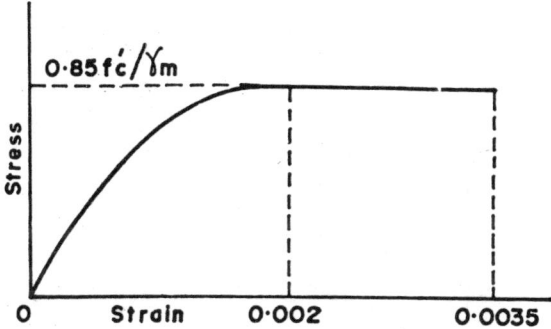

Fig. 1.5 Design curve for concrete recommended by FIP Sixth Congress, Prague, June, 1970

If the relationship between the cylinder compressive strength and the cube compressive strength for normal concrete is taken to be,

$$f_c' = 0.79 \, f_{ck}$$
then
$$0.85 \, f_c' = 0.85 \times 0.79 \, f_{ck}$$
$$= 0.67 \, f_{ck}$$

The design curve for concrete as adopted by the IS Code: 456—1978 is the same as that recommended by the FIP Sixth congress. However, the peak stress is written as $0.67 \, f_{ck}/\gamma_m$ instead of $0.85 \, f_c'/\gamma_m$. The design curve is shown in Fig. 1.6.

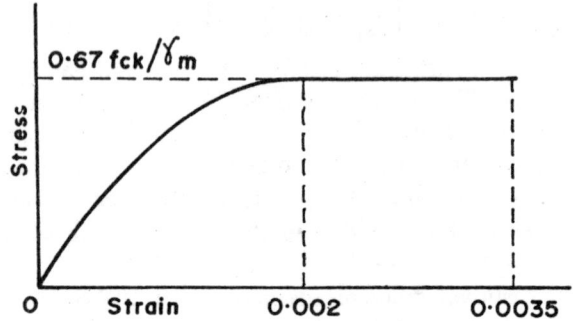

Fig. 1.6 Design curve for concrete adopted by the IS Code: 456—1978

1.2.7 Elastic modulus of concrete. The short term stress-strain curve for concrete as obtained from a compression test on cylindrical specimens is non-linear. See Fig. 1.7. The slope of the tangent drawn to the stress-strain curve is maximum at the origin and reduces to zero at the peak. To adopt this slope as the elastic modulus of concrete would be both erroneous and inconvenient. However, the slope

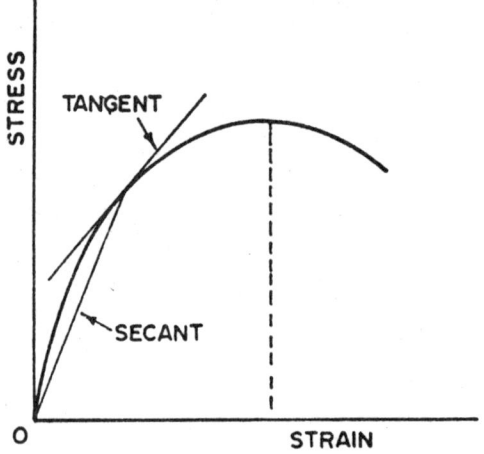

Fig. 1.7

of the secant which is the line joining a point on the curve to the origin, does not vary too widely between the origin and the peak. As the lower portion of the stress-strain curve is relatively straight, an elastic modulus may be conveniently defined in that region.

Thus, the elastic modulus of concrete is taken to be the slope of the secant drawn to the stress-strain curve at a point corresponding to 40% of the maximum stress. It is approximated by the following empirical formula:

$$E_c = 5700 \sqrt{f_{ok}} \quad N/mm^2 \tag{1.8}$$

The elastic modulus of concrete varies with the strength of concrete, its age, properties of aggregates, type of specimen and the speed of loading. Hence, it is almost impossible to predict the value of the elastic modulus of concrete. In the absence of more accurate test data, the short term elastic modulus may be estimated by the above mentioned formula,

1.2.8 Dynamic elastic modulus of concrete. The dynamic modulus for concrete is determined in laboratory by determining the lowest natural frequency of longitudinal vibration of prismatical specimens. The dynamic modulus E is calculated from

$$E = K\eta^2 L^2 \rho \tag{1.9}$$

where K = a constant

η = the lowest natural frequency, cycles/sec

L = length of specimen

ρ = unit weight of concrete

1.2.9 Tensile strength of concrete. A knowledge of the tensile strength of concrete is important in order to estimate the forces which may cause cracking, Direct tension tests on concrete specimens can not be reliably conducted as any accidental eccentricity of loading may result in lower apparent strength. Hence, indirect tension tests are applied in practice. Flexural test on plain concrete beam specimens for determining the tensile strength of concrete is the most popularly used laboratory method. The specimens are 700 mm in length and 150 × 150 mm in cross-section. These are tested in flexure over a simple span of 600 mm by applying two point loads as shown in Fig. 1.8.

However, if the maximum size of the aggregate in concrete does not exceed 20 mm, specimens 500 mm in length and 100 mm × 100 mm in cross-section may be tested in flexure over a span of 400 mm in a similar manner. If the specimen fractures in the portion between the two point loads, the flexural tensile stress at failure f_{or}, assum-

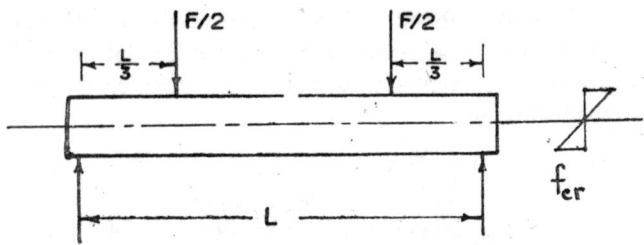

Fig. 1.8 Flexural tension test

ing concrete to be homogeneous and elastic, is given by

$$f_{or} = \frac{M}{Z}$$

where M = bending moment at failure

$$= \frac{F}{2} \times \frac{L}{3}$$

$$= \frac{FL}{6}$$

Z = section modulus

$$= \frac{1}{6} bD^2$$

Hence, $f_{or} = \frac{FL}{bD^2}$

The flexural tensile stress in concrete at failure is termed as modulus of rupture of concrete. An estimate of the same can be made from the known compressive strength of concrete by means of the empirical relationship,

$$f_{or} = 0.7\sqrt{f_{ck}} \quad N/mm^2 \tag{1.10}$$

IS Code: 456-1978 requires that the modulus of rupture of concrete at 7 days should not be less than the values given in Table 1.3 below :

Table 1.3

MODULUS OF RUPTURE REQUIRED AT 7 DAYS

(BASED ON TABLE 5 IS CODE 456—1978)

Concrete grade	M10	M15	M20	M25	M30	M35	M40
N/mm^2		1.7	2.1	2.4	2.7	3.0	3.2	3.4

Another indirect method of determining the tensile strength of concrete is by loading cylindrical specimens of concrete along the diameter. See Fig. 1.9 Length of specimen = 300 mm and diameter = 150 mm. The specimen is placed with its axis horizontal, between the platens of a testing machine. Under increasing load the cylinder fails by slitting along its vertical diameter.

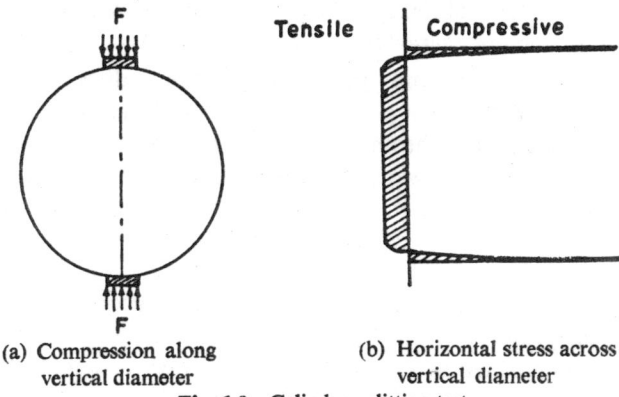

(a) Compression along   (b) Horizontal stress across
vertical diameter   vertical diameter

Fig. 1.9   Cylinder splitting test

In small vicinity of the applied load   high compressive stress acts in horizontal direction. However, failure in compression does not take place. For the major portion of the vertical diameter the horizontal stress is tensile and of uniform magnitude given by

$$f_{ot} = \frac{2F}{\pi LD} \tag{1.11}$$

where   F = compressive force
L = length of cylinder
D = diameter of cylinder

The cylinder splits under the tensile stress. The tensile strength of concrete thus determined from the splitting test is believed to be closer to the true tensile strength of concrete, than the modulus of rupture.

1.2.10 Shrinkage of concrete.   Withdrawal of water by evaporation from concrete, effected by dry atmospheric conditions, causes shrinkage in its volume. A considerable part of shrinkage can be recovered by subjecting concrete again to wet condition.

During drying of concrete the free water which is lost from the capillaries does not cause shrinkage but the adsorbed water which is removed from the surface of particles does cause it. In general, the water content of a mix would indicate the order of shrinkage to be expected. A wet cement paste undergoes volumetric contraction which may be nearly 1% of the volume of dry cement. This is known as plastic shrinkage. Evaporation of water from the surface of concrete aggravates the plastic shrinkage.

By itself the size and grading of concrete does not affect shrinkage. However, well graded aggregates of larger size permit a leaner mix and hence, the amount of shrinkage produced is reduced. Shrinkage is reduced if water loss from the member is restricted. As such, the size and shape of a member affect shrinkage. The influence may be represented by effective thickness which is the ratio of volume to surface area of concrete member. Apart from shrinkage on account of drying, concrete shrinks due to carbonation. Atmospheric CO_2 in the presence of moisture carbonates $Ca(OH)_2$ to $CaCO_3$. This is accompanied by shrinkage. In short, shrinkage is directly influenced by the amount of water present during mixing, cement content in the mix, effective thickness of member and the humidity and temperature of the atmosphere. It is indirectly affected by the size and gradation of the aggregates.

As moisture evaporation takes place from the surface of concrete, a moisture gradient is established within the concrete member. Hence, tensile stresses are created on the surface and balancing compressive stresses are created within the core. This results in surface cracking of concrete. A curve showing shrinkage strain against time after commencement of drying is shown in Fig. 1.10.

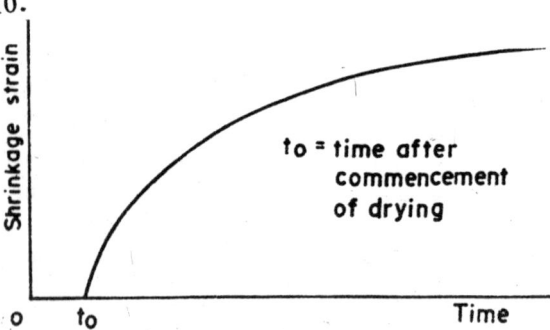

Fig. 1.10 Shrinkage curve for concrete

Evidently, the rate of shrinkage decreases with time. The total shrinkage strain may be in the range of 0.0002 to 0.0006. The IS Code: 456-1978 recommends a value for shrinkage to be 0.0003 for the purpose of design.

1.2.11 Creep of concrete. Creep is defined as the increase in strain under sustained stress. This increase in strain can be several times as large as the instantaneous elastic strain which takes place as soon as the concrete is loaded. Creep hardly affects the strength of a structure, though, it results in the increase of deflections. It causes redistribution of stresses within the reinforced concrete members. Concrete stresses set up due to differential settlement in a structure are partly relieved on account of creep. Tensile cracking of concrete on account of shrinkage is delayed

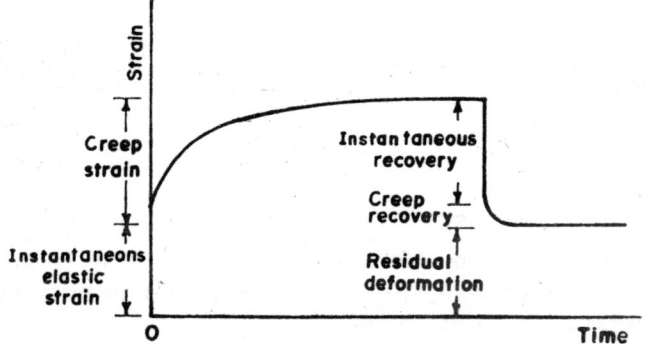

Fig. 1.11 Typical creep curve for concrete

because of creep effect. An illustration of the creep deformation in a specimen subjected to constant axial compression is given in Fig. 1.11.

Creep strain increases with time, though, the rate of increase becomes smaller and smaller. If the specimen which is maintained under a constant stress for a long time is finally unloaded, there is an instantaneous partial recovery of the elastic strain followed by the recovery of a small proportion of the total creep strain. A very large amount of creep is irreversible.

It has been experimentally observed that creep is directly proportional to the stress within the stress limit of one-half of the ultimate strength. Considering the constituents of concrete, creep depends upon the type of aggregate and the aggregate/cement ratio, cement content and type, water/cement ratio and the additives. An increase in the aggregate/cement ratio, a decrease in the water/cement ratio and a decrease in the cement content, all decrease the creep. With a high relative humidity in the environment, creep strains tend to be low. Creep is reduced if the loss of free water from concrete is restricted. Thus, with a high volume to surface area in a concrete member, creep will be less. Creep also depends upon the age of concrete at loading as well as upon the duration of loading. According to the IS Code:456—1978, creep strain may be estimated from the values of the creep coefficient which are given in Table 1.4 below.

Table   1.4

Creep coefficient $= \dfrac{\text{Total creep strain}}{\text{Elastic strain}}$,

for the given age of concrete at first loading

Age at loading	Creep coefficient
7 days	2.2
28 days	1.6
1 year	1.1

In order to estimate the deformation at some stage before the total creep takes place, the values recommended by the IS Code: 456—1978 may be used. These are:

(i)   About half of total creep may be assumed to take place in the first month.

(ii) About three-quarters of total creep may be assumed to take place in the first six months.

1.3   STEEL REINFORCEMENT

1.3.1   General.   The steel reinforcement for concrete is generally round in cross-section. It may be in the form of mild steel plain bars, high yield strength deformed bars or welded wire fabric. Sometimes rolled steel sections are also used.

1.3.2   Mild steel plain round bars.   Mild steel bars are manufactured by hot rolling process. The stress-strain curve is obtained by testing the bars in uniaxial tension. The behaviour in compression is assumed to be the same as that in tension. A typical stress-strain curve for mild steel is shown in Fig. 1. 12. The curve exhibits a linearly elastic behaviour up to the yield point stress f_y. Beyond this point the strains become plastic and increase considerably without increase of stress. This region of the curve is known as the yield plateau. The plastic strain within the yield plateau. may be as much as 10 times the elastic strain occurring before the yield stress. On further loading, the stress again increases with strain on account of strain hardening. It then reaches the peak value known as the ultimate stress and then decreases until the bar finally fractures.

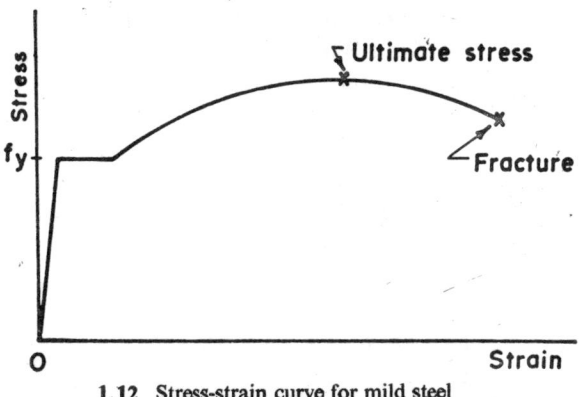

1.12 Stress-strain curve for mild steel

Once the mild steel bars in a loaded reinforced concrete specimen are stressed to the yield point, they considerably stretch, causing excessive cracking of concrete and permanent deformation in the

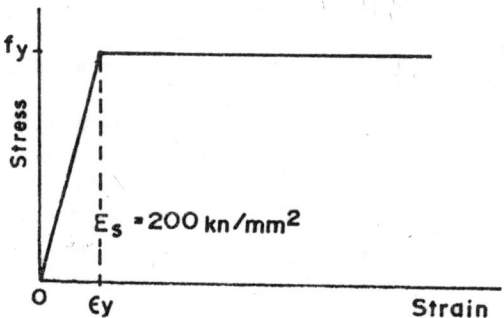

Fig. 1.13 Idealized stress-strain curve for mild steel

member. The yield point is indicative of the limit of usefulness of mild steel reinforcement, and as such, it is an important limit.

An idealized stress-strain diagram for mild steel as adopted by the IS Code: 456-1978 is shown in Fig. 1.13.

Plain mild steel bars have smooth surface. Load transference from concrete to steel takes place on account of bond strength between the two materials at the interface.

1.3.3 High yield strength deformed bars. If a mild steel bar is strained beyond the yield plateau and then unloaded, it is found on reloading that the yield plateau has completely vanished. This process is known as Cold Working. A typical stress-strain curve for a cold worked bar is shown in Fig. 1.14. Cold working may be done either by stretching or twisting of the bar. By this process greater stresses can be attained in the reinforcement bar for smaller strains, than the yield point stress of a mild steel bar.

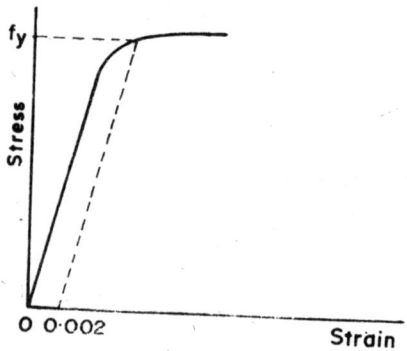

Fig. 1.14 Stress-strain curve for cold worked steel

Cold worked bars do not show a well defined yield point. As such, their strength is fixed up in terms of proof stress which is the stress level causing a definite residual strain upon unloading. IS Code: 456—1978 has accepted 0.2% proof stress as the standard value for the cold worked bars in lieu of yield point stress. It is the stress level corresponding to a residual strain of 0.002. An idealised stress-strain curve for high yield strength deformed bars (HYSD bars) as adopted by the IS Code: 456—1978 is shown in Fig. 1.15.

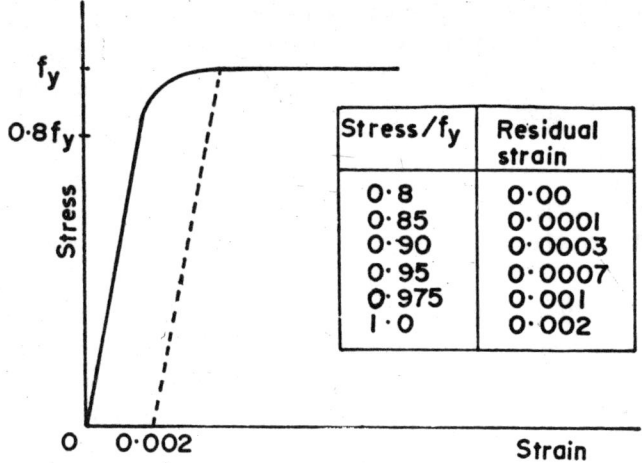

Stress/f_y	Residual strain
0·8	0·00
0·85	0·0001
0·90	0·0003
0·95	0·0007
0·975	0·001
1·0	0·002

Fig. 1.15 Idealized stress-strain curve for HYSD bars

High yield strength deformed bars have protrusions and indentations rolled on to the surface, which produce mechanical interlocking with concrete in addition to plain bonding. This enhanced gripping of the bar in concrete makes it possible to develop high magnitude of axial stress in the reinforcement without the risk of slipping.

1.3.4 Welded wire fabric. Welded wire fabric is made from steel wires running in two orthogonal directions and welded at their intersections. The wire for the fabric is obtained by cold drawing mild steel bars through a series of dies of successively reduced sizes. The wire fabric may be of square mesh type. In square mesh the wire diameter ranges from 3 mm to 10 mm, whereas the pitch ranges from 50 mm to 200 mm. In oblong mesh the main wire diameter ranges from 4 mm to 10 mm and the pitch ranges from 75 mm to 150 mm, whereas, at right angles the wire diameter ranges from 3 mm to 6.5 mm and the pitch ranges from 150 mm to 400 mm.

Welded wire fabric is used for reinforcing slabs, pavements and shell structures.

1.3.5 Characteristic strength of steel reinforcement. Characteristic strength means that value of the strength below which not more than 5% of the test results are expected to fall. The strength referred to is the yield strength or 0.2% proof stress found from testing bars in uniaxial tension. The characteristic strengths of of steel reinforcements specified by the IS Code 465-1978 are given in Table 1.5 below.

Table 1.5

CHARACTERISTIC STRENGTH OF REINFORCEMENTS, f_y

Type of reinforcement	Characteristic strength, N/mm²	
	Yield stress	0.2% Proof stress
Mild steel	250	
HYSD Bars (Fe 415)		415
HYSD Bars (Fe 500)		500

1.3.6 Nominal diameter. The nominal diameter is the diameter calculated from the volume of unit length of bar assuming it to be perfectly cylindrical. The manufactured bar sizes in terms of the standard nominal diameter, and the corresponding sectional areas are mentioned in Table 1.6 below. The bar sizes apply both to plain and the deformed bars.

Table 1.6

NOMINAL BAR DIAMETER AND SECTIONAL AREA

Nominal diamater, mm	Sectional area, mm²	Nominal diameter, mm	Sectional area, mm²
5	19.63	22	380.13
6	28.27	25	490.87
8	50.26	28	615.75
10	78.54	32	804.24
12	113.1	36	1017.87
16	201.06	40	1256.62
18	254.47	45	1590.41
20	314.16	50	1963.48
		—	

1.4 STEEL-CONCRETE BOND

1.4.1 Bond strength. In reinforced concrete members the reinforcement bars receive their share of the load from surrounding concrete due to shear stress developing at the steel-concrete interface. This is termed as bond stress which provides the resistance to reinforcement bars against slipping through the hardened conerete. In plain mild steel bars which have smooth surface, bond strength develops due to adhesion of cement mortar to steel surface and due to the friction on the interface on account of pressure exerted by shrinking concrete. High yield strength deformed bars have lugs or protrusions rolled on to the bar surface. Longitudinal ribs are also present on the surface of the bar. The indentations and the protrusions on the bar surface produce mechanical interlock between concrete and steel, which enhance the resistance of bar against slipping through concrete. Thus, deformed bars have higher bond strength in concrete than the plain bars. In reinforced concrete members HYSD bars rely on the enhanced bond strength in concrete for developing high magnitude of stress of which they are capable.

1.4.2 Development length of bar. A certain length of bar is needed beyond a section to transmit the bar force to the surrounding concrete by bond action. In other words, a certain length of bar is needed to be embedded in concrete in order to develop the required force in bar by bond. This length is called the development length of bar. See Fig. 1.16,

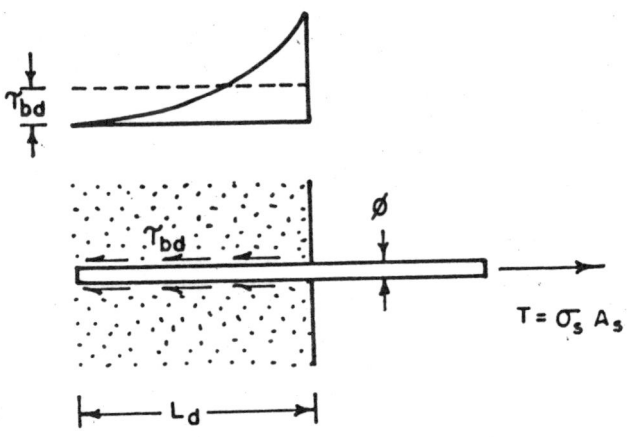

Fig. 1.16 Development of stress in a reinforcement bar

Development length of a bar is given by

$$L_d = \frac{\phi \sigma_s}{4\, \tau_{bd}} \tag{1.12}$$

where ϕ = the nominal diameter of bar

    σ_s = the design axial stress in bar

    τ_{bd} = design bond stress between concrete and reinforcement bar

  Actual distribution of bond stress along the bar length is non-uniform.   However, for the purpose of estimating the development length, an average value is used.

2

Cement and Concrete Technology

2.1 CEMENT

Cements have the property of setting and hardening on chemical reaction with water and are, therefore, called hydraulic cements. On hydration, cements acquire adhesive and cohesive properties, and are then able to hold the mineral aggregates together in a compact mass known as concrete. Hydrated cement is stable in water.

2.1.1 Portland cement. The name Portland cement was originally given because of the resemblance of set cement, in colour and quality, to Portland Stone which was quarried in Dorset, United Kingdom. The principal raw materials used in the manufacture of cement are:

(i) Calcareous meterial i.e. calcium carbonate. in the form of limestone or chalk.

(ii) Argillaceous materials i.e. silicates of alumina, in the form of clays and shales.

Marl, which is a mixture of calcareous and argillaceous materials is also used.

The ingredients are mixed approximately in the proportion of 2 parts of calcareous material to 1 part of argillaceous material. They are then crushed and ground in ball mills in dry state, or in wet state if water is premixed. In the wet process the chemical composition of slurry can be checked easily. The dry powder or the slurry, as the case may be, is then passed into a rotary kiln and burnt at a temperature of 1400° to 1500° C. The fuel may be pulverised coal, gas or oil. The material sinters and fuses into roundish lumps known as klinker. The latter is cooled and passed on to ball mills, and after adding Gypsum, is ground to the required fineness. The resulting product is Portland cement, which is generally available in market in 50 Kg bulk packed in jute sacks or multi-ply paper bags.

Chemical composition : The chemical constituents of the present-day Portland cements are as mentioned in Table 2.1. The first four oxides mentioned in the Table are generally referred to as major oxides.

Table 2.1

CHEMICAL COMPOSITION OF PORTLAND CEMENT

Oxide		Approximate composition limits (% by weight)
Lime	CaO	60 — 67
Silica	SiO_2	17 — 25
Alumina	A_2O_3	3 — 8
Iron oxide	Fe_2O_3	0.5 — 0·6
Magnesia	MgO	0.1 — 5.5
Sulphur trioxide	SO_3	1 — 3
Alkali oxide	$Na_2O + K_2O$	0.5 — 1.3
Titania	TiO_2	0.1 — 0.4
Phosphorous Pentoxide	P_2O_5	0.1 — 0.2

Compound composition : Cement is a mixture of fine grained minerals which are difficult to separate by chemical means. However, from high-temperature phase equilibrium studies it may be concluded that the four major oxides CaO, SiO_2, Al_2O_3 and Fe_2O_3, when clinkered, would eventually combine to form the main compounds which are as mentioned in Table. 2.2. Actual proportions of these compounds vary from cement to cement. In fact, different types of cements are obtained by suitable proportioning of these main compounds.

Table 2.2

MAIN COMPOUNDS OF CEMENT

Name of compound		Abbreviation	% eviation
Tricalcium silicate	$(3CaO.SiO_2)$	C_3S	30 - 50
Dicalcium silicate	$(2CaO.SiO_2)$	C_2S	20 - 45
Tricalcium aluminate	$(3CaO.Al_2O_3)$	C_3A	C_3A
Tetracalcium Aluminoferrite	$(4CaO.Al_2O_3.Fe_2O_3)$	C_4AF	6 - 10

These compounds are known as Bogue compounds as they have been identified mainly due to him. These compounds are also referred to as Major Constituents of cement. Their combined content is approximately 90% of the weight of cement. The remaining 10% weight consists of MgO, Na_2O and/or K_2O, TiO_2, P_2O_5 and $CaSO_4$, which are referred as Minor Constituents.

C_3S rapidly hydrates on adding water, giving out heat (500 J/g). After setting, it hardens within a few hours. Its strength development is rapid and takes a few days. Its strength is high. It is the characteristic constituent of Portland cements. C_2S hydrates slowly over a few days, giving out smaller amount of heat (250 J/g). Its strength development is steady but slow, and takes a few weeks. Its final strength is of the same order as that of C_3S. C_3A instanteously reacts with water giving a flash set. It gives out a large amount of heat (850 J/g). Its strength rapidly develops within 24 hours, though, it is low. It is unstable in water and is subject to sulphate attack. C_4AF very rapidly reacts with water and sets within minutes. It gives out heat on hydration (420 J/g). Its strength rapidly develops within 24 hours, though, it is low. It is responsible for the characteristic grey colour of Portland cement.

C_3S and C_2S constitute 70 to 80% of all Portland cements. they are the most stable and contribute most to the final strength. They make concrete resistant to corrosive salts, acids and alkalis. C_3S contributes more to the early strength. whereas, C_2S contributes to strength after 7 days for upto 1 year. C_3A contributes but very little to the strength principally within 24 hours. It is the least stable of the four main compounds. C_3A may decompose to $Ca(OH)_2$ and $Al(OH)_3$ and is easily attacked by salts and alkalis. As such, for hydraulic or marine works its presence is undesirable. C_4AF also contributes little to the strength, though, it is more stable than C_3A.

Gypsum ($CaSo_4 . 2H_2O$) is added, during grinding of klinker, to regulate the setting time of cement. An optimum gypsum content imparts to the cement maximum strength and minimum shrinkage. An excess amount may cause cracking and deterioration in set cement.

Free or uncombined lime (CaO) may be present in cement if the raw materials contain more lime than can combine with SiO_2, Al_2O_3

and Fe_2O_3.   Free lime may also occur because its reaction with
the oxides are not complete during the clinkering process. Free
lime can be greatly avoided in cement by restricting the amount
of lime in the raw materials. For this purpose, the Lime
Saturation Factor (LSF) for the raw materials should be restricted
between O.66 and 1.02.

$$LSF = \frac{CaO - 0.7\ SO_3}{2.8\ SiO_2 + 1.2\ Al_2O_3 + 0.65\ Fe_2O_3}$$

Even under controlled conditions a small amount of free lime
(less than 1 %) is always present in Portland cements. Free lime in
cement is 'hard burnt' and, as such, it hydrates very slowly. Its
hydration, which takes place after the cement has set, is accompanied by volume expansion. Hence, cracking and deterioration
of concrete may result. This is known as unsoundness of cement
which is caused due to free lime.

Free Magnesia (MgO) which exists in cement in crystallized form
is known as periclase. Hydration of periclase takes place over a
period of years and is accompanied by volume expansion. As such,
it may cause unsoundness.

The alkali oxides (Na_2O and K_2O) are introduced into the
cement through the raw materials, and exist in cement in solid
solution form. Some aggregates used in making concrete are
alkali-reactive as they contain a form of silica which combines
with the alkali oxides released from cement, resulting in the
formation of alkali - silica gel. This combination takes place
slowly and is accompanied by volume expansion, which may cause
cracking and deterioration of concrete.

Titanium oxide (TiO_2) is introduced into cement through the
raw material clay or shale. Its content is very small. Phosphorus
pentoxide (P_2O_5) is generally introduced into the cement through
the raw material limestone. It exists in solid solution form and
slows down hardening of cement. If present in large amount it
indirectly causes unsoundness by releasing free lime (CaO).

2.1.2 Different kinds of Portland cements. Mainly, the
relative proportions of the four major constituents and the degree
of fineness of cement determine the properties of the different
varieties of Portland cements. The latter are of the following
verieties:

(a) *Ordinary Portland cement.* The cement of this variety shall comply with the requirements of chemical and physical test as mentioned in Tables 2.3 and 2.4 respectively.

Table 2.3

CHEMICAL REQUIREMENTS FOR ORDINARY PORTLAND CEMENT
(BASED ON IS CODE : 269 — 1976)

(i)	Lime saturation factor	0.66 to 1.02
(ii)	Ratio of Al_2O_3 to Fe_2O_3	$\not<$ 0.66
(iii)	Insoluble residue	$\not>$ 2%
(iv)	MgO content	$\not>$ 6%
(v)	SO_3 content	$\not>$ 2.75% if C_3A content is $< 7\%$
		$\not>$ 3% if C_3A content is $> 7\%$
(vi)	Loss on ignition	$\not>$ 5%

Note: C_3A content is calculated by the formula :
$$C_3A = 2.65 \, Al_2O_3 - 1.69 \, Fe_2O_3$$

Table 2.4

PHYSICAL REQUIREMENT FOR ORDINARY PORTLAND CEMENT
(BASED ON IS CODE : 269 — 1976)

(i)	Fineness of grinding	$\not<$	2250 cm²/g.
(ii)	Soundness :		
	Unaerated cement		Expansion $\not>$ 10 mm
	Aerated cement		Expansion $\not>$ 5 mm
(iii)	Setting time :		
	Initial	$\not<$	30 minutes
	Final	$\not>$	600 minutes
(iv)	Compressive strength :		
	At 3 days	$\not<$	15.69 N/mm²
	At 7 days	$\not<$	21.57 N/mm²

(b) *High strength ordinary Portland cement.* This variety of Portland cement is particularly suitable in specialized works such as prestressed concrete, and itemes of precast concrete requiring high compressive strength. The cement of this variety shall comply with the chemical test requirement (i) to (v) as mentioned in Table 2.3 for ordinary Portland cement. The only exception is in loss on ignition which shall not exceed 4%.

Physical test requirement regarding soundness and setting time are the same as mentioned in Table 2.4 for ordinary Portland cement. The remaining requirements are as follows:

Fineness of grinding $\not< \ $ 3500 cm²/g

Compressive strength :

At 3 days $\qquad \not< \ $ 22.56 N/mm²

At 7 days $\qquad \not< \ $ 32.37 N/mm²

At 28 days $\qquad \not< \ $ 42.18 N/mm²

(c) *Rapid hardening Portland cement.* This variety of cement shall comply with the same chemical test requirements as mentioned in Table 2.3 for ordinary Portland cements.

Physical test requirements regarding soundness and setting time are the same as mentioned in Table 2.4 for ordinary Portland cement. The remaining requirements are as follows:

Fineness of grinding $\qquad \not< \ $ 3250 cm²/g

Compressive strength

At 1 day $\qquad \not< \ $ 15.69 N/mm²

At 3 days $\qquad \not< \ $ 26.97 N/mm²

(d) *Low heat Portland cement.* This variety of cement shall comply with the chemical test requirements (ii) to (vi) as mentioned in Table 2.3 for ordinary Portland cement. The only exception is in the lime content which will be as follows :

$$CaO - 0.7\, SO_3 \not> 2.4\, SiO_2 + 1.2\, Al_2O_3 + 0.65\, Fe_2O_3$$
$$\not< 1.9\, SiO_2 + 1.2\, Al_2O_3 + 0.65\, Fe_2O_3$$

The physical test requirements for this cement are mentioned in Table 2.5

Table 2.5

PHYSICAL REQUIREMENTS FOR LOW HEAT PORTLAND CEMENT
(BASED ON IS CODE : 269 — 1976)

(i)	Fineness of grinding	$\not<$	3200 cm²/g
(ii)	Soundness:		
	Unaerated cement		Expansion $\not>$ 10 mm
	Aerated cement		Expansion $\not>$ 5 mm
(iii)	Setting time:		
	Initial	$\not>$	60 minutes
	Final	$\not>$	600 minutes
(iv)	Compressive strength:		
	At 3 days	$\not<$	9.81 N/mm²
	At 7 days	$\not<$	15.70 N/mm²
	At 28 days	$\not<$	34.34 N/mm²
(v)	Heat of hydration:		
	At 7 days	$\not>$	65 cal/g
	At 28 days	$\not>$	75 cal/g

This cement is for use where low heat of hydration is required, as in mass concrete work in dams.

(e) *Portland slag cement.* This variety of cement is obtained by mixing Portland cement clinker and grannulated blast furnace slag and then intimately grinding the mixture to required fineness.

The proportion of the slag shall be between 25% to 65% of the total bulk. Its physical properties are similar to those of ordinary Portland cement. In addition, it has low heat of hydration. It is more resistant to soils and water containing sulphates of alkali metals, alumina and iron, and to acidic waters. For hydraulic and marine structures it is advantageous.

Portland cement clinker used in the manufacture of Portland slag cement shall comply with the chemical test requirements mentioned in Table 2·3. Chemical composition of the grannulated blast furnace slag shall be as follows :

Constituent	Percent
Silicon dioxide (SiO_2)	27 — 32
Alumunium oxide (Al_2O_3)	17 — 31
Ferrous oxide (FeO)	0 — 1
Calcium oxide (CaO)	30 — 40
Magnesium oxide (MgO)	0 — 17
Sulphide sulphur (S)	0 — 2

Portland slag cement shall comply with the chemical test requirements as mentioned in Table 2.6.

Table 2.6

CHEMICAL REQUIREMENTS FOR PORTLAND SLAG CEMENT
(BASED ON IS CODE : 456 — 1976)

Constituent	Percent, Maximum
Magnesium oxide (MgO)	8.0
Sulphur trioxide (SO_3)	3.0
Sulphide sulphur (S)	1.5
Loss on Ignition	4.0
Insoluble residue	2.5

Portland slag cement shall comply with exactly the same physical test requirements as for ordinary Portland cement which are mentioned in Table 2.4.

(f) *Portland — Pozzolana cement.* This variety of cement is produced by intimately grinding together Portland cement clinker and pozzolana with addition of gypsum in suitable proportions.

Pozzolana is a silicious material which, possessing little or no cementitious properties, will in finely divided form react with calcium hydroxide in presence of water at ambient temperature, to form compounds possessing cementitious properties.

Volcanic ashes or pumicites, diatomaceous earth, opaline cherts and shales are some of the natural pozzollanas. Calcined clay, precipitated silica and fly ash are the artificial pozzolanas.

Portland - pozzolana cement releases lesser heat of hydration and is more resistant to aggressive waters than ordinary Portland cement. It reduces leaching of $Ca(OH)_2$ during setting and hydration of cement. It can be used in all situations where ordinary Portland cement is used. It is particularly useful for hydraulic and marine construction as well as in mass concrete work.

The Portland cement clinker constituent should comply with the chemical test requirements mentioned in Table 2.3. For the pozzolana constituent the Lime Reactivity test should give a compressive strength not less than 50 kg/cm². The amount of pozzolana constituent shall be between 10 to 25% of the whole bulk.

Portland - Pozzolana cement shall satisfy the chemical test requirements as mentioned in Table 2.7

Table 2.7

CHEMICAL REQUIREMENTS OF PORTLAND-POZZOLANA CEMENT
(BASED ON IS CODE : 1489 — 1976)

(i)	Loss on ignition	≯	5.0%
(ii)	Magnesia (MgO)	≯	6.9%
(iii)	Insoluble material	≯	$x + \dfrac{2\,(100 - x)}{100}$
(iv)	SO_3 content	≯	2.75 if C_3A content is ≮ 7%
		≯	3.0 if C_3A content is > 7%

Note :- x is the percentage of pozzolana in Portland - Pozzolana cement. C_3A is calculated by the formula :

$$C_3A = 2.65\ Al_2O_3 - 1.69\ Fe_2O_3$$

The physical test requirements for this cement as regards soundness and setting time are the same as laid down in Table 2.4 for ordinary Portland cement. The remaining requirements as per IS Code : 1489--1976 are as follows:-

Fineness of grinding	≮	3200 cm²/gm
Compressive strength :		
At 7 days	≮	21.58 N/mm²
At 28 days	≮	30.41 N/mm²
Drying shrinkage :	≯	0. 15%

(g) *Hydrophobic Portland cement.* This variety of cement is obtained by intergrinding Portland cement clinker with hydrophobic agents such as oleic acid, stearic acid, naphthenic acid or pentachlorophenol etc. Hydrophobic agent may be 0.1 to 0.5% of the weight of clinker. Hydrophobic properties are imparted because of the formation of water repellent film around the particles of cement. This film breaks during wet attrition in a concrete mixer, and then, mormal hydration takes place. This cement does not deteriorate on prolonged storage in wet climatic conditions.

The chemical test requirements of hydrophobic cement shall be as laid down in Table 2.3 for the ordinary Portland cement.

The physical test requirements of hydrophobic cement for soundness and setting time shall be as laid down in Table 2.4 for ordinary Portland cement. The remaining requirements as per IS Code : 8043 — 1978 are as follows :

Fineness of grinding	≮	3500 cm²/gm
Compressive strength:		
At 3 days	≮	15.69 N/mm²
At 7 days	≮	21.57 N/mm²
At 28 days	≮	30.40 N/mm²
Loss on ignition at 550°C	≯	30% of that for ordinary Portland cement
Floatation on water surface	≮	24 hours

(h) *White portland cement.* This variety of cement is produced in the same way as ordinary Portland cement except that the amount of iron oxide is restricted to 1%. Suitable raw materials are chalks and lime stones with low iron contents, and white clays. In the kilning process oil fuel may be used in place of pulverised coal.

White cements are not as strong as ordinary Portland cement, and their price is considerably higher.

(i) *Coloured Portland cements.* Most coloured cements are obtained by adding pigments to Portland cements. Strong pigments up to 10% can be added to ordinary Portland cement. For lighter shades white Portland cement should be used. Pigments are best added during the process of grinding. Pigments should be permanent and chemically inert with cement.

2.1.3 Physical tests on cements. The procedures for these tests are described in the following paras:

(a) *Fineness by air permeability method.* The method consists of drawing a definite quantity of air through a bed of cement of given porosity. See Fig. 2.1. The apparatus is first calibrated using a NBS sample No. 114. The volume of compacted bed of powder is found by mercury displacement method. Weight of the standard sample required for calibration is given by

$$W = P_s \, V(1 - e_s)$$

where P_s = specific gravity of sample (=3.15)

V = volume of bed of sample

e_s = desired porosity of sample (0.5 ± 0.005)

The standared sample is placed in the permeability cell A and compacted to the standard volume. Air is drawn off through tube D untill the liquid reaches mark E in the manometer. The valve is then closed. Time required for the liquid to fall from F to G is noted and a mean of three tests is taken.

Sample of cement of which specific surface is to be found is then tested, taking the same weight as that of the standard sample used for calibration. The time of flow of air is noted as before.

Then, the specific surface of the sample of cement is given by

$$S = \frac{S_s \, P_s (1 - e_s) \sqrt{n_s e^3 T}}{P(1 - e) \sqrt{n e^3_s \, T_s}}$$

where S = specific surface of test sample cm^2/gm

S_s = specific surface of standard sample used for calibration

T = time of flow of air through test sample

T_s = time of flow of air through calibration sample

n = viscosity of air while testing test sample

n_s = viscosity of air while testing calibration sample

e = porosity of test sample

e_s = porosity of calibration sample

ρ = specific gravity of test sample
ρ_s = specific gravity of calibration sample

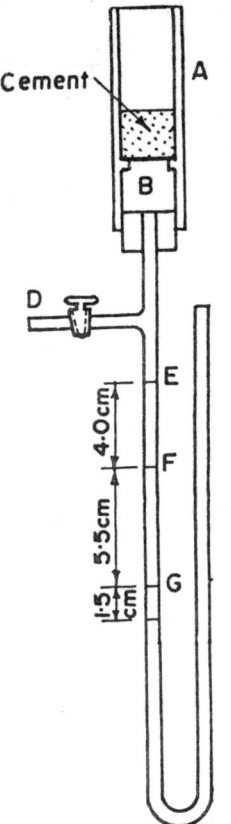

Fig. 2.1 Blaine Air permeability apparatus

The number and size of pores in a bed of cement of given porosity is a function of the size of the particles, which determine the rate of flow of air through it.

(b) Consistency of standard cement paste. By this test is determined the quantity of water required to produce a cement paste of standard consistency. The Vicat apparatus, shown in fig. 2.2 is used with the plunger G. A cement paste, placed in the Vicat mould E, is said to be of standard consistency if it permits penetration by the Vicat plunger up to a point 5 to 7 mm above the bottom. Trial cement pastes are prepared keeping gauging time between 3 to 5

minutes. The gauging time is counted from the moment water is added to the dry cement and before filling up the mould E with cement paste. After filling up, the top surface of the paste is struck smooth. The plunger is brought in contact with the top surface of the cement paste and quickly released. The test is repeated with varying percentages of water untill the quantity of water required for standard consistency is found. During the test the temperature shall be maintained at $27 \pm 2°C$.

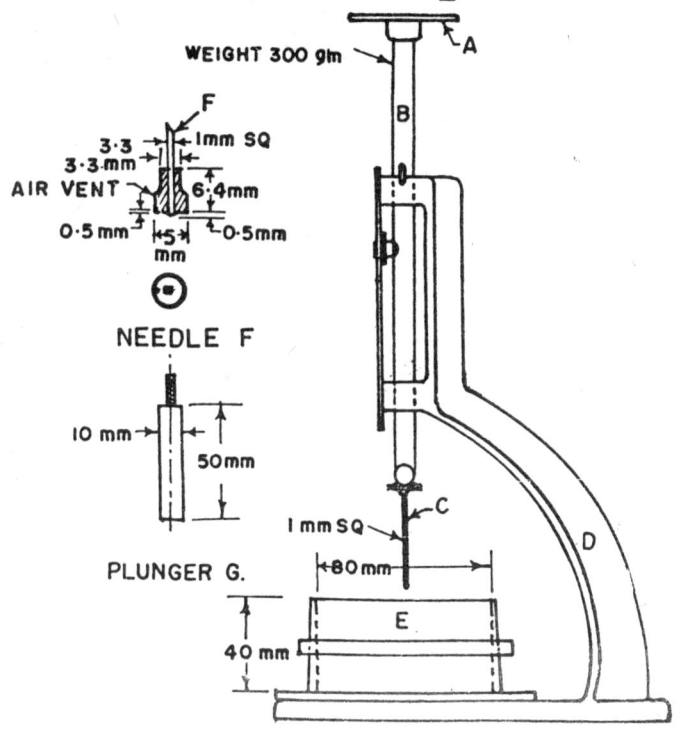

Fig.   2.2   Vicat apparatus

(c)   Initial and final setting times. A neat cement paste is prepared by gauging cement with 0.85 times the water reguired for standard consistency. A stop watch is started when the water is added. The Vicat mould E is filled up with the cement paste and its top surface is struck smooth. Thus, a test block is obtained.

For initial setting time the Vicat needle C is brought in contact with the top surface of the test block and quickly released. In the beginning the needdle may completely penetrate the test block. The test is repeated seveial times untill the needdle fails to completely penetrate the test block and remains short of the bottom of the mould by 5 $\pm$ 0.5 mm. The time noted on the stop watch upto this moment since the adding of water is known as the initial setting time.

For final setting time the Vicat needle F which has an annular attachment is brought in contact with the top surface of the test block and quickly released. The cement is considered to be finally set when the neddle makes an impression on the surface while the attachment fails to do so. The time noted on the stop watch up to this moment since the adding of water is known as the final setting time.

(d) Soundness by 'Le Chatelier' test. The apparatus consists of a split cylinder mould of brass with two 165 mm long pointers as shown in Fig. 2.3. The mould is placed on a glass sheet. Cement paste is prepared by ganging it with 0.78 times the water required for standard consistency. The mould is filled with the paste and covered with another sheet of glass. The whole

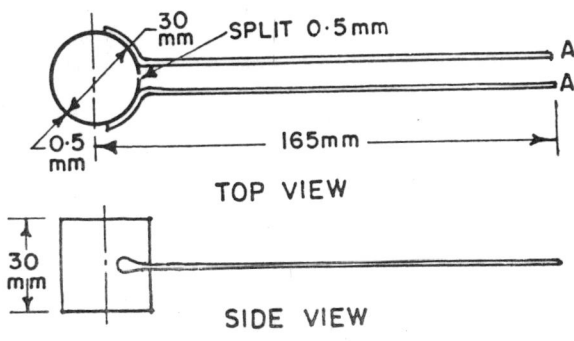

Fig. 2.3 Le Chatelier test apparatus.

assembly is submerged in water for 24 hours at a temperature of 27° $\pm$ 2° C. Distance between the pointers is noted. The water is then brought to boiling in 25 to 30 minutes and kept boiling for 3 hours. Then, the mould is rewoved from water and allowed to cool. Distance between the pointers is measured. The defference between the two measurements represents the expansion of cement.

(e) *Compressive strength of cement*   strength of cement is determined by comdressive strength tests on mortar cubes. Test specimens are in the form of cubes having area of face $= 50$ cm². The meterial for each cube shall be mixed separately in the following proportion;

Cement	200 g
Standard sand	600 g
Water	$(\dfrac{P}{4} + 3.0)$ percent of combined weight of cement and sand.

where P is the percentage of water required for standard consistency. Cement and sand are mixed dry for one minute. Then, water is added till the mixture is of uniform colour within 3 to 4 minutes. The mixture is placed in cube mould and piodded with poking rod 20 times in 8 seconds. The remaining quantity of mortar is placed in the hopper of the mould which is placed in the vibration machine. The mould is vibrated for 2 minutes in the machine running at 12000 ± 400 r.p.m. The mould is then removed and the top surface finished smooth with a trowel. It is then kept at a temperature of $27° \pm 2°C$ at 90% relative humidity for 24 hours. Then the cube specimen is removed and placed in clean water for curing untill tested.

Three cube specimens are tested for compressive strength at the end of period relevant to the cement being tested. The load shall be steadily and uniformly applied, starting from zero at a rate 34.34 N/mm²/min., Compressive strength is calculated from the crushing load.

(f) *Heat of hydration.* This test is conducted to determine the heat of hydration of low heat Portland cement. A calorimeter is used for this purpose. The method consists of noting the temperature rise as cement sets and hardens under adiabatic conditions, The heat of hydration in calories per g is

$$= \frac{\text{heat evolved (cal.)}}{\text{weight of cement (g)}}$$

(g) *Specific gravity of cement.* For this purpose a Le Chatelier flask, as shown in Fig. 2.4, is used. The flask is filled

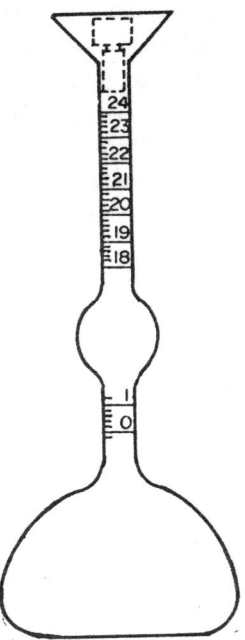

Fig. 2.4 Le Chatelier flask

with kerosene upto a point on the stem between zero and the 1 — ml mark. The reading is recorded. A weighed quantity of cement is then introduced in the flask. The stopper is placed and the flask rolled in inclined position to expel any air from cement. The final reading is then taken in the upper series of graduations. The difference between the final and the first readings represents the volume of liquid displaced by cement. The specific gravity is given by

$$\text{sp gr} = \frac{\text{weight of cement in g}}{\text{volume in ml}}$$

Duplicate determinations shall agree within 0.01.

2.2 AGGREGATES

Aggregates, crushed or uncrushed, are derived from natural sources such as river terraces and river beds, glacial deposits, rocks, boulders and gravels, for use in the production of concrete for structural purposes and mass concrete work.

Fine aggregate is the aggregate most of which passes 4.75 mm IS Sieve and contains no more coarser material than permitted. Natural sand is the fine aggregate which may result from natural disintegration of rock or deposited by rivers or glaciers. Crushed stone sand is the fine aggregate obtained by crushing stone. Crushed gravel sand is the fine aggregate obtained by crushing natural gravel.

Coarse aggregate is the aggregate most of which is retained on 4.75 mm IS Sieve and contains no more finer material than permitted. Uncrushed gravel is the coarse aggregate which results from natural disintegration of rock. Crushed gravel is the coarse aggregate obtained by crushing of gravel or stone. Partially crushed gravel is obtained by blending of crushed and uncrushed gravels.

All-in-aggregate is the mixture of fine and coarse aggregates.

Aggregate shall comprise of naturally occuring, crushed or uncrushed stones, gravel or sand, or a mixture of these. It should be hard, strong, dense, clear and free of veins and adherent coating. It should be free of vegetable matter, alkali and other deleterious substances. Flaky, scoraceous and elongated pieces should be avoided. Deleterious materials such as pyrites, coal, lignite, mica,

Table 2.8
LIMITS OF DELETERIOUS MATERIALS, % AGE
(Based on Table 1 IS Code: 383-1970)

Description of material	Fine aggregate		Coarse aggregate	
	Uncrushed	Crushed	Uncrushed	Crushed
(a) Coal and lignite	1.0	1.0	1.0	1.0
(b) Clay lumps	1.0	1.0	1.0	1.0
(c) Passing 75-micron IS Sieve	3.0	15.0*	3.0	3.0
(d) Soft fragments			3.0	
(e) Shale	1.0			
Total =	5.0	2.0	5.0	5.0

*Not to be counted in the total.

shale, clay, alkali, soft fragments, sea shells and organic impurities shall not be in such quantity as to affect the strength and durability of concrete. Limits of deleterious substances are given in Table 2.8.

The strength and hardness requirement for the aggregates shall be as laid down in Table 2.9.

If the concrete is liable to be exposed to the action of frost, then coarse and fine aggregates shall pass an accelerated soundness test with sodium or magnesium sulphate solution as per IS Code: 2386 (Part V) — 1963. The limits of test are set by the purchaser and the supplier.

Table 2.9

(Based on IS Code 383 — 1970)

Type of test	Concrete use in	
	structural work	wearing surfaces
Aggregate crushing value, %	45	30
Aggregate impact value, %	45	30
Aggregate abrasion value, %	50	30

2.2.1 Sieve analysis. Particle size distribution of fine, coarse and all-in-aggregate is determined by sieve analysis. The sieve sizes to be used are mentioned in Table 2.10.

Table 2.10

IS SIEVE FOR SIEVE ANALYSIS OF AGGREGATES

(Based on IS Code: 2386 (Part 1) — 1963)

Sieve type	Aperture size designation	
	mm	micron
Perforated plate with square holes	80,63,50,40 31.5,25,20,16, 12.5,10,6.3,4.75	
Wire mesh	3.35,2.36, 1.18	600,300,150,75

(a) *For coarse and fine aggregates.* The sample for sieving is prepared from the larger sample either by quartering or by using a sample divider. The sample is first brought to an air dry condition and then weighed. It is then sieved successively through the appropriate IS Sieves, beginning with the largest size. If done

Table 2.11
SIZES OF COARSE AGGREGATES FOR STRUCTURAL CONCRETE
(Based on Table 2 IS Code: 383-1970)

IS Sieve Designation	Percentage Passing for Single-Sized Aggregate of Nominal Size						Percentage Passing for Graded Aggregate of Nominal Size			
	63 mm	40 mm	20 mm	16 mm	12.5 mm	10 mm	40 mm	20 mm	16 mm	12.5 mm
(1)	(2)	(3)	(4)	(5)	(6)	(7)	(8)	(9)	(10)	(11)
80 mm	100	—	—	—	—	—	100	—	—	—
63 mm	85 to 100	100	—	—	—	—	—	—	—	—
40 mm	0 to 30	85 to 100	100	—	—	—	95 to 100	100	—	—
20 mm	0 to 5	0 to 20	85 to 100	100	—	—	30 to 70	95 to 100	100	100
16 mm	—	—	—	85 to 100	100	—	—	—	90 to 100	—
12.5 mm	—	—	—	—	85 to 100	100	—	—	—	90 to 100
10 mm	0 to 5	0 to 5	0 to 20	0 to 30	0 to 45	85 to 100	10 to 35	25 to 55	30 to 70	40 to 85
4.75 mm	—	—	0 to 5	0 to 5	0 to 10	0 to 20	0 to 5	0 to 10	0 to 10	0 to 10
2.36 mm	—	—	—	—	—	0 to 5	—	—	—	—

manually, each sieve shall be shaken separately at least for 2 minutes. However, if sieving is done on an electrical sieve shaker through a nest of IS Sieves, the sieving time shall be at least 10 minutes. The weight of the material passing each sieve is determined aud expressed as a percentage of the sample weight.

(b) *For all-in-aggregates.* The procedure for this material is the same as that for the coarse and fine aggregates. However, heavy overloading of the finer sieves frequently results. As such, it may be neccessary to make a preliminary separation of the all-in-aggregate into coarse and fine aggregates by using a 4.75 mm sieve. The weight of the material passing each sieve is determined and expressed as a percentage of the sample weight.

2.2.2 Determining materials finer than 75 microns. The test sample is first dried at a temperature of 110° ± 5° C and then weighed. It is placed in a container, covered with water, and then vigourously agitated to bring all particles flner than 75 microns into suspension. The wash water is poured over IS Sieves 1.18 mm and 75 micron, with the coarses sieve on the top. The procedure is repeated till the wash water is clear. The material retained on the nested sieves is returned to tne washed sample. After drying it at 110° C it is weighed. The'loss of weight represents the material finer than 75 micron and is expressed as a percentage of the weight of the sample.

Table 2.12

SIZES OF COARSE AGGREGATES FOR MASS CONCRETE
(Based on Table 3 IS Code: 383-1970)

Class and Size	IS Sieve Designation	Percentage Passing
Very large, 150 to 80 mm	160 mm*	90 to 100
	80 mm	0 to 10
Large, 80 to 40 mm	80 mm	90 to 100
	40 mm	0 to 10
Medium, 40 to 20 mm	40 mm	90 to 100
	20 mm	0 to 10
Small, 20 to 4.75 mm	20 mm	90 to 100
	4.75 mm	0 to 10
	2.36 mm	0 to 2

*There being no IS Sieve having an aperture larger than 100 mm a perforated plate having a square aperture of 160 mm may be used.

2.2.3   Size and grading. For structural concrete work single sized coarse aggregates as well as the graded aggregates shall conform to the nominal sizes specified in Table 2.11 For mass concrete work the coarse aggregates shall conform to the nominal sizes specified in Table 2.12.

The grading of fine aggregates shall lie within the limits specified in Table 2.13. Hence, the grading zones I, II, III and IV. If the grading of a material falls outside the limits of any particular grading zone of sieves other than 600 micron sieve by a total of not more than 5%, it shall be regarded to be falling within that zone. This tolerance shall not be applied to percentage of material passing 600 micron sieve, or sieves on the coarser limit of grading zone I, or sieves on the finer limit of grading zone IV.

Table 2.13

SIZES OF FINE AGGREGATES

(Based on Table 4 IS Code 383-1970)

IS Sieve Designation	Percentage Passing for			
	Grading Zone I	Grading Zone II	Grading Zone III	Grading Zone IV
10 mm	100	100	100	100
4.75 mm	90-100	90-100	90-100	95-100
2.36 mm	60-95	75-100	85-100	95-100
1.18 mm	30-70	55-90	75-100	90-100
600 micron	15-34	35-59	60-79	80-100
300 micron	5-20	8-30	12-40	15-50
150 micron	0-10	0-10	0-10	0-15

Note 1 — For crushed stone sands, the permissible limit on 150-micron IS Sieve is increased to 20 percent. This does not affect the 5 percent allowance applying to other sieve sizes.

Note 2 — Fine aggregate complying with the requirements of any grading zone in this table is suitable for concrete but the quality of concrete produced will depend upon a number of factors including proportions.

Note 3 — Where concrete of high strength and good durability is required, fine aggregate conforming to any one of the four grading

The grading of all-in-aggregates shall conform to that specified in Table 2.14.

zones may be used, but the concrete mix should be properly designed. As the fine aggregate grading becomes progressively finer, that is, from Grading Zones I to IV, the ratio of fine aggregate to coarse aggregate should be progressively reduced. The most suitable fine to coarse ratio to be used for any particular mix will, however, depend upon the actual grading, particle shape and surface texture of both fine and coarse aggregates.

Note 4 — It is recommended that fine aggregate conforming to Grading Zoze IV should not be used in reinforced concrete unless tests have been made to ascertain the suitability of proposed mix proportions.

Table 2.14
SIZES OF ALL-IN-AGGREGATE
(Based on Table 5 IS Code 383-1970)

IS Sieve Designation	Percentage Passing for All-in-Aggregate of	
	40 mm Nominal Size	20 mm Nominal Size
80 mm	100	—
40 mm	95 to 100	100
20 mm	45 to 76	95 to 100
4.75 mm	25 to 45	30 to 50
600 micron	8 to 30	10 to 35
150 micron	0 to 6	0 to 6

2.2.4 Flakiness index. This test is applicable to aggregates of size larger than 6.3 mm. Flakiness index is the percentage by weight of particles whose least dimension is less than 0.6 times their mean dimension. The material is divided into 9 fractions of different mean sizes by sieving through 18 IS Sieves according to Table 2.15. A thickness gauge of metal having 9 slots is used. Widths of slot openings equal 0.6 times the mean sizes of the 9 fractions. At least 200 pieces of aggregate in each fraction shall be taken out. Thus, each fraction is gauged for thickness and the amount passing is weighed. Total weight of the material passing the various thickness gauges is expressed as a percentage of the total weight of the sample. This is the flakiness index.

2.2.5 Elongation index. This test is applicable to aggregates of size larger than 6.3 mm. Elongation index is the percentage by weight of particles whose greatest dimension is greater than 1.8 times their mean dimension. The aggregate sample is prepared exactly in the same manner by sieving as that for the flakiness index test. See Table 2.15. A metallic length gauge is used for this test. Each fraction is gauged separately and the total amount retained on the length gauge is weighed. The elongation index is the total weight of material retained on various gauges expressed as a percentage of the total weight of the sample.

2.2.6 Specific gravity and water absorption.

(a) *For aggregate larger than 10 mm.* The aggregate is cleaned of finer particles and dust, placed in a wire basket and immersed in distilled water with a cover of 50 mm of water, at a temperature of

Table 2.15

THICKNESS AND LENGTH GAUGE DIMENSIONS
(Based on Table V IS Code 2386 (Part 1)-1963)

Size of Aggregate		Thickness Gauge*	Length Gauge**
Passing Through IS Sieve	Retained on IS Sieve	mm	mm
63-mm	50-mm	33.90	---
50-mm	40-mm	27.00	81.0
40-mm	25-mm	19-50	58.5
31.5-mm	25-mm	16.95	---
25-mm	20-mm	13.50	40.5
20-mm	16-mm	10.80	32.4
16-mm	12.5-mm	8.55	25.6
12.5 mm	10-mm	6.75	20.2
10-mm	6.3-mm	4.89	14.7

*This dimension is equal to 0-6 times the mean sieve size.
**This dimension is equal to 1.8 times the mean sieve size.

22° to 32° C. The sample is raised by 25 mm and then dropped inside water 25 times to remove entrapped air. It is kept immersed for 24 hours. It is then weighed under water (weight $= A_1$) The wire basket is then taken out and emptied of the aggregates. The wire basket alone is then weighed under water (weight $= A_2$). The aggregate is surface dried with the help of dry cloth, kept in unheated air for 10 minutes, and then weighed (weight $= B$). The

aggregate is then placed in oven at a temperature of 100° to 110° C for 24 hours. It is finally weighed (weight = C).
Then

$$\text{specific gravity} = \frac{C}{B-A}$$

$$\text{Apparent specific gravity} = \frac{C}{C-A}$$

$$\text{Water absorption} = \frac{100(B-C)}{C}$$

where

A = the weight of the saturated aggregate in water
$= A_1 - A_2$
B = weight of the saturated aggregate in air
C = weight of the oven-dried aggregate in air.

(b) *For aggregate smaller than 10 mm* The aggregate is placed in a tray and covered with distilled water at a temperature of 22° to 32°C. Entrapped air is removed by gentle agitation with a rod. It is kept immersed for 24 hours. Then, the water is drained from the aggregate by decantation and passed through filter paper, any material retained being returned to the sample. The aggregate is exposed to a gentle current of warm air while being stirred till it is surface dry. Then, it is weighed (weight = A). The aggregate is then placed in a pycnometer which is filled with distilled water to the top. It is then weighed (weight = B). The pycnometer is emptied completely, refilled with distilled water and weighed (weight = C). The aggregate is dried in an oven at a temperature of 100° to 110° C for 24 hours. After cooling it is finally weighed (weight = D). Two tests should be made.

Then

$$\text{specific gravity} = \frac{D}{A-(B-C)}$$

$$\text{Apparent specific gravity} = \frac{D}{D-(B-C)}$$

$$\text{Water absorption} = \frac{100(A-D)}{D}$$

2.2.7 Bulk density and voids. For this purpose a cylindrical metallic measure is used. The capacity of the measure depends upon the maximum nominal aggregate size. See Table 2.16.

Table 2.16

Max. aggregate size, mm.	Capacity of measure, litres
$\leqslant 4.75$	3
> 4.75 but $\leqslant 40$	15
> 40	40

The measure is filled to overflowing by discharging the aggregate from a height not exceeding 50 mm above the top of the measure. The surface of the aggregate is then levelled off. The net weight of the aggregate is determined and the bulk density determined:

$$\text{Bulk density } \gamma = \frac{\text{Net weight of aggregate}}{\text{Volume of the measure}}$$

Hence

$$\text{percentage voids} = \frac{100\,(s - \gamma)}{s}$$

where

$$s = \text{specific gravity of aggregate.}$$

2.2.8 Bulking of sand. This is the increase in volume of a given weight of sand as the particles are pushed apart on account of thin films of water coating them. The extent of bulking depends on the percentage of moisture in the sand and its fineness. With the increase in moisture content bulking increases. For a moisture content of 5 to 8% the bulking may be as much as 20 to 30%. On further addition of water the films of moisture merge and the water moves into the voids resulting in a decrease of volume. The volume of sand when fully saturated is nearly the same as when it is dry.

The bulking of a moist sample of sand can be found by means of a cylindrical container. The container is two-thirds filled up loosely with the sample with top surface levelled off. A steel rule is vertically inserted through the sample and its height measured (height = h). The sample is then taken out and the container is half filled up with water. The sample is again poured into the container and its top surface levelled off. The steel rule is again inserted vertically through the sample and its height measured (height = h').

Then

$$\text{percentage bulking} = \left(\frac{h - h'}{h'}\right)100$$

On a construction site if the sand is being measured by volume it needs correction on account of bulking. The measured volume of sand should be increased by percentage bulking in order that the mix may contain the right amount of sand.

2.3 CONCRETE MIX DESIGN

In concrete, coarse aggregates provide the main bulk while their voids are filled up by the fine aggregates, the hydrated cement acts as the binder. Proportioning of the ingredients of concrete aims to attain some definite properties for both freshly mixed and hardened concrete. Freshly mixed concrete derives its flowing property on account of cement paste, fine aggregate and admixtures, if any. It can be easily placed and compacted if it is workable to the desired degree. In the absence of proper compaction the performance of the hardened concrete is greatly impaired. The strength and durability of hardened concrete depends on the water/cement ratio in the mix. The objectives of the concrete mix design are as follows:

(i) Attainment of the required strength of hardened concrete mainly by monitoring the water/cement ratio.

(ii) Attainment of the required durability of hardened concrete mainly by ensuring a minimum cement content and limiting the water/cement ratio.

(iii) Attainment of the workability of the freshly mixed concrete mainly by proper proportioning of coarse and fine aggregates and by fixing up the aggregate/cement ratio.

(iv) Attainment of economy by ensuring that the above properties are achieved by using the minimum cement content.

The data which is used in deciding concrete mix proportions should be looked upon as a guide only. It is meant only for the first stage in designing a mix. This stage should be followed by making of trial mixes. The mix proportions are readjusted and refinements done till optimum proportions are obtained and the hardened concrete has the required strength and durability.

Concrete mix design is, therefore, basically a problem of trial and error, and the calculations based on the design data merely provide a means of starting point for conducting the first test.

The essential properties of hardened concrete such as strength, durability and impermeability can be achieved if the concrete mix has adequate cement content, optimum water content and proper workability for placement and compaction in the forms giving maximum density with the available compacting effort.

2.3.1 Strength and the water/cement ratio.

By strength of concrete is meant the compressive strength of hardened concrete which is determined by compressive testing of 150 mm cubical specimens. Compressive strength, characteristic strength and the grades of concrete have been already discussed under the paras. 1.2.2, 1.2.3 and 1.2.4 respectively.

Compressive strength of concrete depends upon the strength of the cement paste, the paste-aggregate bond and the properties of the aggregate. Strength of the cement paste is determined by its porosity which is determined by the water/cement ratio and the degree of hydration. Hence, for the same degree of hydration, the strength of the cement paste depends solely on the water/cement ratio. The latter also mainly determines the strength of the paste-aggragate bond. The compressive strength of concrete is based on the water/cement ratio law which is generally credited to Abrams. According to this law, for the given conditions of test, the compressive strength of a fully compacted concrete depends solely on the water/cement ratio; it is not affected by the type, shape and surface texture of the aggregate and its grading. Made from a given cement and sound aggregates, the compressive strength of fully compacted concrete, for a given method of testing, type and duration of curing, and at a certain age, depends solely upon the water/cement ratio. Abram's law states that

$$S = \frac{A}{B^w} \qquad (2.1)$$

where.

S is the compressive strength of concrete and

w is the water/cement ratio by weight.

A and B are the constants which depend on the age of concrete, duration and type of curing, type of cement and the method of testing.

Trial values of the water/cement ratios, as guidance for obtaining the required compressive strength of concrete is shown in Fig. 2.5.

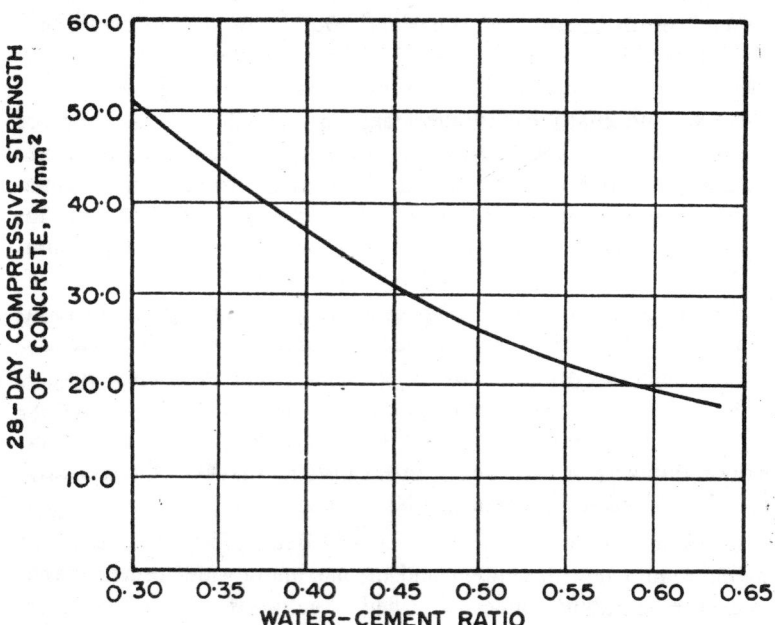

Fig. 2.5 Relationship between Water/Cement Ratio and Compressive Strength of Concrete

The curve is based on a large number of results on Indian Cements. It may need a little modification for a particular type of cement. Hence, it may be used as a guide only. Adjustment in water/cement ratio should be made for the variation in maximum aggregate size as well as for type of aggregate. The water/cement ratio should finally be determined by means of trial mixes.

On an average, cement requires 0.253 times its own weight for complete hydration. Any excess water finally results in increased porosity in cement paste. Hence, an increase in water/cement ratio results in a decrease in the strength of concrete.

2.3.2 Durability of hardened concrete. Durability implies the resistance of concrete to weathering action which is brought about by environmental conditions such as changes of temperature and humidity, abrasion, chemical attack, frost and fire. One of the main factors affecting the durability of concrete is its permeability. The volume of the hardened cement paste is less than the combined volumes of the cement and the water required for chemical reaction. Naturally, voids exist in the paste. Excess water travels through these pores, making continuous minute channels within the mass of concrete, and finally evaporates from the exposed surface. Higher the water/cement ratio, the more permeable and porous will the concrete be. It becomes liable to ingress of environmental elements which cause disintegration by direct chemical attack or by periodical volume changes on account of moisture and temperature variations. Water freezing within the pores causes disintegration of concrete. Unsound aggregates generate alkali-aggregate reaction causing disintegration of concrete by volume expansion. Durable concrete can be produced by using sound and dense aggregates, sufficiently low water/cement ratio and thorough compaction of concrete. Cement content should be adequate to ensure proper workability with a low water/cement ratio so that the concrete may be fully compacted with the available means.

Recommended guide lines as per IS Code 456-1978 in terms of the minimum cement content and the maximum water/cement ratio to ensure durability under specified conditions of exposure and sulphate attack are given in Table 2.17 and 2.18 respectively.

2.3.3 Workability of freshly mixed concrete. Compaction of concrete is achieved by elimination of entrapped air untill it achieves as close a configuration as is possible. Energy is expended in overcoming friction between particles in concrete and also between concrete and the surface of the mould and that of the reinforcement. These are called internal friction and surface friction respectively. The former is the intrinsic property of the mix. Hence, workability is defined as the amount of useful internal work required to overcome internal friction in order to produce full compaction.

The major factor affecting workability is the quantity of water in the mix. An increase in the amount of cement must require an increase in the amount of water to maintain the workability.

Table 2.17

REQUIREMENTS OF DURABILITY AGAINST SPECIFIED CONDITIONS OF
EXPOSURE

(Based on Table 19, IS Code: 456-1978)

Type of exposure	Plain concrete		Reinforced concrete	
	Minimum cement content, kg/m³	Maximum w/c ratio by weight	Minimum cement content, kg/m³	Maximum w/c ratio by weight
Mild—completely protected against weather, or aggressive conditions, except for a brief period of exposure to normal-weather conditions during construction	220	0.70	250	0.65
Moderate—sheltered from heavy wind-driven rain and against freezing, whilst saturated with water; buried concrete in soil, and concrete continuously under water	250	0.60	290	0.55
Severe—exposed to sea-water, alternate wetting and drying and to freezing whilst wet; subject to heavy condensation or corrosive fumes	310	0.50	360	0.45

Notes:-1 When the maximum w/c ratio can be strictly controlled, the cement content may be reduced by 10 per cent

2 Minimum cement content is based on 20-mm aggregate; for 40 mm aggregate reduce it by 10 per cent, and for 12.5-mm aggregate, increase it by 10 percent

Table 2.18

REQUIREMENT OF DURABILITY OF COCRETE EXPOSED TO SULPHATE
ATTACK

(Based on Table 20 IS Code: 456-1978)

Class	Concentration of Sulphates Expressed as SO_3			Type of Cement	Requirements for Dense, Fully Compacted Concrete Made with Aggregates Complying with IS: 383-1970	
	In Soil		In Ground water			
	Total SO_3 (Percent)	SO_3 in 2:1 Water Extract g/L	(Parts per 100 000)		Minimum Cement Content	Maximum Free Water /Cement Ratio
(1)	(2)	(3)	(4)	(5)	(6)	(7)
					kg/m³	
1.	Less than 0.2	—	Less than 30	Ordinary Portland cement or Portland slag cement or Portland pozzolana cement	280	0.55
2.	0.2 to 0.5	—	30 to 120	Ordinary Portland cement or Portland slag cement or Portland pozzolana cement	330	0.50
				Supersulphated cement	310	0.50
3.	0.5 to 1.0	1.9 to 3.1	120 to 250	Supersulphated Cement	330	0.50

Note 1 — This table applies only to concrete made with 20 mm
aggregates complying with the requirements of IS: 383-1970 placed

in near-neutral ground waters of pH 6 to pH 9, containing naturally occurring sulphates but not contaminants such as ammonium salts. For 40 mm aggregate the value may be reduced by about 15 percent and for 12.5 mm aggregate the value may be increased by about 15 percent. Concrete prepared from ordinary Portland cement would not be recommended in acidic conditions (pH 6 or less). Supersulphated cement gives an acceptable life provided that the concrete is dense and prepared with a water/cement ratio of 0.4 or less, in mineral acids, down to pH 3.5.

Note 2 — The cement contents given in Class 2 are the minimum recommended. For SO_3 contents near the upper limit of Class 2, cement contents above these minimum are advised.

Note 3 — Where the total SO_3 in col 2 exceeds 0.5 percent then 2 : 1 water extract may result in a lower site classification if much of the sulphate is present as low solubility calcium sulphate.

Note 4 — For severe conditions such as thin section under hydro-static pressure on one side only and sections partly immersed, considerations should be given to a further reduction of water/cement ratio, and if necessary an increase in the cement content to ensure the degree of workability needed for full compaction and thus minimum permeability.

Note 5 — Portland slag cement conforming to IS: 455-1976 with slag content more than 50 percent exhibits better sulphate resisting properties.

Note 6 — Ordinary Portland cement with the additional requirement that C_3A content be not more than 5 percent and $2 C_3 A + C_4 AF$ (or its solid solution $4 CaO, Al_2O_3, Fe_2O_3 + 2 CaO, Fe_2O_3$) be not more than 20 percent may be used in place of supersulphated cement.

To minimise the possibilities of deterioration of concrete from harmful salts, the levels of such salts in concrete coming from cement, aggregates, water and admixtures as well as by diffusion from environment should be restricted. Generally, total amounts of chlorides (as Cl) and sulphates (as SO_3) in concrete at the time of placing should be restricted respectively to 0.15% and 4% of the weight of cement.

Finely ground cement requires more water to produce the same workability as coarsely ground cement. The grading of agregate which gives the smallest surface area for a given amount of aggregate requires the least amount of water for given workability. The larger the maximum aggregate size, the coarser is the grading resulting in smaller surface area. As such, the amount of water required for workability is also smaller. The advantage of using a large maximum aggregate size is considerable with lean mixes. However, the risk of seggregation is increased. Hence, the mix must be kept dry. Risk of seggregation increases with coarser grading, increase in maximum aggregate size, and with increase in the amount of water. Workability is governed by the aggregate/cement ratio and water/cement ratio for a given type of aggregate and its grading.

In the overall context workability involves ease of placing and compacting concrete under the particular circumstances, absence of seggregation and bleeding, and the degree of cohesion. Workability depends on the amount of cement, fineness of cement and its chemical composition, the amount of water, the grading and shape of the fine aggregate, the grading, shape and surface texture of coarse aggregate, the ratio of coarse to fine aggregate, and the presence of admixtures and entrained air.

Fine aggregate has a greater influence on workability than coarse aggregate. An increase in the proportion of fines requires an increase in the amount of water to maintain workability. A rounded river sand imparts more workability than an angular sand obtained from crushed rock. A rounded coarse aggregate also produces more workable concrete than an angular aggregate. A smaller amount of sand is required in a mix having rounded aggregate, beyond that required to fill the voids, in order to promote free movement between the coarser particles. Air entrainment also promotes workability. The effect is more pronounced with lean and harsh mixes. Air entrainment reduces the tendency for seggregation and bleeding, and as such, helps workability.

Measurement of workability : No test is known to directly measure the workability as defined. Attempts have been made, however to correlate workability with some easily determinable physical measurement. Though not fully satisfactory, these provide useful information within a range of variation in workability. It is important to make the tests for the maintainence of even workability, to ensure uniform quality as well as to effect economies by adjustment of mix proportions. some of these tests are described below.

(a) *The slump Test :* This test is carried out with an apparatus shown in Fig. 2.6. The mould A is in the form of a frusium of a cone having an upper diameter of 100 mm, a lower diameter of 200 mm and a height of 300 mm. The mould is placed on a smooth surface and filled with freshly mixed concrete in four equal layers, each layer being compacted with 25 strokes of a standard punner. The top is struck off level and the cone lifted immediately. The amount by which concrete slumps is measured by a scale. This test is applicable for concretes

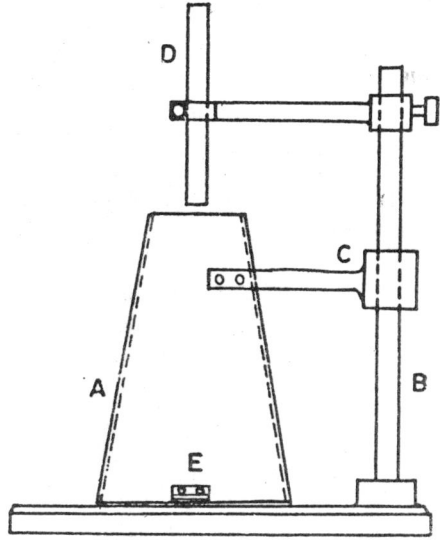

Fig. 2.6 Slump test apparatus

with maximum aggregate size not exceeding 38 mm and for slumps ranging between 13 mm and 150 mm. It is not suitable for either dry or very wet concretes.

There are three main forms of slump which are shown in Fig. 2.7 These are : true slump, shear slump and collapse slump. shear slump is irregular and unsatisfactory from the point of view of measurement. Both shear slump and the collapse slump indicate that concrete is unsatisfactory for placing.

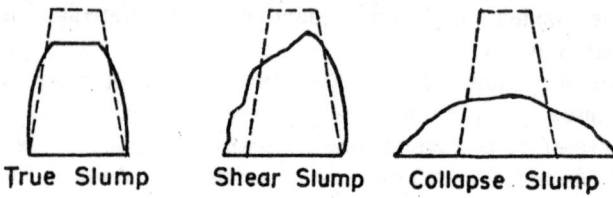

True Slump   Shear Slump   Collapse Slump

Fig. 2.7 Types of slump

The slump may be irregular and may not agree with the observed workability.   It is difficult to see how it is related   to workability. However, this test is useful as a means to control   concrete on the site and to detect   differences in the water content of successive identical mixes.

(*b*) *Compacting factor test:*   In this test a given amount of

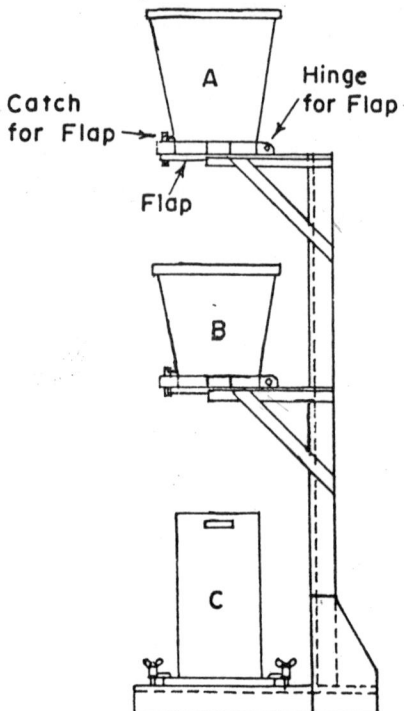

Fig 2.8 Compacting factor apparatus

work is applied to a given quantity of concrete to produce compaction by overcoming internal friction. The work lost in overcoming friction between concrete and the containing surfaces is minimised. This test is a good measure of the workability. It is more sensitive and gives more consistent result than the slump test. Even dry mixes which show no slump can be tested by this method.

The apparatus is illustrated in Fig. 2.8. It can be used for testing concrete having maximum aggregate size upto 38 mm. The apparatus consists of two conical hoppers A and B fitted with doors at their bases, and a cylinder C. The top hopper is first filled with the freshly mixed concrete to be tested and then the door at its bottom is opened to allow concrete to fall into the lower hopper. The bottom door of the lower hopper is then opened, allowing concrete to fall into the cylinder C.

The concrete is struck off level with the top and the cylinder is weighed (weight = w_1). The cylinder is emptied and refilled with concrete taken from the same sample and heavily compacted preferably by vibration. The top is struck off level and the cylinder with fully compacted concrete is weighed (weight = w_2). Weight of empty cylinder is also found. (weight = w).

Then

$$\text{Compacting factor} = \frac{w_1 - w}{w_2 - w}$$

Weight of fully compacted concrete can also be determined from concrete mix proportions and the specific gravities of the constituents.

(c) The V-B test : The apparatus consists of a slump cone placed in a cylindrical pan. Freshly mixed concrete is filled in the cone which is then withdrawn. Concrete slumps. A transparent circular disc is brought in contact with the top of concrete and the cylindrical pan is vibrated. The concrete gets remoulded into cylindrical form Remoulding is considered to be complete when the concrete is in complete contact with the disc. The time· in seconds noted for the complete remoulding operation is quoted as the number of V-B seconds.

The concrete mix proportion should have adequate workability so that it may be easily placed and compacted with the means available. Suggested ranges of the values of workability of concrete for some placing conditions are given in Table 2.19.

Table 2.19

SUGGESTED RANGES OF WORKABILITY FOR SOME PLACING
CONDITIONS OF CONCRETE
(REF : IS CODE : 456 — 1978)

Placing conditions	Degree of workability	Values of workability
Concreting of shallow sections with vibration	Very low	20-10 seconds, Vee-bee time or 0.75-0.80, compacting factor
Concreting of lightly reinforced section with vibration	Low	10-5 seconds, Vee-bee time or 0.80-0-85, compacting factor
Concreting of lightly reinforced sections without vibration, or heavily reinforced section with vibration	Medium	5-2 seconds, Vee-be time or 0.85-0.92, compacting factor or 25-75 mm, slump for 20 mm* aggregate
Concreting of heavily reinforced section without vibration	High	Above 0.92, compacting factor or 75-125 mm, slump for 20 mm* aggregate

*For smaller aggregate the values will be lower.

2.3.4 Proportioning of concrete mix.

(a)   Trial - mix method : This is a simple method which is suggested for both field and laboratory work. It suits all types of aggregate. It is assumed that provided full compaction is achieved an appropriate water/cement ratio will result in concrete of the required characteristic strength. Trial value of water/cement ratio may be chosen from Fig. 2.5. Both coarse aggregate and fine aggregate are graded. The objective is to determine a concrete mix

of the required workability and strength. The cost of a mix depends upon its cement content. Hence, the proportions of coarse and fine aggregates which need the least amount of cement for a given water/cement ratio are found by making trial mixes and testing them for the required workability and strength. Sometimes a single trial mix may be successful. Otherwise, more such mixes may be made and tested.

(b) Maximum density method : This method uses a theoritical approach to determine the complete grading of the ingredients. Fuller's formula is used for arriving at the proportion of different particle sizes beginning with the largest aggregate size and going down to the smallest size including particles of cement. It is written as

$$p = 100 \sqrt{\frac{d}{D}}$$

where

d = a particular particle size

D = maximum size aggregate

p = percentage of particles smaller than size d.

For a mix having 40 mm as the maximum size of aggregates, the percentage of particle sizes smaller than 20 mm is given by

$$p_{20} = 100 \sqrt{\frac{20}{40}}$$
$$= 70.7\%$$

The formula assumes that the unit weights of coarse aggregate, fine aggregate and the cement are the same, which is incorrect. For a given maximum size aggregate the proportion of fine aggregate along with the cement works out to be a fixed value. Also, the aggregate/cement ratio becomes fixed. For a given strength, when water/cement ratio is chosen, the workability of the mix also becomes fixed. Hence, by this method the requirements of both strength and workability cannot be fulfilled simultaneously in most of the cases. This is a serious drawback of the method.

(c) Minimum voids method : This method is based on an experimental approach. The voids in the coarse aggregate should be filled in by the fine aggregate and the voids in the latter should be filled in by the cement paste. A little extra quantity of fine aggregate as well as that of cement paste is needed to impart

flowing property to the mix as well as to make up for the extra voids created by wedging action of the filling material on the material being filled

(d) Fineness modulus method : To determine the fineness modulus of an aggregate the IS Sieves designation used begin from 150 micron, each higher size being twice that of the preceeding designation. The cumulative percentages of material retained on each sieve are added and divided by 100, and thus, the fineness modulus is obtained. Maximum and minimum grading limits expressed in terms of fineness modulus as recommended by Abrams are given in Table 2.20. Aggregates having different fineness modulii can be combined to produce a mixed aggregate of repuired fineness modulus. Let

A = fineness modulus of coarse aggregate

B = fineness modulus of fine aggregate

C = required fineness modulus of the mixed aggregate

p = the percentage of the fine aggregate in the mixed aggregate

Then

$$p = \frac{A-C}{A-B} \times 100$$

Table 2.20

LIMITS OF FINENESS MODULUS RECOMMENDED BY ABRAMS

Type of aggregate	Maximum size of Aggregate mm	Limits of Fineness Modulus	
		Maximum	Minimum
Fine		3.5	2.0
Coarse	20	6.9	6.0
	40	7.5	6.9
	75	8.0	7.5
	150	8.5	8.0
Mixed	20	5.1	4.7
	25	5.5	5.0
	40	5.9	5.4
	75	6.3	5.8
	150	7.0	6.5

(e) Proportioning aggregates to obtain a type grading : Coarse and fine aggregates conforming to their respective gradings specified

in the IS Code : 383 can be mixed in a certain proportion so as to yield a mixed aggregate of the desired grading.

Example :

Given batches of coarse aggregates are to be mixed with a batch of fine aggregate to yied a mixed aggregate. The observed gradings of the coarse and the fine aggregates as well as the required grading of the mixed aggregate are given in Table 2.21. Calculate the proportion in which the four aggregates should be mixed together.

Table 2.21

GRADINGS OF THE AGGREGATES

(% PASSING THE SIEVE SIZE)

IS Sieve Designation	Coarse Aggregate			Fine Aggregate	Mixed Aggregate	
	20 mm	12.5-10 mm	10-6.3 mm		Required	Obtained
40 mm	100	100	100	100	100	100
20 mm	100	100	100	100	100	100
10 mm		18.96	98.45	100	55 - 65	58.98
4.75 mm				100	34 - 41	34.00
2.36 mm				96.54	28 - 35	32.82
1.18 mm				73.75	20 - 28	25.08
600 micron				53.42	12 - 20	18.16
300 micron				9.60	3 - 5	3.26
150 micron				1.96	0 - 1	0.67

Let the four types of aggregate be mixed in the proportion by weight given by a:b:c:d to produce the mixed aggregate of the desired grade. Then, following relationships are obtained considering 1 Kg of the mixed aggregate :

For material passing 20 mm sieve
$$a + b + c + d = 1.0$$

For material passing 10 mm sieve
$$0.1896b + 0.9845c + d = 0.65$$

For material passing 4.75 mm sieve
$$d = 0.34$$

The solution works out to
$$b = 0.428 - 1.239a$$
$$c = 0.232 - 0.239a$$
$$d = 0.34$$

Hence, the proportion of the aggregates can be written as
a ; (0.428 — 1.239a) : (0.232 + 0.239a) : 0.34

Any value of a less than $\dfrac{0.428}{1.239}$ i.e, 0.345 can be chosen to work out the proportions.

Choosing a = 0.2, the proportions work out to
0.2 : 0.18 : 0.28 : 0.34
or   1 : 0.9 : 1.4 : 1.7

The grading of the mixed aggregate obtained with these proportions is shown in the last column of Table 2.21. Evidently, it satisfies the requirement of grading.

(f)   R.R.L method: This method was developed by the Road Research Laboratory United Kingdom. It evolved as a result of experimental study and research based mainly on parameters such as nominal maximum size aggregate viz. 40 mm, 20 mm and 10 mm, grading of aggregate, type of aggregate viz. rounded, irregular gravel or crushed rock, workability, strength, durability and type of cement.

Four type grandings have been adopted for each of the maximum size aggregate i.e. 40 mm, 20 mm and 10 mm. However, these are not the ideal gradings but only the ones which were actually used for the tests. These gradings have been presented in both tabular and graphical forms. Workability has been catagorised as very low, medium and high. The parameters deciding the degree of workability were taken to be maximum size aggregate viz. 40 mm, 20 mm and 10 mm, the grading of aggregate, the type of aggregate viz. rounded, irregular gravel or crushed rock, the water/cement ratio, and the aggregate/cement ratio. Values of the aggregate/cement ratios required to give the four degrees of workability for the three maximum aggregate size of the three types, have been presented in both graphical and tabular form, and can be read out for different water/cement ratios for the four types of gradings of the aggregates. Graphs have also been presented showing the relationship between the compressive strength of concrete and the water/cement ratio for both ordinary and the rapid hardening cement.

The steps in the design of a concrete mix by this method are as follows :-

(i)   Fix up the laboratory design strength required of a mix from the specified characteristic strength.

(ii) Fix up the water/cement ratio from the consideration of both strength and durability.

(iii) Fix up the degree of workability required according to the conditions at the site.

(iv) Perform the sieve analysis of the defferent size aggregates and fix up their gradings.

(v) Determine the proportions of different size aggregates in order to give a combined grading reasonably agreeing with a type grading.

(vi) Determine the aggregate/cement ratio from the appropriate graph or table.

A trial mix is then made and tested to check whether it satisfies the requirements. Otherwise, the mix may be modified by altering the relevant parameters.

For more details reference may be made to a standard text book on Concrete Technology.*

2.3.5 Sampling and test strength : Sampling should provide a reasonable representation of the concrete batch under consideration. Strict adherence to statistical random sampling may not be possible in practice. However, sampling procedure shall be adopted to ensure that every batch of concrete shall have a reasonable chance of being sampled and tested. As such, the sampling should cover all mixing units and should be spread over the entire period of concreting.

According to the IS : 1199 — 1959 the sample should be at least 0.02 m³ in volume made up by collecting concrete from 3 locations. This volume is about twice that required for making three specimens of 150 mm cube. Three test specimens shall be made from each sample for testing at 28 days. For other purposes, additional cubes may be made. According to the IS : 516 — 1959 the individual variation in the compressive strength of cubes in a sample should not be more than $\pm$ 15 percent of the average. Strength of the specimens even if made from the same sample differ from each other on account of the heterogeneous nature of concrete and the human element involved in obtaining the sample and in casting, curing and testing of specimen.

---

*Concrete Technology by D.F.Orchard, vol 1, published by John Wiley & Sons.

The minimum frequency of sampling of concrete according to the IS : 456 — 1978 shall be as follows :-

Volume of concrete in the work, m³	Number of samples
1 to 5	1
6 to 15	2
16 to 30	3
31 to 50	4

Note:—Take out one additional sample for each additional volume of upto 50 m³.

2.3.6 Standard deviation. The standard deviation is an index of the scatter of the test results from the mean value. As such, it reflects the degree of control exercised in the manufacture of concrete. The plot of compressive strength against the frequency (Nos. of samples) ideally follows the Normal distribution. See Fig. 1.2. One can look upon the mean strength as a measure of central tendency and the standard deviation as a measure of dispersion. A curve with a dominant peak reflects a small value of standard deviation, whereas, a flatter curve reflects a large value of the same.

For the purpose of specification and design reference is made to the characteristic strength of concrete. However, considering the inherent variability of concrete strength during production, it is necessary to design the concrete mix for a target mean strength. From Eq. (1.6), one can write

$$\mu = f_{ck} + zs \qquad (2.2)$$

where,

μ = target mean strength

f_{ck} = characteristic strength

s = standard deviation

z = a constant equal to 1.65 meaning that not more than 5% of the test results may fall below the characteristic strength.

The number of test results, required for obtaining a reliable value of the standard deviation, shall not be less than 30. The standard deviation of concrete shall be calculated from the results of individual tests as follows :

$$s = \sqrt{\frac{\Sigma(x - \overline{x})^2}{N - 1}} \qquad\qquad (2.3)$$

where

x = observed strength of a specimen

$$\overline{x} = \frac{\Sigma x}{N}$$

N = total number of test specimens.

where sufficient number of test results are not available for concrete, the values of standard deviation may be assumed as specified in Table 2.22.

Table 2.22

SUGGESTED STANDARD DEVIATION
(BASED ON TABLE 6. IS CODE : 456 — 1978)

Concrete grade	Standard deviation N/mm²
M10	2.3
M15	3.5
M20	4.6
M25	5.3
M30	6.0
M35	6.3
M40	6.6

2.3.7 Acceptance criteria. The concrete shall be considered to comply with the strength requirements if one of the following criteria are satisfied :

(i) For every specimen

$x \not< f_{ok}$

(ii) For any specimen

$x \not< f_{ek} - 1.35s$

$\not< 0.8\, f_{ck}$

provided

$$\overline{x} \not< f_{ek} + \left(1.65 - \frac{1.65}{\sqrt{N}} \right)s$$

However, concrete shall be deemed not to comply with the strength requirements if

$$x < f_{ck} - 1.35\ s$$
$$< 0.8\ f_{ck}$$

or if $\overline{x} < f_{ck} + \left(1.65 - \dfrac{3}{\sqrt{N}} \right) s$

2.3.8 Nominal mix concrete. Nominal mix concrete may be used upto grade M20 only. The proportion of the fine aggregate to coarse aggregate by mass is generally kept 1:2. This proportion is adjusted from an upper limit of $1:1\frac{1}{2}$ to a lower limit of $1:2\frac{1}{2}$ as the maximum size of the coarse aggregate becomes larger and the grading of fine aggregate becomes finer. Proportions by weight for nominal mixes are given in Table 2.23.

Table 2.23

PROPORTIONS FOR NOMINAL MIX CONCRETE
(BASED ON TABLE 3 IS CODE: 456—1978)

Concrete grade	Aggregate/cement ratio	Water/cement ratio
M5	16	1.2
M7.5	12.5	0.9
M10	9.6	0.68
M15	7	0.64
M20	5	0.6

2.4 COMPACTION OF CONCRETE

Concrete should be placed in the form work immediately after the completion of mixing. Dropping from great heights may cause segregation, entrainment of air, displacement of reinforcement and damage to the already placed concrete. However, dropping of concrete from a height of 10 to 15 metres is not uncommon.

Compaction is achieved by removal of the entrapped air from the concrete and thus obtaining as dense a mass as possible. Friction between the concrete particles, between concrete and reinforcement and concrete and the mould surface hampers compaction. To reduce this friction it is necessary to add more water than required to hydrate the cement. Excess water forms voids which reduce the strength of concrete. Concrete can be compacted by the following methods:

(i) Punning with rods.

(ii) Application of vibration internally and/or externally

(iii) Application of centrifugal force by spinning (pre-cast units only)

(iv) Application of pressure on concrete (pre-cast unit only)

(v) Application of shock by dropping the mould through a small height (pre-cast units only)

The method of vibration is the most popular one. Due to oscillation of the particles the fluidity of concrete increases and the friction is reduced. Hence, concrete gets compacted easily.

Higher degree of compaction achieved through vibrations primarily enables a lower water/cement ratio to be used which means an increase in strength of concrete. Further advantages are : reduced shrinkage and creep, greater density and reduced permeability, improved durability, higher elastic modulus and enhanced bond with reinforcement.

(a) Internal vibrators. The interior of a concrete mass can be vibrated by means of an immersion vibrator. It is in the form of a metallic tube of diameter ranging from 38 mm to 102 mm. An eccentric rotating mass within the tube causes vibration. The mass is driven by an electric motor through a flexible shaft. The frequency of vibration is normally 6000 cycles per minute. The normal radius of action of an immersion vibrator is approximately 10 times the tube diameter. The depth of one concrete layer should not exceed 600 mm.

(b) External vibrators. External vibrators are clamped to the mould or to the shuttering. Part of the energy is consumed in vibrating the mould or the shuttering and part is lost due to damping effect of the form work. As such external vibrators consume more power, Compaction can be achieved up to a distance of 450 mm from the face of shuttering. They are extensively used for pre-cast units such as pipes and poles, thin walls and columns. Shutter vibrators are of eccentric mass type and are driven by electric motor with frequency upto 3000 cycles per minute. Electro magnetic type vibrators are also used. With external vibrators the suttering has to be very strong.

(c) *Vibrating table* Vibrating tables may be driven by an electro mangnetic vibrator or by an electric motor utilizing an eccentric mass. Frequency of vibration may vary from 3000 to 6000 cycles per minute. The moulds are kept on the table and filled up progressively after the vibration begins. Sometimes vertical pressure can be applied on the top surface of concrete it so needed.

(d) *Surface vibrators.* Surface vibrators are applied on the exposed top surface of concrete. There are Pan vibrators which consist of steel pan with a motor mounted on its top. An eccentric mass driven by the motor produces vtbration. Frequencies from 3000 to 6000 cycles per minutes are common. They are used for compaction of small slabs. However, the thickness of concrete should not exceed 150 mm. Vibrating beams are also used for compacting concrete in slabs. Normally, one driving unit is needed for every 1.8 m length of vibrating beam. The latter may be upto 5.5 m long. It can be handled by two persons hoding it at the opposite ends. Thickness of concrete to be compacted should not normally exceed 150 mm. High degree of compaction can be achieved since the vibration acts In· the same dtrection as gravity.

2.5 CURING OF CONCRETE

After placing and compaction of concrete it is cured under an environment favourable to setting nnd hardening, for a period of time. Curing involves maintenance of a suitable level of temperature and humidity governing the rate of hardening, prevention of loss of moisture and protection from premature stressing.

Concrete can be cured in various ways depending upon the requirements of the situation. Curing of concrete by ponding of water is the most eflective method applicable to flat horizontal slabs. Where ponding is not possible, the concrete surface may be covered by jute or cotton mats which are kept wet. Continuous spraying of water on uncovered concrete surface may not be verv effective. However, these methods are inconvenient and expensive on account of supervision and labour involved. In regions where there is a shortage of water, these methods may be impossible to apply. Hence, alternative methods of curing have been developed based on the principal of preventing the evaporation of water initially held by concrete. Thus, concrete can be effetively cured by covering its surface with water proof paper. However, any loss of moisture taking place before covering of the surface can not be made good. Sometimes the shuttering is kept on to prevent the loss of moisture by evaporation. If calcium chloride is mixed with concrete it accelerates the rate of hardening. Hence, the curing time is considerably reduced with the normal methods. In a moist environment, if calcium chloride is applied on the exposed surface of concrete, it may keep it moist on account of hygroscopic action. However, in an environment of low humidity, calcium chloride will draw water from concrete itself. Hence, this method is the least effective. Spraying concrete surface with sodium silicate does not effectively inhibit evaporation of water. The rate of laying concrete is high in airfield construction. Hence, evaporation is effectively prevented by applying an impervious membrane on top by spraying. Curing of concrete by application of heat while maintaining moist environment is mainly uied in the manufacture of precast products in a factory. These methods include low pressure steam curing, high piessure steam curing and electrical curing. In a large mass of concrete, curing mainly consists of removal of heat by circulating cooling water through pipes laid in concrete.

2.6 ADMIXTURES

Concrete often incorporates admixtures in addition to cement, aggregates and water. Admixtures are added immediately before or during mixing to improve the properties of freshly mixed or hardened concrete.

Pozzolanas like fly ash and burnt clay particularly improve the workability of concrete and thereby reduce the water required for a given workability. However, because of the pozzolanic reactions these materials can partly replace the cement. As such, they are added in much larger quantities than the admixtures.

Some admixtures are likely to contain water soluble chlorides and sulphates. If present in large quantities, chlorides may cause corrosion of steel reinforcement while sulphates may cause disintegration of concrete by forming of sulphoaluminates. Different types of admixtures covered by the Indian Standards are as follows:

(i) **Accelerators** These materials when added to concrete increase the rate of hydration and the rate of strength development while reducing the setting time. Soluble chlorides, carbonates, silicates fluoro-silicates and hydroxides are some inorganic admixtures. Triethanolamine is an organic admixture.

Acceleiators are particularly useful when concrete is to be placed in cold weather. Calcium chloride up to a maximum of 1.5% by weight of cement is recommended for concreting at low temperature. Not only it accelerates the rate of strength development but also increase the strength and abrasion resistance.

(ii) Retarders These substances delay the setting of cement. As such, they are used in hot weather when the normal setting time of cement is reduced due to high temperature.

Sugar, carbohydrate derivatives, soluble zinc salts and soluble borates etc. exhibit retarding action. Sugar about 0.05% by mass of cement delays the setting time of concrete by 4 hours. However, the action of sugar depends upon the chemical composition of the cement. As such, trial mixing is required before it is used in construction.

(iii) Water-reducing agents These substances increase the workability of concrete. They belong to the following two main groups:

(a) Lignosulphonic acids and their salts
(b) Hydroxylated carbolic acids and their salts.

The setting time is increased by about 2 to 6 hours. As such, the concrete can be vibrated, revibrated and finished. It is particularly advantageous in hot weather and where construction demands a time gap between the successive placing of concrete. The set-retarding property of this substance can be offset by incorporating an accelerator like calcium chloride or triethanolamine.

(iv) Air entraining agents These substances incorporate minute air bubbles in concrete during mixing in order to increase workability and the resistance to freezing and thawing. They are prepared from the following original materials:

(a) Natural wood resins
(b) Animal fats or vegetable oils
(c) Wetting agents like alkali salts of sulphated and sulphonated organic compounds
(d) Water soluble soaps of resin acids and animal or vegetable fatty acids

(e) Sodium salts of petroleum sulphonic acids, hydrogen peroxide and aluminium powder etc.

The entrained air bubbles restrict the effective length of capillary pores in concrete, and as such, reduce the capillary forces. They do not get filled with water from capillaries because of surface tension effects. Under freezing conditions they accommodate the ice formed. When the ice melts the water is drawn off into the capillaries by surface tension. Thus, the con crete is protected against frost damage.

3

Singly Reinforced Beam

3.1 INTRODUCTION

The analysis of structure for the internal stress resultants which are created by any prescribed loading can be performed conveniently by a suitable method. In a structure, a beam is that member which bears bending moments and shear forces on account of transverse loading or otherwise.

In a beam, though concrete can safely bear flexural compressive stress, it is incapable of resisting the flexural tensile stress since the tensile strength of concrete is very low compared with the compressive strength. Tensile strength of concrete is of the order of one-tenth of the compressive strength. As such, steel bars are provided in the tensile zone of the beam to reinforce the concrete there. Thus, a singly reinforced concrete beam is obtained. See Fig. 3.1(a). Compressive force in concrete in the compression zone together with an equal and opposite tensile force in the steel reinforcement form a couple which resists the external bending moment at any section of the beam.

While designing a reinforced concrete beam it is necessary to keep the magnitude of flexural stresses in concrete and steel reinforcement within the permissible limits in order to ensure adequate margin of safety and proper serviceability.

3.2 PERMISSIBLE STRESSES

The permissible stress in a material is given by

$$\text{Permissible stress} = \frac{\text{Limiting strength}}{\text{Factor of safety}}$$

In the case of steel reinforcement the limiting strength is either the yield stress or 0.2 % proof stress, as the case may be. For concrete the limiting strength is the crushing strength in compression. The factor of safety in the case of tensile steel reinforcement is approximately $= 1.82$. Hence, the permissible tensile stress in steel is $\sigma_{st} \approx 0.55f_y$. For concrete the factor of safety is higher than for steel. This is so because concrete suffers from higher degree of variability regarding its strength and properties than steel which is produced under well controlled conditions. The factor of safety for flexural compressive stress in concrete is $= 3$. Thus, the permissible compressive stress in concrete in flexural compression is $\sigma_{cbc} = 0.333f_{ck}$. The permissible stresses in concrete and steel as discussed above, are given in Tables 3.1 and 3.2 below. These tables also give permissible stresses other than those discussed above, which shall be explained at an appropriate place later.

Table 3.1

PERMISSBLE STRESSES IN CONCRETE, N/mm^2

(BASED ON TABLE 15, IS CODE : 456-1978)

Concrete grade	M10	M15	M20	M25	M30	M35	M40
σ_{cbc}	3.0	5.0	7.0	8.5	10.0	11.5	13.0
σ_{cc}	2.5	4.0	5.0	6.0	8.0	9.0	10.0
τ_{bd}		0.6	0.8	0.9	1.0	1.1	1.2

Table 3.2

PERMISSIBLE STRESSES IN STEEL REINFORCEMENT, N/mm^2

(BASED ON TABLE 16, IS CODE: 456—1978)

Type of stress	Grade of Steel		
	Mild steel bars	HYSD bars grade Fe 415	HYSD bars grade Fe 500
(i) Tensile σ_{st} or σ_{sv}			
Bar dia. <20 mm	140	230	275
Bar dia. >20 mm	130		
(ii) Compressive σ_{sc}			
In columns	130	190	190
In beams	130	190	190

Or the calculated compressive stress in surrounding concrete multiplied by $1.5m$, whichever is less.

3.3 BENDING THEORY OF SINGLY REINFORCED RECTANGULAR BEAM

The theory of bending for reinforced concrete beam is based on the elastic theory of bending of a prismatic beam. The following assumptions are made:

(i) Sections of beam which are plane before bending remain plane after bending. The implication is that in a beam section the strain due to bending at any point is proportional to its distance from the neutral axis. Hence, the strain variation along the depth of beam is linear as shown in Fig. 3.1(b). At the neutral axis the strains are zero. Above the neutral axis the strains are compressive and below the neutral axis the strains are tensile.

(ii) Concrete does not bear any tensile stress which is taken up by the steel reinforcement only.

(iii) The stress-strain relationship for concrete as well as for steel is linear within the limits of the permissible stresses. This implies that the stress variation along the depth of beam is also linear as shown in Fig. 3.1(b). At the neutral axis the stress is zero. Above the netural axis the concrete has compressive flexural stress with the maximum value occuring at the top fibres. Below the neutral axis concrete is cracked, and as such, it does not bear tensile stress. This situation is indicated by the dashed line in the stress diagram. All the tension is taken up by steel reinforcement which has a tensile stress $= f_{st}$. In the stress diagram the tensile stress is shown to be $= \dfrac{f_{st}}{m}$ which is the equivalent stress in concrete at the level of steel reinforcement. The symbol m is the modular ratio which is $= \dfrac{E_s}{E_c}$. However, for the purpose of design, the IS Code: 456—1978 gives an empirical value viz., $m = \dfrac{280}{3\sigma_{cbc}}$. Calculated values of modular ratio m for different grades of concrete are given in Table 3.3 below:

Table 3.3

Modular ratio $m = \dfrac{280}{3\sigma_{cbc}}$

Concrete grade	M15	M20	M25	M30	M35	M40
m	18.67	13.33	10.98	9.33	8.12	7.18

It is also assumed that steel bars are perfectly bonded to concrete and do not slip.

Consider the strain diagram in Fig. 3.1(b). From the similarity of triangles,

$$\frac{\epsilon_c}{nd} = \frac{\epsilon_{st}}{(1-n)d}$$

Substituting $\epsilon_c = \dfrac{f_c}{E_c}$ and $\epsilon_{st} = \dfrac{f_{st}}{E_s}$ in the above equations,

$$\frac{f_c}{E_c nd} = \frac{f_{st}}{E_s(1-n)d} \quad \text{and hence,}$$

$$n = \frac{1}{1 + \dfrac{f_{st}}{mf_c}} \tag{3.1}$$

where $\qquad m = \dfrac{E_s}{E_c}$ the modular ratio

Eq. 3.1 gives the neutral axis factor n in terms of the extreme fibre compressive stress in concrete, the tensile stress in steel reinforcement and the modular ratio m.

Now consider the stress diagram in Fig. 3.1.(b). The concrete above the neutral axis has a uniformly varying compressive stress. Hence, the total compressive force is given by

$$C = \frac{1}{2}f_c bnd \tag{3.2}$$

The tensile force in the steel reinforcement is

$$T = f_{st}A_{st} \tag{3.3}$$

Since there is no external axial force on the section, these two forces must balance each other. Hence.

$$T = C$$

or $\qquad A_{st}f_{st} = \dfrac{1}{2}f_c bnd$

or $\qquad \dfrac{A_{st}}{bd} = \dfrac{nf_c}{2f_{st}}$

i.e. $\qquad p = \dfrac{nf_c}{2f_{st}} \tag{3.4}$

where $\qquad p = \dfrac{A_{st}}{bd}$ is the proportion of the tensile reinforcement in the section which is known as the tension reinforcement index.

The equal and opposite forces C and T are parallel to each other and are separated by a distance $=$ jd. They form a couple which resists the bending moment at the section. Hence, the bending strength is

$$M = Cjd$$

$$= \frac{1}{2} f_c \, bndjd$$

$$= \frac{1}{2} f_c \, njbd^2 \qquad (3.5)$$

Also
$$M = Tjd$$

$$= f_{st} \, A_{st} \, jd \qquad (3.6)$$

where
$$jd = d - \frac{nd}{3}$$

$$= \left(1 - \frac{n}{3}\right)d$$

Hence $\quad j = 1 - \frac{n}{3}$, which is known as the lever arm factor.

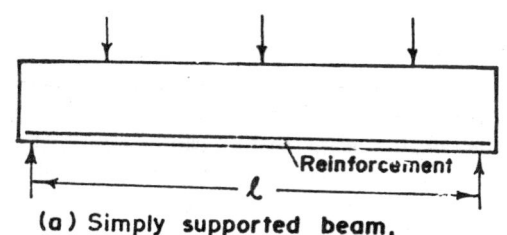

(a) Simply supported beam.

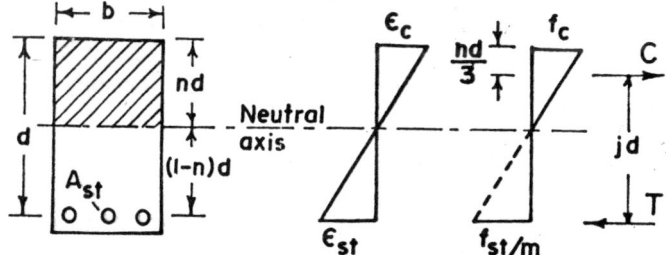

Cross − section   Strain Diag.   Stress Diag.

(b) Strain and Stress diagrams for a Section.

Fig. 3.1 Singly reinforced rectangular beam

On eliminating the ratio $\dfrac{f_{st}}{f_c}$ between the Eq. (3.1) and Eq. (3.4), a quadratic equation in n is obtained:

$$n^2 + 2mpn - 2mp = 0 \qquad (3.7)$$

This equation can also be obtained by considering that the neutral axis passes through the centroid of the effective section. Hence, the moment of the compressive concrete area about the neutral axis. is equated to that of the tensile reinforcement about the same:

$$\frac{bn^2 d^2}{2} = m\, A_{st}\,(1 - n)\,d$$

or $$n^2 = 2\,m\,\frac{A_{st}}{bd}\,(1 - n)$$

or $$n^2 = 2\,mp\,(1 - n)$$

i.e. $$n^2 + 2\,mpn - 2\,mp = 0$$

For the known values of the modular ratio m and the proportion of steel reinforcement p, Eq. (3.7) can be solved for n.

3.4 FORMULATION FOR BENDING STRENGTH

The bending strength of a beam section depends upon factors such as strengths of concrete and steel, the dimensions of section and the proportion of steel reinforcement. For the given grades of concrete and steel, a beam section can be classified into three types depending upon the proportion of steel reinforcement within the section. These three types are discussed below:

3.4.1 Balanced section. A reinforced concrete beam section is said to be balanced if the flexural extreme fibre compressive stress in concrete and the tensile stress in steel reinforcement reach their permissible values simultaneously at the design bending moment. Hence,

$$f_c = \sigma_{obc}$$

and $$f_{st} = \sigma_{st}$$

From Eq. (3.1) the neutral axis factor can be written for the balanced section as:

$$n = \frac{1}{1 + \dfrac{\sigma_{st}}{m\sigma_{cbc}}} \qquad (3.8)$$

From Eq. (3.4), the proportion of steel reinforcement for a balanced section can be written as:

$$p = \frac{n\sigma_{cbc}}{2\sigma_{st}} \qquad (3.9)$$

Evidently, a balanced section has a definite proportion of steel reinforcement which is given by Eq. (3.9).

From Eq. (3.5) the bending strength of a balanced section can be written as:

$$M_r = \frac{1}{2}\,\sigma_{cbc}\,njbd^2 \qquad (3.10)$$

$$= Rbd^2$$

Where, $R = \frac{1}{2}\sigma_{cbc}\,nj$ is known as the moment factor. It has a definite value for a balanced section. The bending strength can also be written from Eq. (3.6) as:

$$M_r = \sigma_{st}\,A_{st}\,jd \qquad (3.11)$$

Ready calculated values of n, p and R for balanced section, for different grades of concrete and steel reinforcement are given in Tables 3.4, 3.5 and 3.6 respectively.

Table 3.4

NEUTRAL AXIS FACTOR n FOR BALANCED SECTION

Concrete grade	Bar dia mm	Steel grade		
		Mild	Fe 415	Fe 500
M15	<20	0.4	0.289	0.253
	>20	0.418		
M20	<20	0.4	0.289	0.253
	>20	0.418		
M25	<20	0.4	0.289	0.253
	>20	0.418		

Table 3.5

PROPORTION OF STEEL REINFORCEMENT $p = \dfrac{A_{st}}{bd}$ FOR BALANCED DESIGN

Concrete grade	Bar dia. mm	Steel grade		
		Mild steel	Fe 415	Fe 500
M15	<20	0.00714	0.00314	0.0023
	>20	0.00804		
M20	<20	0.01	0.0044	0.00322
	>20	0.01125		
M25	<20	0.01214	0.00534	0.00391
	>20	0.01367		

Table 3.6

MOMENT FACTOR $R = \dfrac{1}{2}\,\sigma_{cbc}\,nj$ FOR BALANCED SECTION

Concrete grade	Bar dia. mm	Steel grade		
		Mild steel	Fe 415	Fe 500
M15	<20	0.867	0.653	0.579
	>20	0.9		
M20	<20	1.213	0.914	0.811
	>20	1.259		
M25	<20	1.473	1.11	0.985
	>20	1.529		

3.4.2 Under-reinforced section. An under-reinforced beam section contains less amount of steel reinforcement than a balanced section. In this section the tensile stress in steel reinforcement reaches the permissible value while the extreme fibre flexural compressive stress in concrete remains less than the permissible value, at the design bending moment. Also, the depth of neutral axis is less than that for the balanced section. In this case p $<$ the proportion of steel for balanced section

$f_{st} = \sigma_{st}$

$f_c < \sigma_{cbc}$

For the known values of p and m the neutral axis factor can be calculated from Eq. (3.7), that is

$$n^2 + 2\,mpn - 2\,mp = 0$$

From Eq. (2.6) the bending strength can be written as

$$M_r = \sigma_{st}\,A_{st}\,jd$$
$$= \sigma_{st}\,pbdjd$$
$$= \sigma_{st}\,pjbd^2 \qquad (3.12)$$

If a beam is to be designed as under-reinforced beam, then a value of p less than that for a balanced section is first chosen. Then, the neutral axis factor n is found from the quadratic equation. Finally, the section dimensions are found from the Eq. (3.12).

It would be in order at this stage to describe the behaviour of an under-reinforced beam tested to failure under gradually increasing transverse loading. As the design bending moment the steel attains full permissible stress while concrete has a stress less than the permissible value. On further loading, the steel reinforcement attains yield point stress at which the tensile force becomes constant i.e. $T = f_y A_{st}$, though the tensile strain keeps on increasing considerably. This results in large deflection of the beam and opening up of wide flexural cracks in the tensile zone. The compressive strain in concrete also keeps on increasing. As a result, the compressive stress diagram becomes non-linear. However, the total compressive force C remains constant just as the tensile force T remains constant. The only way the beam can resist increasing bending moment is by increase in the lever arm. Hence, the neutral axis shifts upwards reducing the area of concrete in compression when the strain in the extreme compression fibre of concrete reaches a value = 0.0035 which is the limiting strain at which crushing of concrete commences and the beam eventually collapses. See Fig. 3.2. This type of failure is called a tension failure since it is initiated by yielding of reinforcement. It is characterised by ductile behaviour of beam which is an important feature of an under-reinforced beam.

Such a behaviour gives adequate warning of impending collapse by way of large deflection and opening up of cracks in tension zone. Hence, in practice it is preferable to have an under-reinforced design.

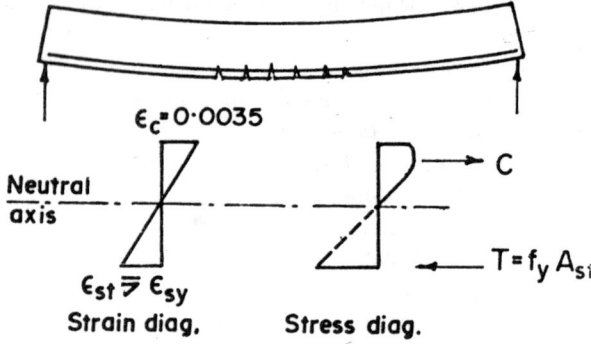

Fig. 3.2   Behaviour of an under-reinforced beam at collapse

3.4.3 Over-reinforced section. An over-reinforced beam section contains more amount of steel reinforcement than a balanced section. In this section the extreme fibre compressive stress in concrete reaches the permissible value while the tensile stress in steel reinforcement remains less than the permissible value, under increasing bending moment. Also, the depth of neutrel axis is more than that for the balanced section. Hence,

p > the proportion of steel for balanced section

$f_c = \sigma_{cbc}$

$f_{st} < \sigma_{st}$

For the known values of p and m the neutral axis factor n can be calculated from Eq. (3.7) i.e.,

$$n^2 + 2mpn - 2mp = 0$$

From Eq. (3.5) the bending strength can be written as,

$$M_r = \frac{1}{2}\,\sigma_{cbc}njbd^2 \tag{3.13}$$

Consider an over-reinforced beam subjected to gradually increasing transverse load. While the tensile stress in steel reinforcement may remain less than the yield point stress the concrete may reach its maximum capacity in compression. In other words, the extreme fibre compressive strain in concrete may reach a value = 0.0035 when the tensile reinforcement may still be stressed within the elastic range only. The beam thus reaches a limiting stage as far as the flexural compression in concrete is concerned. However, the deflections are very small and there is hardly any visible cracking of concrete in the tensile zone. See: Fig. 3.3. The beam suddenly collapses in a violent fashion by crushing of concrete in compression zone. This type of failure is called compression failure. It is characterised by brittle behaviour of beam. An over-reinforced beam does not give any warning of impending failure which may take place in a catastrophic manner. Hence, in practice the design of an over-reinforced beam should be avoided altogether.

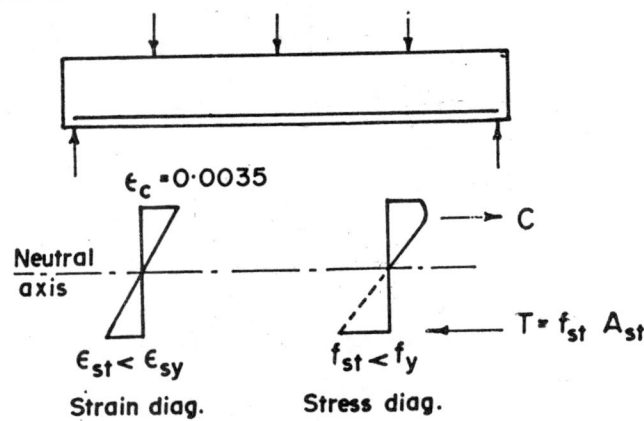

$\epsilon_c = 0.0035$

Neutral axis

$\longrightarrow C$

$\epsilon_{st} < \epsilon_{sy}$

$f_{st} < f_y$

$\longleftarrow T = f_{st} \, A_{st}$

Strain diag.   **Stress diag.**

Fig. 3.3 Behaviour of an over-reinforced beam just before collapse

3.5 DESIGN REQUIREMENTS OF THE IS CODE : 456-1978

The design procedure and the detailing of reinforcement within the cross-section of a beam are governed by the rules given in the IS Code 456-1978. Relevant clauses are mentioned in the following sub-paras.

3.5.1 Effective span. (i) For simply supported beam or slab, see Fig. 3. 4.

Effective span $l \not> l_o + d$
$\not> c/c$ of supports

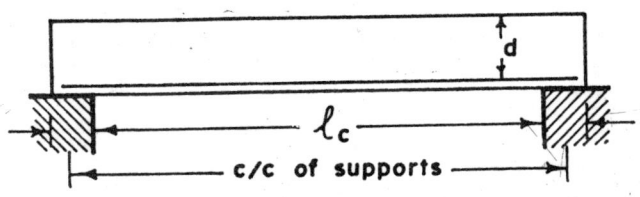

d

ℓ_c

c/c of supports

Fig. 3.4 Simply supported beam or slab

For a cantilever beam or slab,

      l = clear projection from face of support

(ii) Continuous Beam or Slab

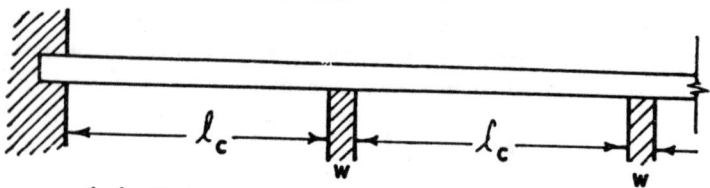

(a) End span having one end fixed

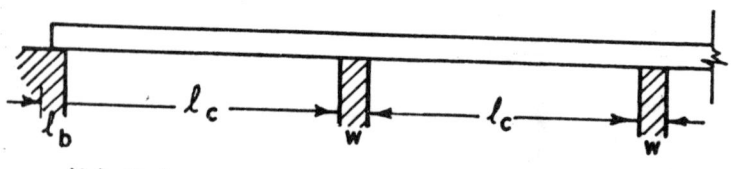

(b) End span having one end simply supported

Fig. 3.5 Continuous beam

Two cases arise depending upon the width of support.

(a) Width of support w $<$ l_c /12. Then

For end span or for intermediate span,

    effective span l $\not>$ l_c + d

                $\not>$ c/c of supports.

(b) Width of support w $>$ l_c/12

      or $>$ 600 mm. Then,

effective spans shall be taken as given in Table 3.7 below:-

Table 3.7
EFFECTIVE SPAN FOR CONTINUOUS BEAM

Condition of beam end in end span	Effective span	
	End span	Intermediate span
Fixed: Fig. 3.5 (a)	l_c	l_c
Simply Supported : Fig. 3.5 (b)	$\not> l_c + 1/2$ d	l_c
	$\not> l_c + 1/2 \ l_b$	

3.5.2 Minimum clear distance between bars. Clear horizontal distance between two parallel main bars shall not be less than the following:

(i) The diameter of the larger bar

(ii) Nominal maximum size of coarse aggregate + 5 mm

The above requirement may be reduced to two-thirds of the nominal maximum size of the coarse aggregate for the bars of a group provided sufficient space is left between groups of bars to immerse a needle vibrator in between.

When the bars are arranged in rows one over the other the bars shall be vertically in line. The clear vertical distance between bars shall not be less than the following :

(i) The diameter of the larger bar

(ii) Two-thirds the maximum size of aggregate

(iii) 15 mm

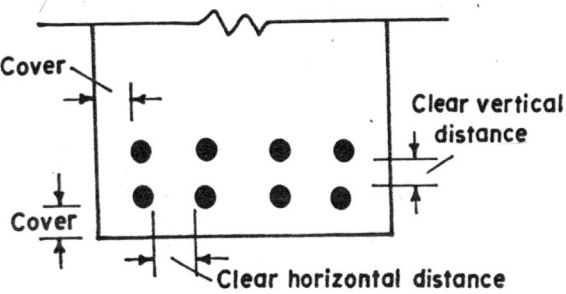

Fig. 3.6 Arrangement of reinforcement bars in a beam

The requirement of minimum distance between the bars ensures free movement of aggregates and the cement mortar around the reinforcement bars when concrete is being compacted in a beam. See Fig. 3.6.

3.5.3 Maximum clear distance between bars. Near the tension face of a beam the clear horizontal distance between the main reinforcement bars, or groups of bars shall not exceed the values given in Table 3.8, which depend upon the amount of redistribution of moment made to and from the section and upon the grade of reinforcement steel.

Table 3.8
MAXIMUM CLEAR DISTANCE BETWEEN BARS, mm

N/mm²	% Redistribution of moment at section				
	−30	−15	0	15	30
250	215	260	300	300	300
415	125	155	180	210	235
500	105	130	150	175	195

Note: The above table is based on Table 10 of IS Code 456-1978

3.5.4 Cover to reinforcement in a beam. A certain minimum thickness of concrete cover has to be provided over reinforcement bars in order to ensure protection of steel against the ingress of atmospheric elements causing corrosion. The cover would also ensure that there is concrete all around the reinforcement bars for proper bonding. See Fig. 3.6. The minimum requirement of cover to reinforcement in a beam according the IS Code: 456-1978 is as follows:

(i) For longitudinal reinforcing bar,
　　　　≮ 25 mm
　　　　≮ the diameter of bar

(ii) At the end of reinforcing bar,
　　　　≮ 25 mm
　　　　≮ twice the diameter of bar

(iii) For any other reinforcement,
　　　　≮ 15 mm
　　　　≮ the diameter of reinforcing bar.

An increase in cover thickness beyond the figures given above may be provided under special circumstances described below:

(a) An increase of 15 to 50 mm when surface of ooncrete is exposed to the action of harmful chemicals, acidic vapour, saline atmosphere, sulphurous smoke etc.

(b) An increase of 40 mm for reinforced concrete members immersed in sea water.

(c) An increase of 50 mm for reinforced concrete members periodically immersed in sea water or subjected to sea spray.

(d) For concrete of grade M25 and above the additional cover described in (a), (b) and (c) above may be reduced to half.

However, for all such cases the total cover should not exceed 75 mm.

3.5.5 Limits on the quantity of reinforcement. The limits on the quantity of tensile reinforcement as per the IS Code: 456-1978 are as follows:

(i) *Minimum reinforcement.* The minimum area of tension reinforcement is given by

$$\frac{A_{st}}{bd} = \frac{0.85}{f_y}$$

i.e. $A_{st} = 0.85\ bd/f_y$ (3.14)

This would avoid the sudden fracture of steel when the concrete in the tensile zone of beam cracks. transferring full tension to steel, when the beam is loaded.

(ii) *Maximum reinforcement.* The maximum area of tension reinforcement shall not exceed 0.04bD. This upper limit would avoid jumbling up of reinforcement and consequent difficulties in compaction.

3.5.6 Side face reinforcement. When the depth of web in a beam exceeds 750 mm, side face reinforcement should be provided along the two faces. Area of side face reinforcement

$$\nless 0.1\% \text{ of } bD$$

and it shall be distributed equally on two faces. Their spacing on each face,

$$s \ngtr b$$
$$\ngtr 300 \text{ mm}.$$

These rules are illustrated in Fig. 3.7.

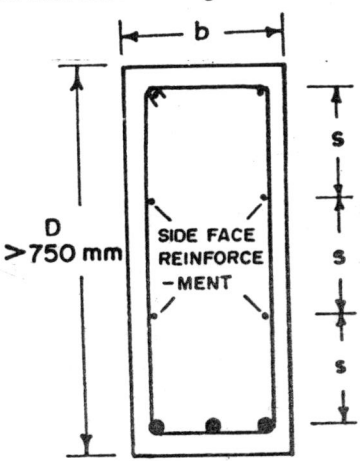

Fig. 3.7 Side face reinforcement in a beam

Example 3.1

A reinforced concrete rectangular beam is to be provided over a clear opening of 4.35 m. Bearing length on each end = 150 mm. The beam is to be singly reinforced having the ratio of width to effective depth = 0.4. The loads on the beam are:

Live load	= 6 kN/m
Superimposed dead load	= 12 kN/m

Concrete grade : M15 and steel grade : Fe 415

Calculate the section dimensions and the reinforcement required.

Solution

First of all the design load is calculated.

Superimposed dead load	=	12 kN/m
Estimated self weight	=	4 kN/m
Live load	=	6 kN/m

Total load w = 22 kN/m

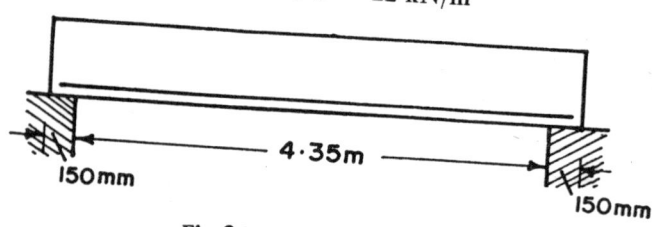

Fig. 3.8 Details for example 3.1

Effective span

$$l = 4.35 + 0.15$$
$$= 4.5 \text{ m}$$

Design bending moment

$$M = \frac{wl^2}{8}$$
$$= \frac{22 \times 4.5^2}{8}$$
$$= 55.69 \text{ kN-m}$$

Permissible stresses are,

$$\sigma_{st} = 230 \text{ N/mm}^2 \text{ from Table 3.2}$$
$$\sigma_{cbc} = 5 \text{ N/mm}^2 \text{ from Table 3.1}$$

Modular ratio is,

$$m = 18.67 \quad \text{from Table 3.3}$$

(a) Balanced design

From Eq. (3.8) the neutral axis factor is

$$n = \cfrac{1}{1 + \cfrac{230}{18.67 \times 5}}$$

$$= 0.289$$

Hence. $\quad j = 1 - \cfrac{0.289}{3}$

$$= 0.904$$

From Eq. (2.10) the moment factor is,

$$R = \tfrac{1}{2} \times 5 \times 0.289 \times 0.904$$

$$= 0.653$$

The values of n and R could have also been read directly from Tables 3.4 and 3.6 respectively. Equating moment of resistance to the bending moment

$$R\, bd^2 = M$$

or $\qquad 0.653 bd^2 = 55.69 \times 10^6$ N-mm

Putting b=0.4d and solving for d,

$$d = 597.4 \text{ mm}$$

Adopt d = 600 mm and b = 240 mm. From Eq. (3.11) the area of tensile reinforcement is given as

$$A_{st} = \cfrac{M}{\sigma_{st}\, jd}$$

$$= \cfrac{55.69 \times 10^6}{230 \times 0.904 \times 600}$$

$$= 446.4 \text{ mm}^2$$

Minimum required quantity of steel reinforcement as per Eq. (3.14) is given as

$$A_{st} = \cfrac{0.85 \times 240 \times 660}{415}$$

$$= 295 \text{ mm}^2$$

Calculated reinforcement is more than the minimum quantity required. Hence O.K.

Provide 12 mm dia. bars 4 nos. giving A_{st}=452 mm².

Required concrete cover on bars

$$\not< \text{ dia. of bar i.e. } 12 \text{ mm}$$
$$\not< 25 \text{ mm}$$

With the arrangement of bars as shown in Fig. 3.9 the cover actually provided on the bottom and the sides is = 29 mm.

Required clear horizontal distance between bars,
  ≮ dia. of bar i.e. 12 mm
  ≮ maximum aggregate size + 5 mm i.e. 25 mm
Clear horizontal distance actually provided between bars
  = (240 − 2 × 35 − 3 × 12)/3
  = 44.67 mm. This is more than 25 mm. Hence O.K.

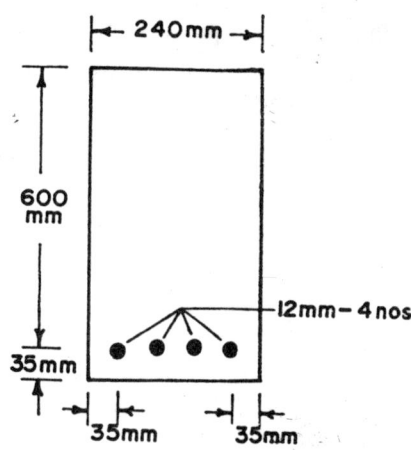

Fig. 3.9 Designed balanced section

(b) *Under-reinforced design*

For balanced design the proportion of steel reinforcement as per Table 3.5 is,

$$p = 0.00314$$

In order to make an under-reinforced design, a proportion of steel reinforcement less than the balanced proportion should be taken.

Take $p = 0.0023$ for this design.

The neutral axis factor is found from Eq. (3.7) i.e.,

$$n^2 + 2 \times 18.67 \times 0.0023n - 2 \times 18.67 \times 0.0023 = 0$$

which gives $n = 0.253$

Hence $j = 1 - \dfrac{0.253}{3} = 0.916$

As the tensile stress in steel reinforcement reaches its permissible value, the design bending strength of the section can be written from Eq. (3.12) as

$$\sigma_{st} \, p \, j \, bd^2 = M$$

Substituting the known values in the above

$$230 \times 0.0023 \times 0.916 \, bd^2 = 55.69 \times 10^6$$

Putting $b = 0.4d$ and solving for d,

$$d = 660 \text{ mm}$$

Adopt $d = 660$ mm and $b = 265$ mm. Area of steel reinforcement, required is given by

$$A_{st} = \frac{M}{\sigma_{st} \, jd}$$
$$= \frac{55.69 \times 10^6}{230 \times 0.916 \times 660}$$
$$= 400.5 \text{ mm}^2$$

Minimum required quantity of steel reinforcement is,

$$A_{st} = \frac{0.85 \, bd}{f_y}$$
$$= \frac{0.85 \times 265 \times 660}{415}$$
$$= 358 \text{ mm}^2$$

Calculated reinforcement is more than the minimum required. Hence O.K.

Provide 16 mm dia. bars, two nos. giving $A_{st} = 402$ mm². The designed cross-section of the beam is shown in Fig. 3.10. Maximum

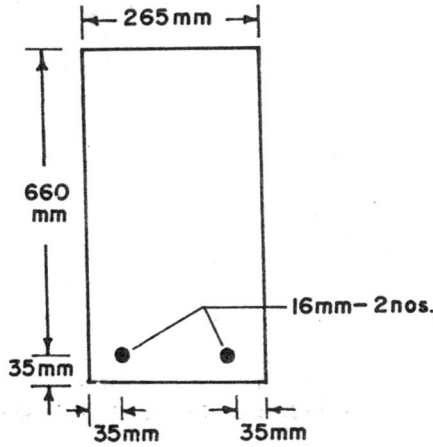

Fig. 3.10 Under-reinforced design

permitted clear distance between bars from Table 3.8 is 180 mm. Actual clear distance provided between the bars = 265 $-2 \times 35-$ 16 = 179 mm.

Hence it is O.K.

Example 3.2

The cross-section of a reinforced concrete beam is shown in Fig. 3.11. Calculate the safe bending strength of the section. Concrete grade is M15 and steel grade is Fe 415.

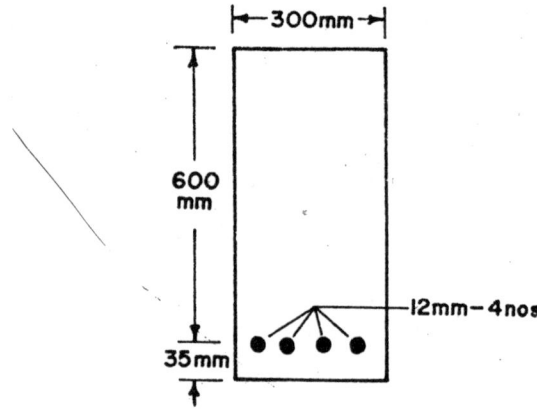

Fig. 3.11 Beam section in Example 3.2

Solution

Proportion of steel reinforcement in the given beam section is

$$p = \frac{4 \times 113.1}{300 \times 600}$$
$$= 0.0025$$

However, the balanced proportion of steel reinforcement from Table 3.5 is

$$p = 0.00314$$

Evidently, the given section has less amount of steel reinforcement than required for a balanced section. Hence, the beam is under-reinforced.

Now

$$\sigma_{st} = 230 \text{ N/mm}^2 \qquad \text{from Table 3.2}$$

and $\qquad m = 18.67 \qquad$ from Table 3.3

Neutral axis factor is found from the quadratic equation Eq. (3.7) i.e.

$n^2 + 2mpn - 2mp = 0$

or, $n^2 + 2 \times 18.67 \times 0.0025 \times n - 2 \times 18.67 \times 0.0025 = 0$

which gives $n = 0.262$

hence $\qquad j = 1 - \dfrac{0.262}{3}$

$\qquad\qquad = 0.913$

From Eq. (3.11) the bending strength is

$\qquad M = \sigma_{st} A_{st} \, jd$

$\qquad\qquad = 230 \times 4 \times 113.1 \times 0.913 \times 600 \times 10^{-6}$

$\qquad\qquad = 57 \text{ KN} - \text{m}.$

3.6 DEVELOPMENT LENGTH OF REINFORCEMENT BAR

As already discussed under Art. 1.4.2 the development length of a bar is the length required to be embedded in concrete in order to develop in it the design axial stress. It is given by

$$l_d = \frac{\phi \sigma_s}{4\tau_{bd}} \qquad (3.15)$$

where $\qquad \phi$ = nominal diameter of bar

$\qquad \sigma_s$ = permissible stress in reinforcement bar, tensile or compressive.

$\qquad \tau_{bd}$ = permissible bond stress between concrete and steel as given in Table 3.1

(a) *Bars in tension*

Here, $\sigma_s = \sigma_{st}$. For calculating the development length of plain bars in tension the values of τ_{bd} and σ_{st} are taken from Tables 3.1 and 3.2 respectively. For deformed bars the value of τ_{bd} may be increased by 40% according to the recommendation of the IS Code: 456-1978. This accounts for the higher degree of bond strength obtained with deformed bars.

For different grades of steel and concrete, calculated development lengths of bars in tension are given in Table 3.9, expressed as multiples of bar diameter.

Table 3.9

DEVELOPMENT LENGTHS OF BARS IN TENSION EXPRESSED AS MULTIPLES
OF DIAMETER

Concrete grade		M15	M20	M25	M30	M35	M40
Plain	$\phi < 20$ mm	58.33	43.75	38.89	35.0	31.82	29.17
Bars	$\phi > 20$ mm	54.17	40.63	36.11	32.5	29.54	27.08
HYSD	Fe 415	68.45	51.34	45.63	41.07	37.34	34.23
Bars	Fe 500	81.85	61.38	54.56	49.11	44.64	40.92

(b) *Bars in compression*

Here, $\sigma_s = \sigma_{so}$. For calculating the development length of plain bars in compression, the value of τ_{bd} and σ_{so} are taken from Tables 3.1 and 3.2 respectively. For bars in compression the value of τ_{bd} may be increased by 25% according to the recommendation of the IS Code: 456-1978. For deformed bars an additional increase of 40% in the value of τ_{bd} is applied. Calculated values of the development length of bar in compression are given in Table 3.10 as multiples of bar diameter.

Table 3.10

DEVELOPMENT LENGTHS OF BAR IN COMPRESSION EXPRESSED AS
MULTIPLES OF DIAMETER

Concrete grade	M15	M20	M25	M30	M35	M40
Plain bars	43.33	32.5	28.89	26.0	23.64	21.67
Deformed bars	48.0	36.0	32.0	28.79	26.17	24.0

3.7 Anchoring reinforcing bars. Deformed bars may be used without end anchorage provided full development length is available beyond the section where the bar carries full design stress.

However, plain bars carrying tension should always be provided with a standard hook at the end.

Anchorage value of standard bends shall be taken as 4 times the bar diameter for each 45° bend subject to a maximum of 16 times the diameter. A standard hook and a standard 90° bend which are most commonly used as end anchorages for tension reinforcement are shown in Fig. 3.12

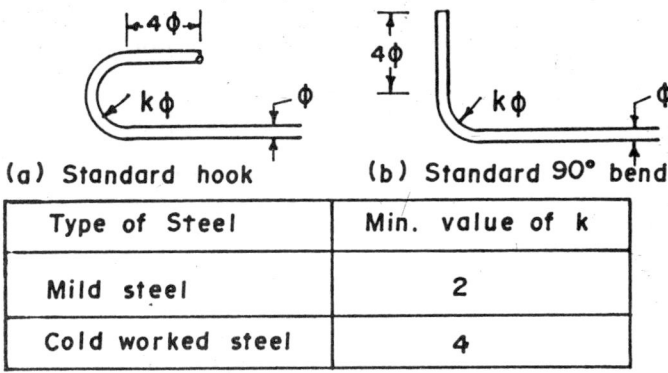

(a) Standard hook   (b) Standard 90° bend

Type of Steel	Min. value of k
Mild steel	2
Cold worked steel	4

Fig. 3.12 End anchorage for tension reinforcement

3.8 TERMINATION AND CURTAILMENT OF TENSION REINFORCEMET

In a simple beam the tension reinforcement is calculated for the maximum bending moment. As the bending moment reduces towards the end of the beam the reinforcement can be curtailed accordingly. The section at which a certain amount of tension reinforcement is no longer required is that where the calculated bending strength considering the continuing bars only is equal to the bending moment. In a continuous beam, at point of inflexion, the bending moment becomes zero and changes sign. As such, the point of inflexion is the termination point for both positive and negative moment reinforcements. Recommendation of the IS Code 456-1978 regarding the termination and curtailment of the reinforce ment bars are as follows :

(i) In simple beams at least one-third of the positive moment reinforcement shall extend along the same face of the member into the support to a length $\frac{l_d}{3}$. See Fig. 3.13 (a).

(ii) In continuous beams at least one-fourth the positive moment reinforcement shall extend along the same face of the member into the support to a length $\frac{l_d}{3}$. See Fig. 3.13 (b).

(iii) The positive reinforcement shall extend beyond the section at which it is no longer required, for a distance

$$\not< \begin{array}{l} d \\ 12 \ \phi \end{array}$$

See Fig. 3.13(a) and (b).

(iv) The diameter of the positive moment reinforcement should

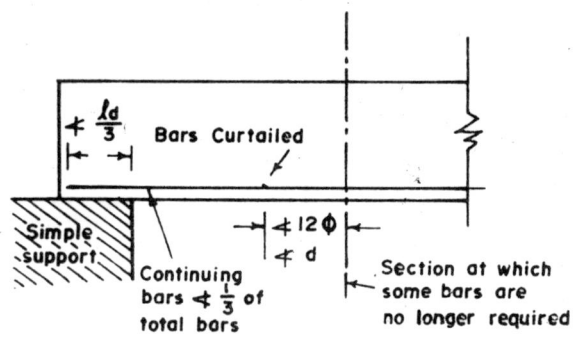

(a) For simple beam

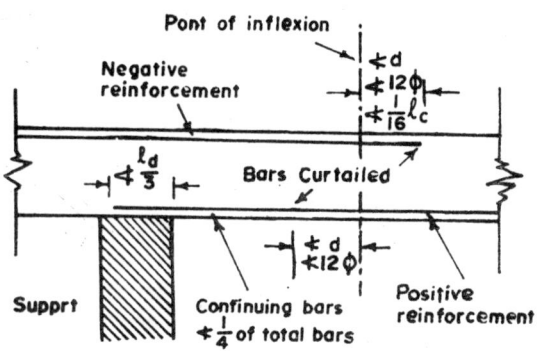

(b) For continuous beam

Fig. 3.13 Curtailment and continuation of tension reinforcement

be so limited, at supports and at points of inflexion, that the calculated development length of bar does not exceed the value

$$\frac{M_1}{V} + L_0 \qquad (3.16)$$

where

$M_1 =$ bending strength available at the section

$V =$ shear force at the section

$L_0 =$ anchorage of bar available beyond the centre of support or the point of inflexion as the case may be.

Note: In the above expression the value of $\frac{M_1}{V}$ may be increased by 30% when the ends of reinforcement are confined by a compressive reaction.

(v) At least one-third of the total negative reinforcement provided over a support in a continuous beam shall extend beyond the point of inflexion for a distance

$$\not< \quad d$$
$$\not< \quad 12\ \phi$$
$$\not< \quad \frac{1}{16} \quad \text{of clear span}$$

See Fig. 3.13 (b).

Example 3.3

A singly reinforced rectangular beam is simply supported over an effective span of 4.5 m. The distributed load is as follows :

Superimposed dead load = 12 kN/m
live load = 6 kN/m

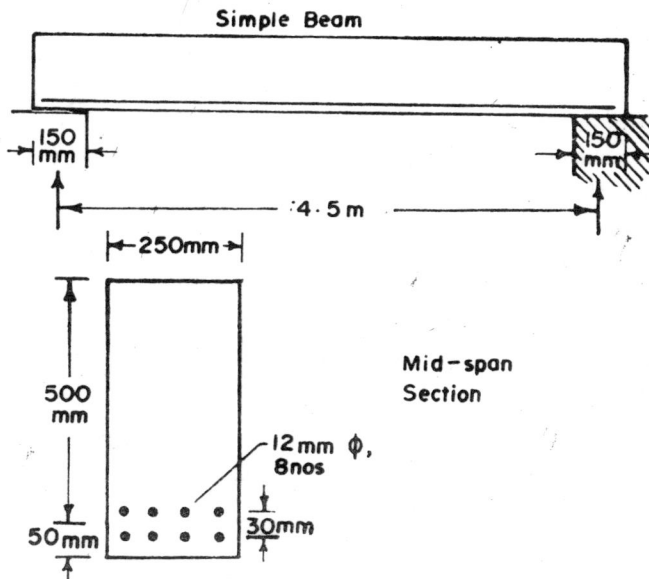

Simple Beam

150 mm

150 mm

4.5 m

250mm

500 mm

12 mm ϕ, 8 nos

50 mm

30 mm

Mid–span Section

Fig. 3.14

The beam section has been designed for balanced conditions, to resist maximum bending moment at mid-span. Concrete grade is M15. The reinforcement consists of 12 mm dia. mild steel bars, 8 nos, laid up in two layers of 4 bars each. See Fig. 3.14. Indicate as to how many bars can suitably be curtailed towards the support and locate their point of cut-off.

Solution

Self weight of beam	=	$0.25 \times 0.55 \times 25$
	=	3.44 kN/m
Superimposed dead load	=	12 kN/m
Live load	=	6 kN/m
Total distributed load	=	21.44 kN/m
Mid-span bending moment	=	$\dfrac{21.44 \times 4.5^2}{8}$
	=	54.27 kN-m

Permissible stresses are

$\sigma_{obc} = 5$ N/mm^2, $\sigma_{st} = 140$ N/mm^2

Modular ratio $m = 18.67$.

As per the IS Code: 456-1978 at least one-third of the tension reinforcement should continue into the supports. Keeping the symmetry of the reinforcement bars in the cross-section, it is proposed to curtail 4 bars of the upper layer, two at a time, towards the support.

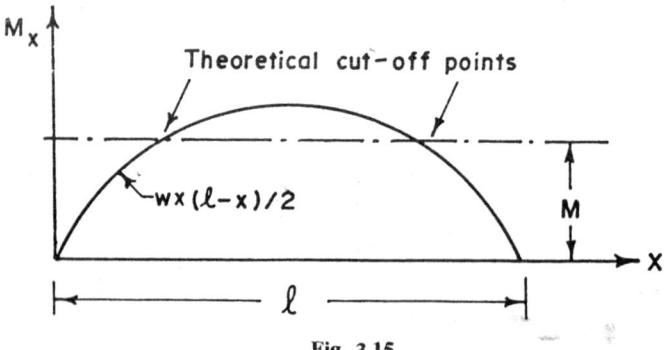

Fig. 3.15

Bending moment at any section distant x from the left hand support is given by

$$M_x = wx\,(l-x)/2$$

Let the bending strength of the section after a certain number of reinforcement bars have been cut off be $== M$. Then, the point of cut-off can be found by equating,

$$w \times (l\text{-}x)/2 = M$$

This quadratic equation gives two values of x corresponding to two cut-off points. See Fig. 3.15

(i) Location of point where 2 bars from the upper layer are curtailed:

Beyond the point of curtailment, only 6 bars will continue. Distance of the centroid of the continning bars from top concrete fibres is

$$= \frac{2 \times 485 + 4 \times 515}{6}$$

$$= 505 \text{ mm}$$

This is the effective depth of the section now.

Area of reinforcement is

$$A_{st} = 6 \times 113.1 = 678.6 \text{ mm}^2$$

Therefore

$$p = \frac{A_{st}}{bd} = \frac{678.6}{250 \times 505}$$

$$== 0.005375$$

Neutral axis factor is found from the quadratic equation,

$$n^2 + 2mpn - 2mp = 0$$

i.e. $n^2 + 2 \times 18.67 \times 0.005375 \, n - 2 \times 18.67 \times 0.005375 = 0$

which gives

$$n = 0.358$$

$$j = 1 - \frac{0.358}{3}$$

$$= 0.881$$

Bending strength of section is

$$= A_{st} \, \sigma_{.t} \, j \, d$$

$$= 678.6 \times 140 \times 0.881 \times 505 \times 10^{-6}$$

$$= 42.27 \text{ kN}-\text{m}$$

Hence. theoretical cut-off point is found by equating

$$21.44 \, x \, (4.5 - x)/2 = 42.27$$

Hence,

$$x = 1.19 \text{ m and } 3.31 \text{ m}$$

As per the IS Code : 456—1978 the bars should extend beyond the point at which they are no longer required by a distance not less than 12 times the diameter of bar or the effective depth, which ever is greater. Hence, the actual point of curtailment of the bars will be at $1.19 - 0.505 = 0.685$ m from the centre of support. See Fig. 3.16 (a).

(ii) Location of point where 4 bars from the upper layer are curtailed:

Effective depth of the remaining 4 bars is
$$= 515 \text{ mm}$$
Area of reinforcement
$$A_{st} = 4 \times 113.1 = 452.4 \text{ mm}^2$$
$$p = \frac{452.4}{250 \times 515}$$
$$= 0.00351$$

Neutral axis factor is found from the quadratic equation
$$n^2 + 2 \times 18.67 \times 0.00351 \ n - 2 \times 18.67 \times 0.00351 = 0$$
which gives
$$n = 0.303$$

and
$$j = 1 - \frac{0.303}{3}$$
$$= 0.899$$

Bending strength of the section
$$= 452 \ 4 \times 140 \times 0.899 \times 515 \times 10^{-6}$$
$$= 29.32 \text{ kN-m}$$

Theoretical point of cut-off is found by equating
$$21.44 \ x \ (4.5 - x)/2 = 29.32$$
which gives
$$x = 0.725 \text{ m and } 3.775 \text{ m}$$

As before, the actual point of curtailment will be at $0.725 - 0.515 = 0.21$ m from the centre of the support. See Fig. 3.16(b).

Continuation of bars into support :

First of all the development length of 12 mm dia. mild steel bar should be found for M 15 grade of concrete. From Table 3.9
$$l_d = 58.33 \times 12 = 700 \text{ mm}$$
$$\therefore \quad l_d/3 = 700/3 = 233.33 \text{ mm}$$

The tension reinforcement should extend beyond the face of support by a distance $l_d/3$ i.e. 233.33 mm. However, the length of support is only 150 mm. Hence, it is proposed to provide a standard 90° bend at the end as shown in Fig. 3.16 (b). Thus, the

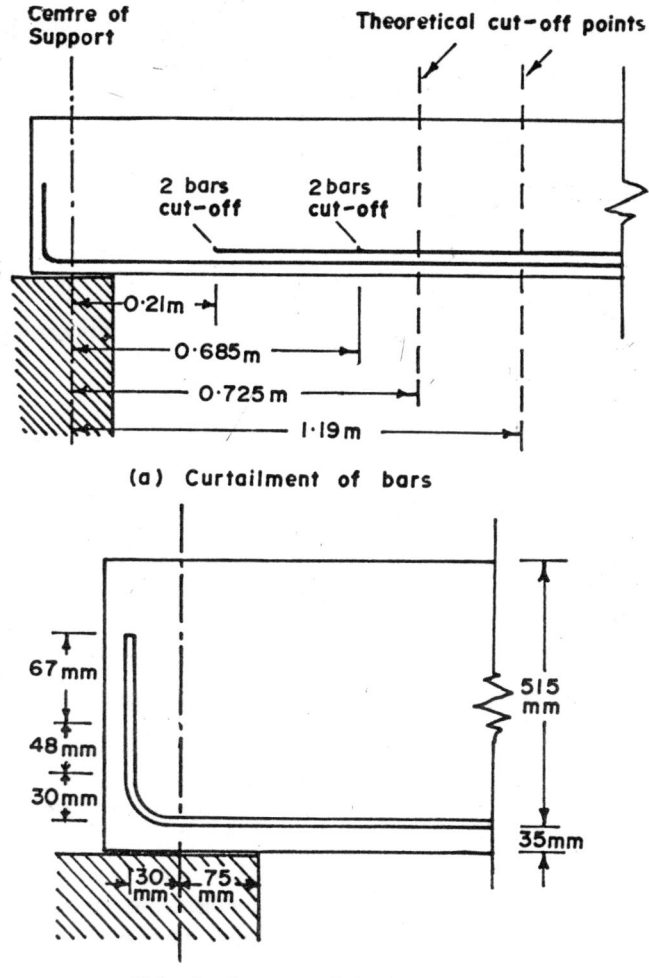

(a) Curtailment of bars

(b) Anchorage of bar at the end

Fig. 3.16

total anchorage of bar available beyond the face of support

$= 75 +$ value of $90°$ bend $+ 67$ mm

$= 75 + 12 \times 8 + 67$

$= 238$ mm

Which is greater than the required length i.e. 233.33 mm. Hence, it is alright.

As four bars have been cut-off, only four bars continue into the support. Moment of resistance of beam section over the support, there fore, as calculated earlier is

$$M_1 = 29.32 \quad kN\text{-}m$$

Shear force at the support is

$$V = \frac{1}{2} \times 4.5 \times 21.44$$

$$= 48.24 \; kN$$

Anchorage of bar available beyond the centre of support is

$$L_0 = 8 \times 12 + 67$$

$$= 163 \; mm$$

Now

$$\frac{M_1}{V} + L_0 = \frac{29.32 \times 10^6}{48.24 \times 10^3} + 163$$

$$= 770.8 \; mm$$

It is evident that l_d is not greater than $\frac{M_1}{V} + L_0$. Hence, the bar diameter provided is alright and does not require to be reduced.

3.9   DESIGN FOR SHEAR

In structural members shear forces seldom act on their own. They generally exist in combination with bending moment and sometimes also with twisting moment. Theoretical analysis for shear stresses within a section applicable to homogeneous isotropic elastic bodies, after suitable modification, can give acceptable predictions regarding crack formation and the strength. However, after development of cracks, the pattern of stresses and the resisting mechanism within a section becomes quite complex and the elastic analysis is hardly relevant to the actual behaviour.

3.9.1   Shear stress and diagonal tension. Consider a simple beam made of homogeneous and elastic material, which is transversely loaded. An element below the neutral axis is subjected to a combination of shear stress and bending stress as shown in Fig. 3.17.

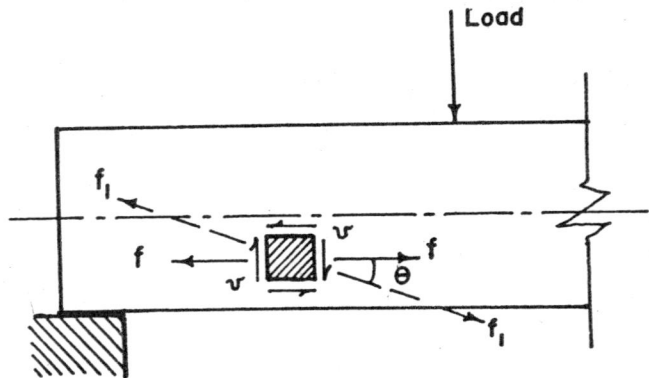

Fig. 3.17 Principal tension in a beam

The resulting principal tensile stress is given by

$$f_1 = \frac{f}{2} + \sqrt{\left(\frac{f}{2}\right)^2 + v^2} \tag{3.17}$$

and its inclination with respect to the beam axis is given by

$$\text{Tan } 2\theta = \frac{2v}{f} \tag{3.18}$$

Near the beam axis the shear stress is maximum whereas the bending stress is zero. Hence, $2\theta = 90°$ or $\theta = 45°$. This implies that the principal tensile stress is inclind at 45° to the axis in the region close to the beam axis. Towards the extreme fibres of beam, the shear stress is zero whereas the bending stress is not zero. Hence, $\theta = 0°$. This implies that the principal tensile stress is parallel to the axis near the extreme fibres of beam.

Since concrete is very weak in tension it cracks under the principal tensile stress. A typical cracking pattern of a reinforced concrete beam is shown in Fig. 3.18. In short span beams carrying heavy loads, diagonal tension cracks may open up in the end region close to the beam axis and inclind to it at 45°. See Fig. 3.18 (a). In beams of comparative longer span, bending moments are predominant. Hence, vertical flexural cracks open up in bottom fibres and may join up with the diagonal cracks in the web of the beam. Towards the mid-span vertical flexural cracks only may appear in the bottom fibres. See Fig. 3.18 (b).

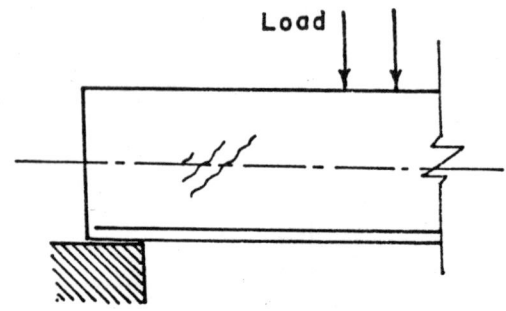

(a) Web cracking in heavily loaded short span beams.

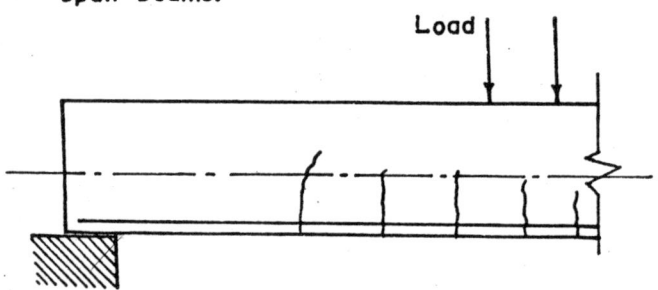

(b) Combination of shear and flexural cracks in long span beams.

Fig. 3.18 Shear cracking in beams

Principal tension is also known as diagonal tension in reinforced concrete beams.

Shear stress distribution within the cross-section of a singly reinforced section can be worked out by a little modification of that for an elastic homogeneous beam. The concrete above the neutral axis bears linearly varying flexural compressive stress. Hence, the shear stress distribution from top fibres down to the neutral axis is parabolic, being maximum at the neutral axis. See Fig. 3.19. Below the neutral axis concrete is cracked and as such, it is unable to bear flexural tensile stress. Hence, the shear stress would remain constant up to the tensile reinforcement where it suddenly drops to zero. Magnitude of the shear stress can be found by equating the applied shear force to the shear resistance of the section.

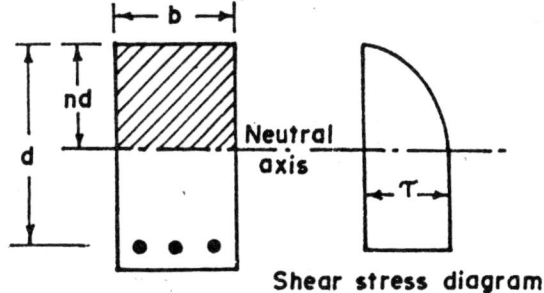

Fig. 3.19   Shear stress distribution in a reinforced concrete beam section

Now,

$$V = b \times \text{area of shear stress diagram}$$

$$= \frac{2}{3}\, \tau b n d + \tau b\,(d - nd)$$

$$= \tau b \left(d - \frac{nd}{3} \right)$$

$$= \tau b d \left(1 - \frac{n}{3} \right)$$

$$= \tau b j d$$

$$\therefore \qquad \tau = \frac{V}{bjd} \qquad\qquad (3.19)$$

This concept presumes that the concrete below the neutral axis, though cracked in flexural tension, is capable of taking up the shear stress. However, Eq. (3.19) is useful only as a convenient index and does not give the value of shear stress at a point in the cracked region of the beam.

3.9.2 Nominal shear stress. A convenient index of shear stress as adopted by the IS Code: 456-1978 is the simple equation:

$$\tau_v = \frac{V}{bd} \qquad\qquad (3.20)$$

This is applicable to uniform beams of rectangular section.

However, for beams of varying depth the equation in modified form is written as :

$$T_v = \frac{V \pm \frac{M}{d} \tan \beta}{bd} \qquad (3.21)$$

where β = angle between top and bottom edges of beam

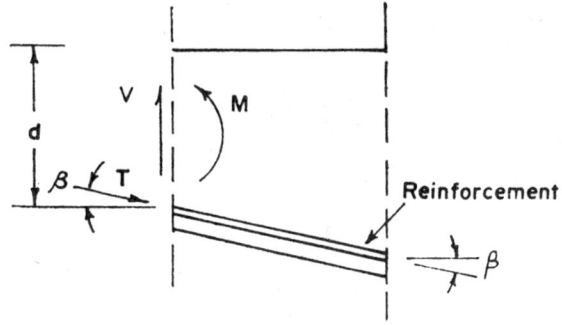

Fig. 3.20 Beams of varying depth

The negative sign applies when the bending moment M increases in the same direction as the depth d increases. Otherwise, the positive sign applies. Consider the beam shown in Fig. 3.20. Tension in the inclind reinforcement is approximately given by

$$T \cos \beta = \frac{M}{d} \qquad (3.22)$$

Vertical component of T opposes the applied shear force. Hence, resultant shear force $= V - T\text{-}\sin \beta$

$$= V - \frac{M}{d} \tan \beta \qquad (3.23)$$

However, if the depth of the beam decreases in the direction of increasing bending moment, the vertical component of T would be additive to V. Hence,

resultant shear force $= V + \frac{M}{d} \tan \beta \qquad (3.24)$

Expressions (3.23) and (3.24) appear in the numerator of Eq. (3.21). for the appropriate cases,

3.9.3 Reinforcement against shear. Consider a diagonal tension crack due to shear in a beam, inclind to the axis at 45° as shown in Fig. 3.21. The reinforcement against shear is in the form of steel bars inclined at an angle α to the beam axis and spaced uniformly at s_v mm c/c horizontally. Now, the number of steel bars crossing the diagonal crack is given by,

$$N = \frac{d \sin (180° - 45° - \alpha)}{s_v \cos 45° \sin \alpha}$$

$$= \frac{\sqrt{2} \; d \; \sin (45° + \alpha)}{s_v \sin \alpha}$$

$$= \frac{d \; (\sin \alpha + \cos \alpha)}{s_v \sin \alpha}$$

Total tensile force in the bar is given by

$$T = \sigma_{sv} A_{sv} N \tag{3.25}$$

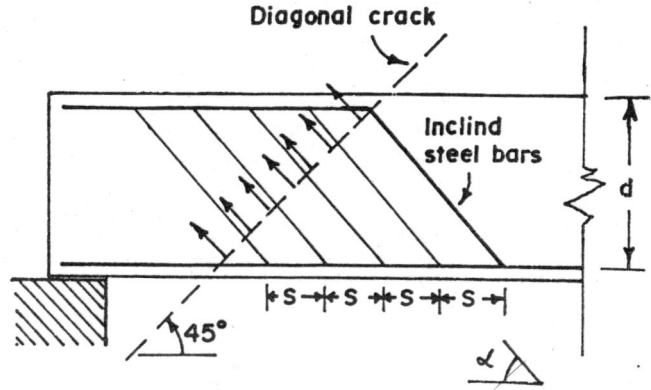

Fig. 3.21 Shear reinforcement

Evidently, the vertical component of the tension in the inclind bars is the strength of the reinforcement in shear. Hence,

$$V_s = T \sin \alpha$$

$$= \sigma_{sv} A_{sv} N \sin \alpha$$

$$= \frac{\sigma_{sv} A_{sv} \; d \; (\sin \alpha + \cos \alpha)}{s_v} \tag{3.26}$$

For vertical shear reinforcement, $\alpha = 90°$ and the above expression reduces to

$$V_s = \frac{\sigma_{sv} A_{sv} \; d}{s_v} \tag{3.27}$$

For a number of bars all bent-up at the same cross-section,

$$V_s = \sigma_{sv} A_{sv} \sin \alpha \tag{3.28}$$

3.9.4 Shear strength of a beam. The shear force at a section is assumed to be partly resisted by concrete and partly by shear reinforcement. Concrete by itself can bear shear force within the limit given by the permissible shear stress. Excess shear force beyond this limit is taken up by the shear reinforcement. Thus, the shear carrying capacity of a beam is written as

$$V = V_0 + V_s \qquad (3.29)$$

where $\quad V_0 = \tau_0\, bd \qquad (3.30)$

τ_0 = permissible shear stress in concrete. See Table 3.11.

V_s = shear taken by shear reinforcement.

Table 3.11

PERMISSIBLE SHEAR STRESS τ_0 IN CONCRETE BEAMS WITHOUT SHEAR REINFORCEMENT, N/mm^2

(BASED ON TABLE 17, IS CODE : 456-1978)

$\dfrac{100\,A_s}{bd}$	Concrete grade					
	M15	M20	M25	M30	M35	M40
0.25	0.22	0.22	0.23	0.23	0.23	0.23
0.50	0.29	0.30	0.31	0.31	0.31	0.32
0.75	0.34	0.35	0.36	0.37	0.37	0.38
1.00	0.37	0.39	0.40	0.41	0.42	0.42
1.25	0.40	0.42	0.44	0.45	0.45	0.46
1.50	0.42	0.45	0.46	0.48	0.49	0.49
1.75	0.44	0.47	0.49	0.50	0.52	0.52
2.00	0.44	0.49	0.51	0.53	0.54	0.55
2.25	0.44	0.51	0.53	0.55	0.56	0.57
2.50	0.44	0.51	0.55	0.57	0.58	0.60
2.75	0.44	0.51	0.56	0.58	0.60	0.62
3.00 and more	0.44	0.51	0.57	0.60	0.62	0.63

Note: In the above table A_s represents the sectional area of tension reinforcement which continues at least one effective depth beyond the section considered. However, at supports full area of tension reinforcement may be used provided it is terminated and anchored according to the provisions described under para. 3.8.

3.9.5 Form of shear reinforcement in a beam

(i) *Stirrups.* The transverse reinforcement should be taken around the tension reinforcement and anchored properly in the compression zone. Two legged and four legged stirrups are illustrated in Fig. 3.22 (a). Vertical stirrups are the most popular, though, inclined stirrups also can be provided, as shown in Fig. 3.22(b). The stirrups should also go around the supporting longitudinal bars which are nominally provided in the compression zone of singly reinforced beams. However, in doubly reinforced beams the stirrups should go around the compression reinforcement.

(ii) *Bent-up tension reinforcement.* Near the support any tension reinforcement no longer required from consideration of flexure, and which is excess over the quantity required to go into the support, can be bent up and properly anchored within the compression zone of the beam, See Fig. 3.22(c). The inclined portion of the bar effectively checks up the opening up of diagonal tension cracks in concrete. This type of reinforcement is generally provided in combination with the stirrups.

3.9.6 Design recommendations as per IS Code: 456-1978.

For a given shear force V the nominal shear stress τ_v is calculated from Eq. (3.20). Three cases can now be considered as follows :

(i) If $\tau_v > \tau_c$ shear reinforcement should be provided.

Shear taken up by the shear reinforcement is given by

$$V_s = V - V_c$$

$$= V - \tau_c \, bd$$

For a chosen diameter and shape of shear reinforcement, A_{sv} is calculated. Then, the spacing of shear reinforcement is calculated from the appropriate expression:

$$V_s = \frac{\sigma_{sv} \, A_{sv} \, d \, (\sin \alpha + \cos \alpha)}{s_v} \quad \text{For inclined shear reinforcement.}$$

$$V_s = \frac{\sigma_{sv} \, A_{sv} \, d}{s_v} \quad \text{For vertical shear reinforcement.}$$

(ii) If $\tau_v < \tau_c$ shear reinforcement need not be designed and provided. However, as per the IS Code : 456 certain minimum shear reinforcement should always be provided as concrete alone

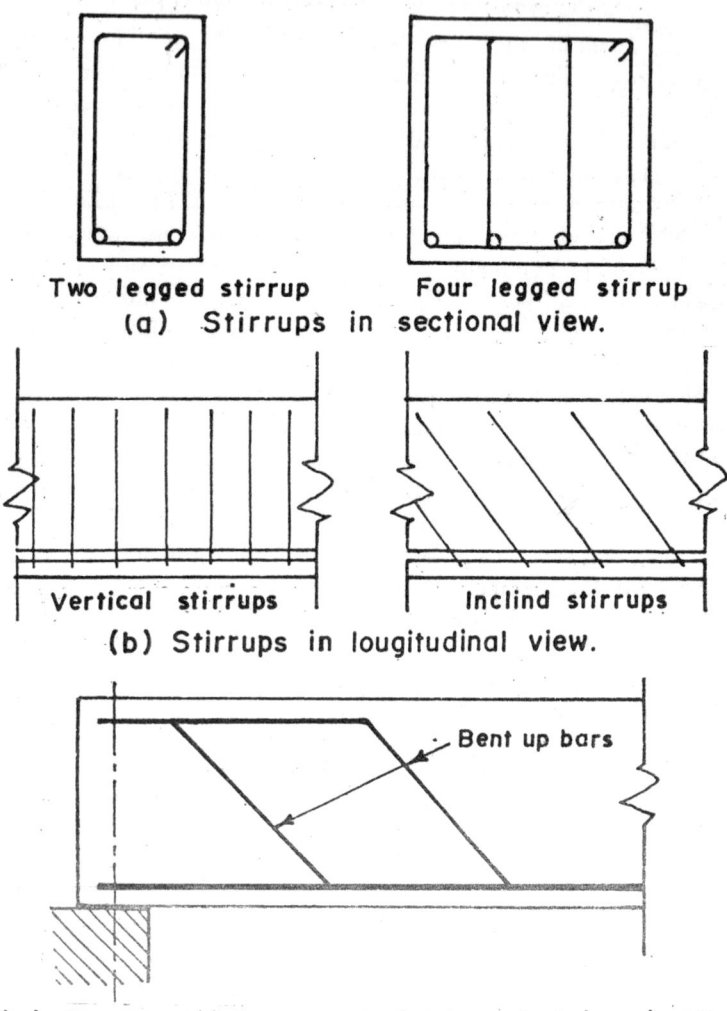

Two legged stirrup   Four legged stirrup

(a) Stirrups in sectional view.

Vertical stirrups   Inclind stirrups

(b) Stirrups in lougitudinal view.

Bent up bars

(c) Tension reinforcement bent up to take shear.

Fig. 3.22

can not be relied upon to take up shear stresses safely. This shear reinforcement is nominal and is given by

$$\frac{A_{sv}}{b\, s_v} \geqslant \frac{0.4}{f_y} \tag{3.31}$$

(iii) If $\tau_v < \frac{1}{2}\tau_c$ no shear reinforcement may be provided in members of minor structural importance such as lintels.

(iv) If $\tau_v > \tau_{omax}$ then the beam section dimensions should be increased. Values of the maximum permissible limits τ_{omax} of the calculated shear stress in a reinforced concrete beam are given in Table 3.12.

Table 3.12
MAXIMUM SHEAR STRESS τ_{cmax}
PERMITTED IN A REINFORCED CONCRETE BEAM, N/mm^2

Concrete grade	M15	M20	M25	M30	M35	M40
N/mm^2	1.6	1.8	1.9	2.2	2.3	2.5

Note: The above table is based on Table 18 of IS Code: 456.

Maximum spacing of shear stirrups. The spacing of shear stirrups measured along the axis of the member shall be restricted to the following limits:

> ≯ 0.75d for vertical stirrups
> ≯ d for inclined stirrups at 45°
> ≯ 450 mm in any case

3.9.7 Critical section for shear. If the support reaction induces compression in the end region of a member, then the

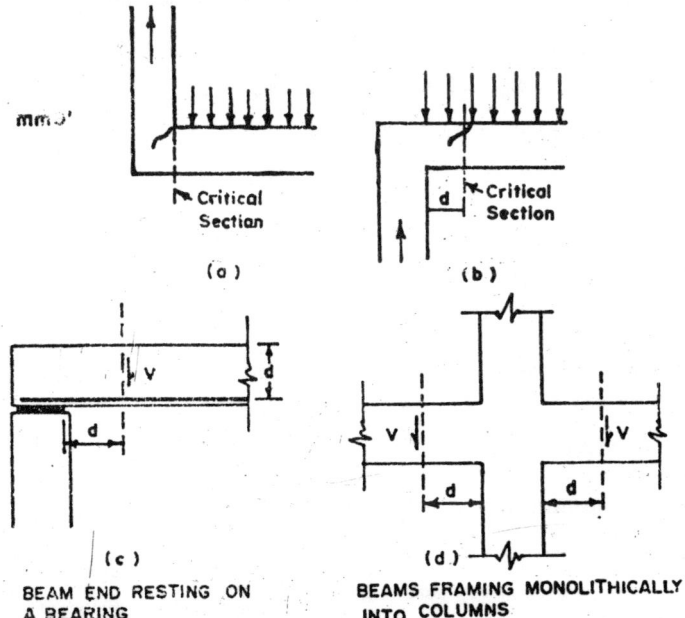

(a)

(b)

(c)

BEAM END RESTING ON
A BEARING

(d)

BEAMS FRAMING MONOLITHICALLY
INTO COLUMNS

Fig. 3.23 Critical section for shear

critical section is considered to be located at a distance d away from the face of the support, See Fig. 3.23 (b). Otherwise, the critical section is taken at the face of support. See Fig 3.23(a). Two situations for the critical sections for beams are illustrated in Fig. 3.23 (c) and Fig. 3.23 (d)

Example 3.4

A reinforced concrete uniform rectangular beam is simply supported over a clear span of 5.7 m. Bearing at each end = 200 mm. The beam supports a uniformly distributed superimposed dead plus live load = 29 kN/m. Concrete grade: M20 and steel grade: Fe 415. Design the beam section for flexure and provide the shear reinforcement as required.

Solution

Clear span + effective depth = 5.7 + 0.75

$$= 6.45 \text{ m}$$

Distance c/c of supports $= 5.9$ m

Effective span is the lesser of the above two values, i.e.

$$= 5.9 \text{ m}$$

Estimated self weight of beam = 6 kN/m

Design load intensity on the beam

$$= 6 + 29$$
$$= 35 \text{ kN/m}$$

Mid-span bending moment $= \dfrac{35 \times 5.9^2}{8}$

$$= 152.3 \text{ kN-m}$$

Permissible stresses are $\quad \sigma_{cbc} = 7 \text{ N/mm}^2$

$$\sigma_{st} = 230 \text{ N/mm}^2$$

Modular ratio $\qquad m = 13.33$

Design constants are

$$n = \cfrac{1}{1 + \cfrac{230}{13.33 \times 7}}$$

$$= 0.289$$

$$j = 1 - \frac{0.289}{3}$$

$$= 0.904$$

$$R = \tfrac{1}{2} \times 7 \times 0.289 \times 0.904$$

$$= 0.914$$

Equating $\quad Rbd^2 = M$

Substituting all the values and keeping $b \approx 0.4\,d$ in the above expression

$$0.914 \times 0.4\,d^3 = 152.3 \times 10^6$$

$$d = 746.85 \text{ mm}$$

Adopt $\qquad d = 750 \text{ mm and}$

$$b = 0.4 \times 750$$

$$= 300 \text{ mm.}$$

Tension reinforcement $A_{st} = \dfrac{152.3 \times 10^6}{230 \times 0.904 \times 750}$

$$= 976.7 \text{ mm}^2$$

Provide 16 mm dia. bars-5 nos. For details See Fig. 3.24

Self weight of beam $\qquad = 0.3 \times 0.8 \times 25$

$$= 6 \text{ kN/m}$$

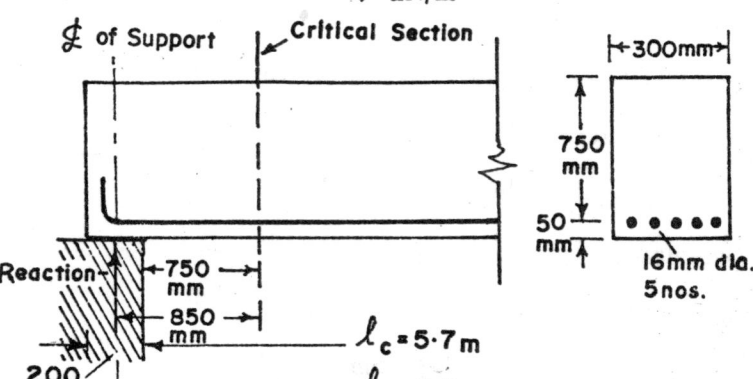

Fig. 3.24 Tension reinforcement for beam in example 3.4

Critical section for shear occurs at one effective depth i.e. 750 mm away from the face of support. Hence, the distance of the same from the centre of support will be $100 + 750 = 850$ mm away. Thus, shear force at the critical section

$$V = \tfrac{1}{2} \times 5.9 \times 35 - 0.85 \times 35$$
$$= 73.5 \text{ kN}$$

Nominal shear stress $\tau_v = \dfrac{73.5 \times 10^3}{300 \times 750}$

$$= 0.327 \text{ N/mm}^2$$

$$\frac{100 \, A_{st}}{bd} = \frac{100 \times 5 \times 201.06}{300 \times 750}$$

$$= 0.447$$

Permissible shear stress from Table 3.11, for grade: M20 concrete is

$$\tau_c = 0.283 \text{ N/mm}^2$$

As $\tau_v > \tau_c$ the beam is unsafe in shear.
Shear resisted by concrete alone

$$V_c = \tau_c \, bd$$
$$= 0.283 \times 300 \times 750 \times 10^{-3}$$
$$= 63.68 \text{ kN}$$

Hence, shear to be resisted by the shear reinforcement is

$$V_s = V - V_c$$
$$= 73.5 - 63.68$$
$$= 9.82 \text{ kN}$$

It is proposed to provide 6 mm dia. two legged stirrups of grade: Fe415 steel.
The shear strength of vertical stirrups is given by

$$V_s = \frac{\sigma_{sv} \, A_{sv} \, d}{S_v}$$

Substituting all the known values in the above expression

$$9.82 \times 10^3 = \frac{230 \times 2 \times 28.27 \times 750}{S_v}$$

$$\therefore \qquad S_v = 993.2 \text{ mm}$$

Minimum shear reinforcement as per IS Code: 456 is given by

$$\frac{A_{sv}}{bS_v} \geqslant \frac{0.4}{f_y}$$

Substituting all the known values in the above expression

$$\frac{2 \times 28.27}{300 \ S_v} \geqslant \frac{0.4}{415}$$

∴ $S_v \leqslant 195.5$ mm

Hence, adopt a stirrup spacing of 195 mm c/c throughout the beam length.

Maximum permitted spacing of stirrups

$$= 0.75 \ d$$
$$= 0.75 \times 750$$
$$= 562.5 \ \text{mm.}$$

Actual spacing is less than the maximum permitted value.
Hence, O.K. Details are shown in Fig. 3.25

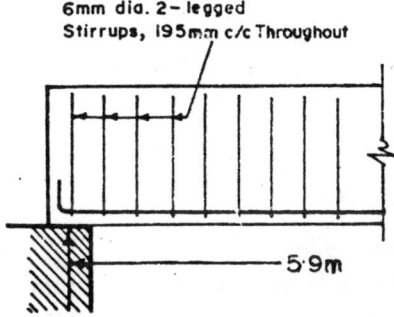

Fig. 3.25 Shear reinforcement for beam in Example 3.4

Example 3.5

A beam of uniform rectangular section is simply supported over a clear apan of 3.6 m. Length of bearing on each side = 150 mm, Width of beam = 400 mm and effective depth = 625 mm. Tension reinforcement is 12 mm bars, 10 nos. Total 4 nos. bars have been be͞ up at 45°, in two steps, as shown in Fig. 3.26. Shear reinforcement consists of 6 mm dia. two legged stirrups at 145 mm c/c throughout. Concrete grade: M 20 and steel grade; Fe 415. Calculate the shear strength of the beam,

    (i) At the critical section

    (ii) At a section 1.3 m away from the face of support.

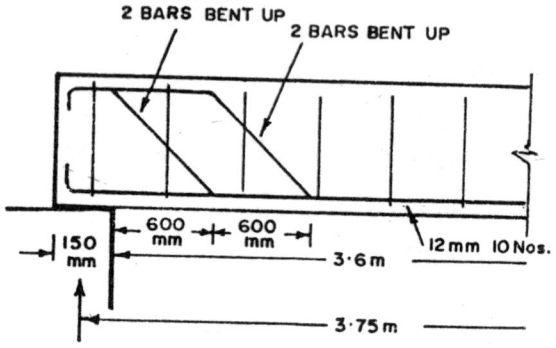

Fig. 3.26 Details for Example 3.5

Solution

For the given beam

$$\frac{100 \; A_{st}}{bd} = \frac{100 \times 10 \times 113.1}{400 \times 625}$$

$$= 0.452$$

From Table 17 of IS Code: 456

$$\tau_c = 0.285 \; N/mm^2$$

(i) At the critical section.

The critical section is at one effective depth i.e. 625 mm away from the face of support. Two bent up bars cross this section. Vertical component of the tensile strength of these two bars

$$= 2 \times 113.1 \times 230 \times \frac{1}{\sqrt{2}} \times 10^{-3}$$

$$= 36.79 \; kN$$

Shear strength of the beam concrete

$$V_c = \tau_c \; bd$$

$$= 0.285 \times 400 \times 625 \times 10^{-3}$$

$$= 71.25 \; kN$$

Shear strength of vertical stirrups

$$V_s = \frac{\sigma_{sv} \; A_{sv} \; d}{S_v}$$

$$= \frac{230 \times 2 \times 28.27 \times 625}{145} \times 10^{-3}$$

$$= 56.05 \ kN$$

Hence, shear strength of beam

$$V = 36.79 + 71.25 + 56.05$$

$$= 164.09 \ kN$$

(ii) At 1.3 m away from the face of support.

At this point no bent-up bars cross the section of beam. Hence, the shear strength of beam

$$= 71.25 + 56.05 = 127.3 \ kN$$

3.9.8 Shear consideration when tension reinforcement is curtailed in tension zone. For the tension reinforcement in a

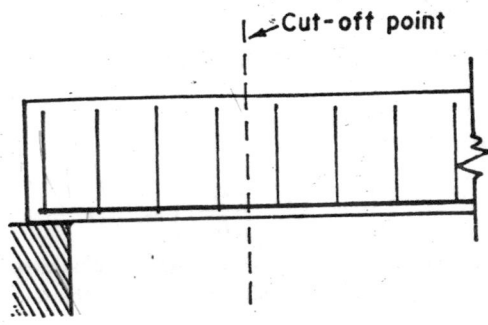

Fig. 3.27

beam, to be curtailed in the tension zone, one of the following conditions should be satisfied as per the requirement of the IS Code: 456-1978 para . **25.2.3.2.**

(i) Shear force at the cut-off point is not more than $\frac{2}{3}$ of the permitted shear at that section. See Fig. 3.27. $V \ngtr \frac{2}{3} (V_c + V_s)$ at cut-off point.

(ii) Stirrups in addition to those required for shear should be provided along the terminated bar from the point of cut-off for a distance equal to $\frac{3}{4}$ of the effective depth. See Fig. 3.28

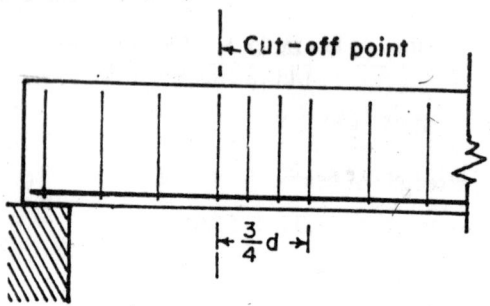

Fig. 3.28

The additional stirrup area is given by

$$= \frac{0.4 b s_v}{f_y}$$

Stirrup area for shear reinforcement as per Eq. (3.28) can be written as

$$= \frac{V_s s_v}{\sigma_{sv} d}$$

Combining the two areas together, the total stirrup area can be written as

$$A_{sv} = \frac{V_s s_v}{\sigma_{sv} d} + \frac{0.4 b s_v}{f_y}$$

$$= \left(\frac{V_s}{\sigma_{sv} d} + \frac{0.4 b}{f_y} \right) s_v \qquad (3.32)$$

For a given shear stirrup, the spacing can be calculated from Eq. (3.32). This spacing should be maintained over the terminated bar from the cut-off point for a distance equal to $\frac{3}{4}$ of the effective depth. However, this spacing should not exceed a minimum limit which is

$$= \frac{d}{8 \beta_b}$$

where $\quad \beta_b = \dfrac{\text{Area of bars cut-off}}{\text{Total area of bars}}$

(iii) At the cut-off point, the continuing bars should provide double the area required for flexure and the shear force should not exceed $\frac{3}{4}$ of that permitted.

Example 3.6

Tension reinforcement was curtailed in the beam based on bending moment consideration, in Example 3.3. Let the beam be reinforced in shear with 6 mm dia. two legged stirrups of Mild Steel at a spacing of 180 mm c/c throughout. Since the main bars have been curtailed in the tension zone check whether the provisions of para. 25.2.3.2 of the IS Code: 456-1978 are satisfied or not. Otherwise, provide the remedial measures.

Solution

For the given beam

$$\frac{100 \, A_{st}}{bd} = \frac{100 \times 8 \times 113.1}{250 \times 500}$$
$$= 0.723$$

From Table 17 of the IS Code 456 the permissible shear stress

$$\tau_c = 0.335 \ \text{N/mm}^2$$

Shear strength of the concrete in the beam

$$= 0.335 \times 250 \times 500 \times 10^{-3}$$
$$= 41.88 \ \text{kN}$$

Strength of the shear reinforcement is

$$V_s = \frac{\sigma_{sv} A_{sv} d}{S_v}$$
$$= \frac{140 \times 2 \times 28.27 \times 500 \times 10^{-3}}{180}$$
$$= 22 \ \text{kN}$$

Hence, the shear strength of the beam section

$$V_c + V_s = 41.88 + 22$$
$$= 63.88 \ \text{kN}$$

The beam can now be checked for the provisions of the para. 25.2.3.2 of the IS Code: 456 at the points of cut-off of the bars.

(i) Section distant 0.685 m from the centre of support, where 2 bars are cut-off.

Shear force at this section is

$$V = (\tfrac{1}{2} \times 4.5 - 0.685) \times 21.44$$
$$= 33.55 \text{ kN}$$

Now $\dfrac{2}{3}(V_c + V_s) = \dfrac{2}{3} \times 63.88$

$$= 42.59 \text{ kN}$$

Evidently $V < \tfrac{2}{3}(V_c + V_s)$. Hence, the criterion of shear strength is satisfied and the 2 bars can be safely cut-off in the tension zone.

(ii) Section distant 0.21 m from centre of support, where 4 bars are cut-off.

Shear force at this section is

$$V = (\tfrac{1}{2} \times 4.5 - 0.21)$$
$$= 43.74 \text{ kN}$$

Evidently $V > \tfrac{2}{3}(V_c + V_s)$

Additional stirrup area should be provided along the terminated bars for a distance $= \dfrac{3}{4}$ of effective depth i.e.

$$= \dfrac{3}{4} \times 500$$

$$= 375 \text{ mm}$$

Over this length, the spacing of 6 mm dia. 2 legged stirrups of Mild Steel can now be found out by substituting the known values in Eq. (3.32). Hence

$$2 \times 28.27 = \left(\frac{22 \times 10^3}{140 \times 500} + \frac{0.4 \times 250}{250} \right) S_v$$

or $S_v = 79.16 \text{ mm}$

Adopt a spacing of 75 mm c/c over a length of 375 mm as shown in Fig. 3.29.

$$\beta_b = \frac{4}{8} = 0.5$$

Maximum limit on spacing of these stirrups is

$$\frac{d}{8\,\beta_b} = \frac{500}{8 \times 0.5}$$

$$= 125 \text{ mm}$$

Actual spacing is less than this limit. Hence O.K.

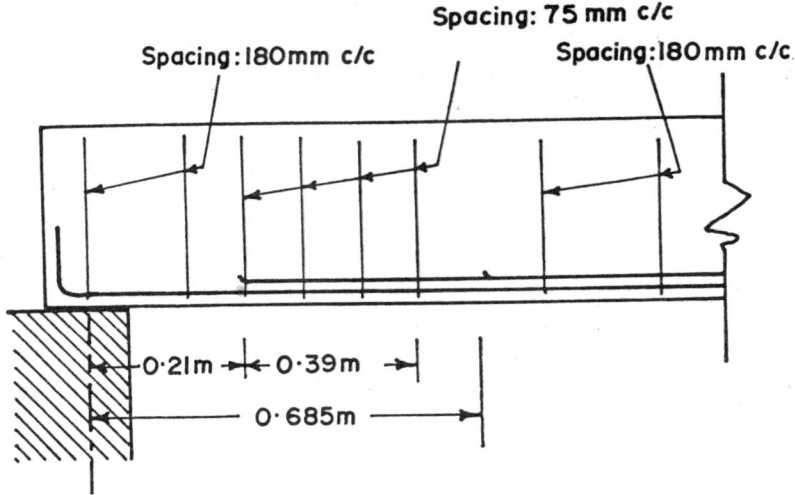

Fig. 3 29 Details for Example 3.6

3.10 One-way slab. A reinforced concrete slab supported along two opposite edges may be structurally treated just like a beam spanning in one direction, as it supports the load one-way only. The main reinforcement is provided along the direction along which the load is supported. Hence, the name one-way slab.

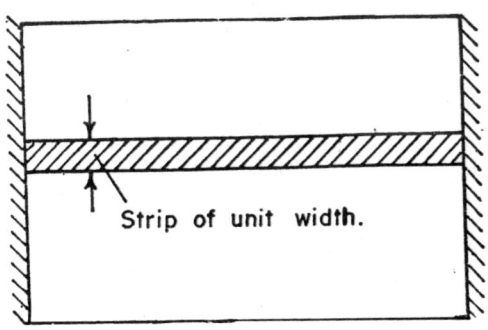

Fig. 3.30 Slab spanning in one direction

If a slab is supported along all the four edges it spans in two directions and supports the load by two-way action. As such, tension reinforcement is required along the two directions to take up the bending moments. Thus, the slab is known as two-way slab. However, if the length to width ratio of the slab in plan is $\geqslant 2$, the slab has a predominantly one-way action along the shorter span. As such, it may be designed as one-way slab only. In one-way simply supported slabs the effective span may be calculated in the same way as for the simply supported beam. It is best to consider a strip of 1 m width as shown in Fig. 3.30. This may be treated as a beam of width equal to unity, for which the effective depth and the reinforcement may be calculated in the usual manner. The main bars are uniformly spaced. c/c spacing of the bars is

$$= 1000 \times \frac{a_{st}}{A_{st}} \text{mm}$$

where a_{st} = sectional area of one bar

A_{st} = area of reinforcement in 1 m width of slab

3.10.1 Temperature and shrinkage reinforcement. At right angles to the main reinforcement, temperature and shrinkage reinforcement is provided. The purpose of this reinforcement is to prevent opening up of any cracks in the slab on account of shortening of length which may occur due to shrinkage or fall of temperature. It also helps in distributing the concentrated load on slab to adjacent areas .in lateral direction. The quantity of temperature and shrinkage reinforcement shall not be less than the following limits which are expressed as % age of the cross-sectional area of concrete:

M.S. bars $\not< 0.15$ %

H.Y.S.D. bars $\not< 0.12$ %

Even the tension reinforcement along the main span should never be less than the above limits. As such, these limits are also known as the minimum reinforcement limits.

3.10.2 Maximum limit on spacing of reinforcement bars. According to the recommendation of the IS Code : 456, the limits on the spacing of the bars are as follows:

(i) For tension reinforcement

$\not> 3d$

$\not> 450$ mm

(ii) For temperature and shrinkage reinforcement

    ≯ 5d

    ≯ 450 mm

3.10.3 Permissible shear stress in slab concrete As per IS Code: 456-1978 the permissible shear stress in concrete in solid reinforced concrete slab shall be taken to be $k \tau_c$. The values of k depend upon the thickness of slab and are given below in Table 3.13.

Table 3.13

(REF: IS CODE: 456—1978, CLAUSE 47.2.1.1)

Thickness of slab, mm	300 or more	275	250	225	200	175	150 or less
k	1.00	1.05	1.10	1.15	1.20	1.25	1.30

Note: The values of τ_c are given in Table 3.11.

Example 3.7

Design a reinforced concrete slab over a clear opening in plan of 3.75 × 9 m. The slab is to be provided with a concrete tile flooring on the top, of thickness = 50 mm. Live load on slab = 2 kN/m². Bearing of slab along the edges = 150 mm. Concrete grade : M15 and steel grade : Fe415.

Solution

Length to width ratio for slab $= \dfrac{9.15}{3.9}$

$$= 2.35$$

This ratio being greater than 2, the slab may be designed spanning one-way over the clear span of 3.75 m. For the purpose of estimating the self weight of slab, its thickness is assumed to be 180 mm. Hence, the load per unit area of slab would be

Self weight, $0.18 \times 1 \times 1 \times 25$ = 4.5 kN/m²

Weight of tiles, $0.05 \times 1 \times 1 \times 24$ = 1.2 kN/m²

Live load   = 2.0 kN/m²

               Total   w ·   = 7.7 kN/m²

Consider a strip of slab of 1 m width, spanning over 3.75 m. This strip may be considered to act as a simply supported beam carrying a load of intensity = 7.7 kN/m.

Effective span

$$= c/c \text{ of supports}$$
$$= 3.75 + 0.15$$
$$= 3.9 \text{ m}$$

Bending moment at mid-span

$$= \frac{wl^2}{8}$$
$$= \frac{7.7 \times 3.9^2}{8}$$
$$= 14.64 \text{ kN-m/m}$$

Permissible stresses are

$$\sigma_{cbc} = 5 \text{ N/mm}^2, \ \sigma_{st} = 230 \text{ N/mm}^2$$

Modular ratio m = 18.67

From Tables 3.4 and 3.6

$$n = 0.289$$

Hence

$$j = 1 - \frac{0.289}{3} = 0.904$$

Moment factor R = 0.653

For a strip of 1 m width, equating

$$Rbd^2 = M$$

i.e.

$$0.653 \times 1000 \text{x} d^2 = 14.64 \times 10^6$$

$$\therefore \quad d = 149.7 \text{ mm.}$$

Adopt an effective depth = 150 mm.

Tension reinforcement is

$$A_{st} = \frac{M}{\sigma_{st} jd} = \frac{14.64 \times 10^6}{230 \times 0.904 \times 150}$$
$$= 469.4 \text{ mm}^2/\text{m}$$

Provide 12 mm dia. bars at a spacing $= \dfrac{1000 \times 113.1}{469.4}$

$$= 240.95 \text{ mm, say } 240 \text{ mm c/c.}$$

Check: Maximum permitted spacing

$$\not> 3d \text{ i.e. } 3 \times 150 = 450 \text{ mm}$$
$$\not> 450 \text{ mm}$$

Actual spacing is within the above limit.

Cover on reinforcement

$\not< 15$ mm

$\not< $ diameter of bar i.e. 12 mm

Adopting an overall thickness of 175 mm for the slab will ensure a cover of 19 mm on the bar. This will be alright.

Main reinforcement bars will run along the short span. At right angles to these bars, temperature and shrinkage reinforcement shall be provided. It is given by

$$= \frac{0.12 \times 175 \times 1000}{100}$$

$$= 210 \text{ mm}^2/\text{m}$$

Provide 10 mm dia. bars at a spacing $= \dfrac{78.54}{210} \times 1000$

$$= 374 \text{ mm c/c, say } 370 \text{ mm c/c.}$$

Check : Maximum permitted spacing

$\not> 5d$ i.e. $5 \times 150 = 750$ mm

$\not> 450$ mm

Actual spacing is within the above limit. A section of the slab showing the details of reinforcement is shown in Fig. 3.31. As a good practice alternate main bars may be bent up at a point 0.15 of span short of the support. In this case alternate bars have been bent up at 0.5 m from the face of the support.

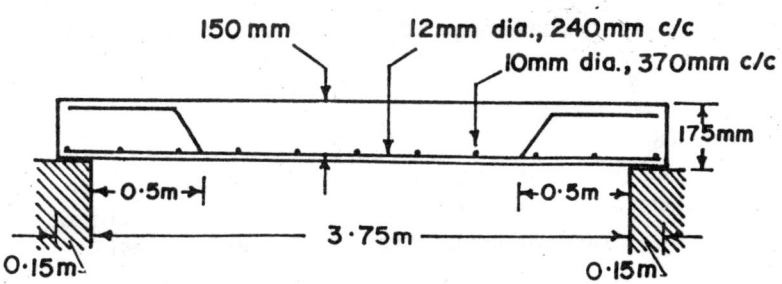

Fig. 3.31 Reinforcement details for one-way slab vide Example 3.7

The slab can now be checked for shear. Critical section occurs at an effective depth i.e. 150 mm from the face of support. Therefore, shear force in a strip of unit width is

$$V = (\tfrac{1}{2} \times 3.75 - 0.15) \times 7.7$$
$$= 13.28 \text{ kN/m}$$

$$\tau_v = \frac{13.28 \times 1000}{1000 \times 150}$$

$$= 0.089 \text{ N/mm}^2$$

Tension reinforcement area per unit width

$$A_{st} = \frac{113.1 \times 1000}{240}$$

$$= 471.25 \quad \text{mm}^2/\text{m}$$

$$\frac{100 A_{st}}{bd} = \frac{100 \times 471.25}{1000 \times 150}$$

$$= 0.314$$

From Table 3.11 for concrete grade: M15

$$\tau_c = 0.238 \text{ N/mm}^2$$

From Table 3.13 for slab thickness $= 175$ mm

$$k = 1.25$$

Hence, permissible shear stress in slab concrete is

$$k \tau_c = 1.25 \times 0.238$$

$$= 0.298 \text{ N/mm}^2$$

Evidently τ_v is very small compared to $k \tau_c$. Hence, the slab is safe in shear.

Example 3.8

A sunshade projects out 1.25 m from the lintel provided over a window opening of 1.5 m. See Fig. 3.32. Design the sunshade and the lintel. Concrete grade: M15 and steel grade : Fe415.

Solution

Permissible stresses are,

$$\sigma_{cbc} = 5 \text{ N/mm}^2, \ \sigma_{st} = 230 \text{ N/mm}^2$$

Modular ratio $= 18.67$

Design constants for balanced condition are:

$$n = 0.289$$
$$j = 0.904$$
$$R = 0.653$$

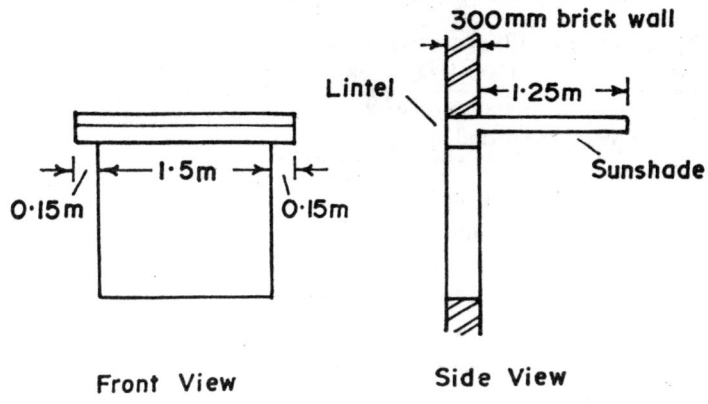

Fig. 3.32 Details for Example 3.8

Design of sunshade:

Consider a strip of width = unity projecting out 1.25 m from the lintel as a cantilever. The loading on the strip is

Self weight + Finishing = 2.5 kN/m (assumed)
Occasional live load   = 0.75 kN/m

Total w   = 3.25 kN/m

Bending moment

$$M = \frac{Wl^2}{2}$$
$$= \frac{3.25 \times 1.25^2}{2}$$
$$= 2.54 \text{ kN-m}$$

Equating

$$Rbd^2 = M$$

i.e. $0.653 \times 1000 \times d^2 = 2.54 \times 10^6$

∴ $d = 62.4$ mm. Adopt $d = 65$ mm.

Tension reinforcement

$$A_{st} = \frac{M}{\sigma_{st}jd}$$
$$= \frac{2.54 \times 10^6}{230 \times 0.904 \times 65}$$
$$= 187.9 \text{ mm}^2$$

Provide 6 mm dia. bars at a spacing

$$= \frac{1000 \times 28.27}{187.9} = 150.45 \text{ mm c/c.}$$

Maximum permitted spacing is

$$\not> 3d \text{ i.e. } 3 \times 65 = 195 \text{ mm}$$
$$\not> 450 \text{ mm}$$

Hence, provide 6 mm dia. bars at a spacing of 150 mm c/c.

Cover on bar

$$\not< 15 \text{ mm}$$
$$\not< \text{ dia. of bar i.e. 6 mm.}$$

Adopt an overall depth of 85 mm. This will ensure a cover of 17 mm on the bar.

Check:

Actual self weight of slab $= 0.085 \times 25 = 2.125 \text{ kN/m}^2$

Finishing, 15 mm thick $= 0.015 \times 24 = 0.36 \text{ kN/m}^2$

Total w $= 2.485 \text{ kN/m}^2$

This is very close to the assumed value of 2.5 kN/m². The main reinforcement in sunshade requires to be embedded into the concrete beyond the side face of lintel in order to develop the design tensile stress. Development length

$$l_d = \frac{\phi \sigma_s}{4 \tau_{bd}}$$

Here $\phi = 6 \text{ mm}$

$\tau_{bd} = 0.6 \text{ N/mm}^2$ from Table 3.1. This may be increased by 40% for deformed bars. Hence

$$l_d = \frac{6 \times 230}{4 \times 0.6 \times 1.4}$$

$$= 410.7 \text{ mm, say } 415 \text{ mm.}$$

The main reinforcement should be embedded by at least 415 mm into the lintel.

Temperature and shrinkage reinforcement

$$= \frac{0.12}{100} \times 85 \times 1000$$

$$= 102 \text{ mm}^2/\text{m}$$

Provide 6 mm dia. bars at a spacing $= \dfrac{1000}{102} \times 28.27$

$\qquad = 277.2$ mm c/c, say 275 mm c/c.

Maximum permitted spacing

$\qquad \not> 5d$ i.e. $5 \times 65 = 325$ mm

$\qquad \not> 450$ mm

Actual spacing is within the limit given above.

Design of Lintel:

Effective span $= $ c/c of supports

$\qquad = 1.5 + 0.15$

$\qquad = 1.65$ m

Due to the arching action in wall the masonry within a triangular region over the lintel, with sides sloping at 60°, imparts its load to the lintel. See Fig. 3.33.

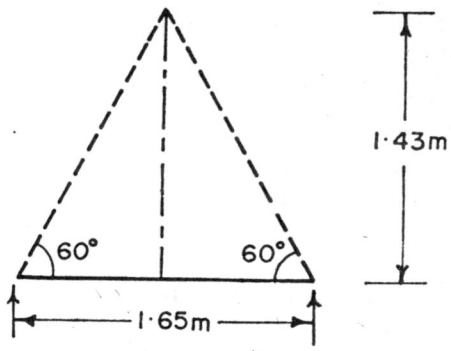

Fig. 3.33 Triangular load on lintel due to arching action in the masonry

For the purpose of estimating the self weight, the thickness of lintel is assumed to be 160 mm.

Self weight of lintel $= 0.16 \times 0.3 \times 25 = 1.2$ kN/m

Load due to sunshade $= 1.25 \times 2.5 \qquad = 3.125$ kN/m

$\qquad\qquad\qquad$ Total u.d.l. $= 4.325$ kN/m

Weight of triangular masonry $= \frac{1}{2} \times 1.65 \times 1.43 \times 0.3 \times 21.7$

$\qquad\qquad\qquad = 7.68$ kN

Hence, end reaction

$$= \tfrac{1}{2} \times 1.65 \times 4.325 + \tfrac{1}{2} \times 7.68$$

$$= 7.41 \text{ kN}$$

Mid-span bending moment

$$= 7.41 \times \frac{1.65}{2} - \tfrac{1}{2} \times 7.68 \times \frac{1.65}{6} - 4.325 \times \frac{1.65^2}{4} \times$$

$$= 6.11 - 1.06 - 1.47$$

$$= 3.58 \text{ kN--m}$$

Equating

$$\text{Rbd}^2 = \text{M}$$

i.e. $\quad 0.653 \times 300 \times d^2 = 3.58 \times 10^6 \text{ N-mm}$

$\therefore \ d = 135 \text{ mm}$

Tension reinforcement

$$A_{st} = \frac{3.58 \times 10^6}{230 \times 0.904 \times 135}$$

$$= 127.5 \text{ mm}^2$$

However, 2 bars of 12 mm dia. may be provided
Cover on bars

$$\nleqslant 25 \text{ mm}$$

$$\nleqslant \text{ dia. of bar i.e. 12 mm.}$$

Adopt an overall depth of 170 mm. This will ensure a cover on bars of 29 mm. Details of reinforcement are shown in Fig. 3.34.

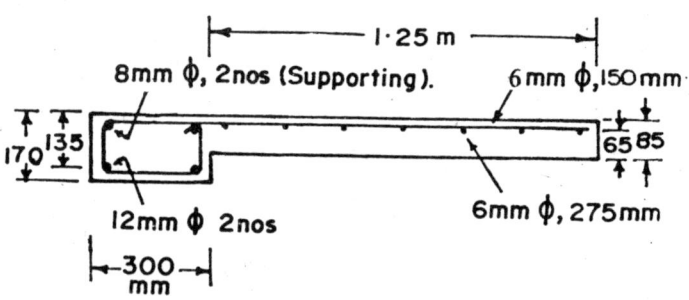

Fig. 3.34. Reinforcement details for sunshade

4

Doubly Reinforced Beam

4.1 INTRODUCTION

There are instances when the dimensions of a beam section are restricted on account of architectural reasons or for gaining more headroom. In such cases, moment capacity of beam section, which may be otherwise inadequate, can be enhanced by adding steel reinforcement in the compression zone of concrete and balancing it with an additional tensile reinforcement. The beam, thus reinforced both in tension and compression, is known as a doubly reinforced beam.

There may also be instances when a beam may be required to bear reversal of bending moment. An example is the brace connecting the columns of the supporting structure of an overhead tank subjected to wind force. Such a beam has to be designed with equal reinforcement in tension and compression zones. It is a special case of a doubly reinforced beam.

4.2 PERMISSIBLE STRESS

Permissible bending stress in compression in concrete and tension in steel reinforcement are the same as for the singly reinforced section. However, permissible bending stress in compression in steel is lower than that in tension, in view of the possibility of buckling. IS Code: 456-1978 specification for the permissible stress

σ_{sc} in bending compression in steel reinforcement is as follows:
σ_{sc} shall be taken to be the lesser of
(i) The value specified in Table 3.2.
(ii) 1.5 m times the calculated compressive stress in the surrounding concrete.

4.3 BENDING THEORY AND FORMULATION OF BENDING STRENGTH

The assumptions in the theory of bending of doubly reinforced section are the same as for the singly reinforced section which have been discussed in Chapter 3.   Consider a beam section shown in Fig. 4.1 in which the concrete in compression and steel in tension reach their permissible stresses simultaneously. Then, the calculated stress in concrete at the level of compression steel is

$$\sigma'_{cbc} = \left(\frac{nd - d_c}{nd} \right) \sigma_{cbc}$$

Hence, the permissible stress in compression steel is

$$\sigma_{sc} = 1.5\, m \left(\frac{nd - d_c}{nd} \right) \sigma_{cbc}$$

or   = the value specified in Table 3.2, whichever is
the lower.   Generally, the former value is the lower of the two.

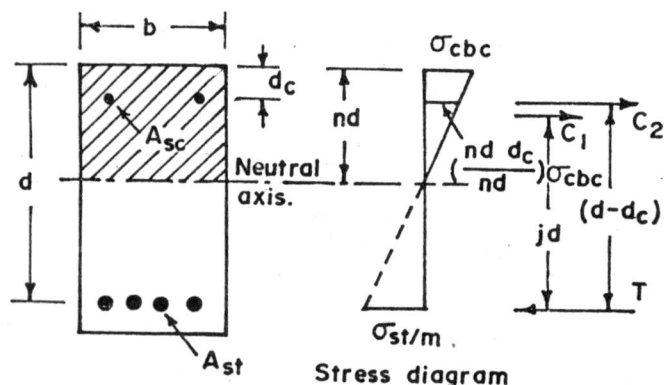

Fig. 4.1 Doubly reinforced section.

Total compression = $C_1 + C_2$

Where

$$C_1 = \text{Compressive force in concrete}$$
$$= \tfrac{1}{2}\,\sigma_{cbc}\,bnd$$
$$C_2 = \text{Compressive force in steel}$$
$$= (\sigma_{sc} - \sigma'_{cbc})\,A_{sc}$$
$$= (1.5\,m - 1)\left(\frac{nd - d_c}{nd}\right)\sigma_{cbc}A_{sc}$$

The moment of resistance can be found by taking moment of the compressive forces about the centre of tension reinforcement:

$$M_r = C_1\,jd + C_2\,(d - d_c)$$
$$= \tfrac{1}{2}\,\sigma_{cbc}\,bndjd + (1.5\,m - 1)\left(\frac{nd - d_c}{nd}\right)\sigma_{cbc}A_{sc}\,(d - d_c)$$
$$= \tfrac{1}{2}\,\sigma_{cbc}njbd^2 + (1.5\,m - 1)\left(\frac{nd - d_c}{nd}\right)\sigma_{cbc}A_{sc}\,(d - d_c)$$
$$= Rbd^2 + (1.5\,m - 1)\left(\frac{nd - d_c}{nd}\right)\sigma_{cbc}A_{sc}\,(d - d_c) \quad (4.1)$$

The first term on the r.h.s. of Eq. (4.1) is the moment of resistance ($= M_b$) of the singly reinforced section and the second term is the additional moment of resistance ($= M'$) offered by the compression reinforcement. Hence, Eq. (4.1) can be rewritten as

$$M = M_b + M' \quad (4.2)$$

This situation can be represented in Fig. 4.2. The tension reinforcement may be considered to be composed two parts A_{st1} and A_{st2}. The part A_{st1} is that required for a singly reinforced balanced section, whereas, the part A_{st2} is that required for balancing the compression steel.

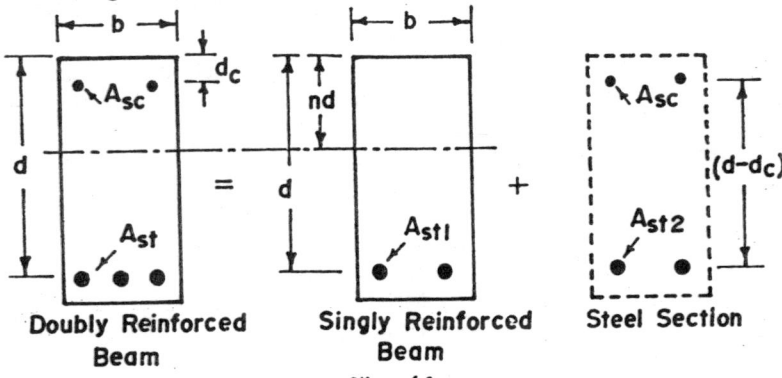

Doubly Reinforced   Singly Reinforced   Steel Section
Beam   Beam

Fig. 4.2

Hence

$$A_{st1} = \frac{M_b}{\sigma_{st} \, jd}$$

$$= \frac{R b d^2}{\sigma_{st} \, jd} \tag{4.3}$$

$$A_{st2} = \frac{M'}{\sigma_{st} \, (d - d_c)} \tag{4.4}$$

and $$A_{st} = A_{st1} + A_{st2} \tag{4.5}$$

Equating the tensile and compressive forces in steel section

$$A_{st2} \, \sigma_{st} = A_{sc} \, (1.5m - 1) \left(\frac{nd - d_c}{nd} \right) \sigma_{cbc}$$

$$\therefore \qquad A_{sc} = \frac{A_{st2} \, \sigma_{st}}{(1.5m - 1) \left(\dfrac{nd - d_c}{nd} \right) \sigma_{cbc}} \tag{4.6}$$

Note: The permissible stress in compression steel is taken to be $1.5m \left(\dfrac{nd - d_c}{nd} \right) \sigma_{cbc}$ assuming that it is less than σ_{sc} which is specified in Table 3.2. However, should σ_{sc} from Table 3.2 be the lesser it should be used instead.

Design Procedure

As there is a restriction on the dimensions of the section, these are known initially.

The tension and compression reinforcement can now be calculated as follows:

(i) The design bending moment is first calculated from the given span and loading.

(ii) From the known dimensions of the section the bending strength of the singly reinforced section is calculated i.e. $M_b = R b d^2$.

(iii) Hence, $M' = M - M_b$ is found.

(iv) Then, tensile reinforcements are calculated:

$$A_{st1} = \frac{M_b}{\sigma_{st} \, j \, d}$$

$$A_{st2} = \frac{M'}{\sigma_{st} \, (d - d_c)}$$

$$A_{st} = A_{st1} + A_{st2}$$

(v) Finally

$$A_{sc} = \frac{A_{st2}\, \sigma_{st}}{(1.5\, m - 1)\left(\dfrac{nd - d_c}{nd}\right)\sigma_{cbc}}$$

(vi) Check whether the value

$$\sigma_{sc} = 1.5\, m\left(\frac{nd - d_e}{nd}\right)\sigma_{cbc}$$

is lower than that specified in Table 3.2. Otherwise, replace the denominator of the r.h.s. of the expression for A_{sc} by the value

$$\sigma_{sc} - \left(\frac{nd - d_e}{nd}\right)\sigma_{cbc}$$

Example 4.1

A beam of rectangular section is to be designed to take up a bending moment of 69 kN-m. The width and overall depth of the beam are to be restricted to 300 mm and 550 mm respectively due to architectural reasons. Concrete grade: M15 and steel grade: Fe 415.

Solution:

Knowing the limits on the dimensions of the section the effective depth may be chosen to be $= 515$ mm. For the given grades of concrete and steel, the permissible stresses are:

$\sigma_{cbc} = 5$ N/mm^2, $\sigma_{st} = 230$ N/mm^2 from Tables 3.1 and 3.2 respectively Modular ratio $m = 18.67$ from Table 3.3. Design constants from Tables 3.4 and 3.6 are:

$$n = 0.289$$

$$j = 1 - \frac{0.289}{3} = 0.904$$

$$R = 0.653$$

Moment of resistance of the singly reinforced section

$$\begin{aligned} M_b &= Rbd^2 \\ &= 0.653 \times 300 \times 515^2 \times 10^{-6} \\ &= 51.96 \text{ kN-m} \end{aligned}$$

This is less than the required moment capacity of 69 kN-m. Hence, compression reinforcement should be provided. The section will therefore, be doubly reinforced. Keeping $d_c = 35$ mm, the calculated value of the permissible stress in compression reinforcement is

$$\sigma_{sc} = 1.5 \times 18.67 \times \left(\frac{0.289 \times 515 - 35}{0.289 \times 515}\right) \times 5$$

$$= 107.1 \ N/mm^2$$

Permissible compressive stress from Table 3.2 is

$$\sigma_{sc} = 190 \ N/mm^2$$

Evidently, the permissible value of the stress lower of the two is = 107.1 N/mm^2 which should be used. Now, the reinforcements can be calculated.

$$A_{st1} = \frac{M_b}{\sigma_{st} \ j \ d}$$

$$= \frac{51.96 \times 10^6}{230 \times 0.904 \times 515}$$

$$= 485.3 \ mm^2$$

$$M' = M - M_b$$

$$= 69 - 51.96$$

$$= 17.04 \ kN\text{-}m$$

$$A_{st2} = \frac{M'}{\sigma_{st} \ (d - d_c)}$$

$$= \frac{17.04 \times 10^6}{230 \times (515 - 35)}$$

$$= 154.35 \ mm^2$$

Hence

$$A_{st} = A_{st1} + A_{st2}$$

$$= 485.3 + 154.35 = 639.65 \ mm^2$$

Finally

$$A_{sc} = \frac{A_{st2} \ \sigma_{st}}{(1.5 \ m - 1) \left(\dfrac{nd - d_c}{nd}\right) \sigma_{cbc}}$$

$$= \frac{154.35 \times 230}{(1.5 \times 18.67 - 1) \ \dfrac{(0.289 \times 515 - 35) \times 5}{(0.289 \times 515)}}$$

$$= 343.77 \ mm^2$$

The designed section is shown in Fig. 4.3.

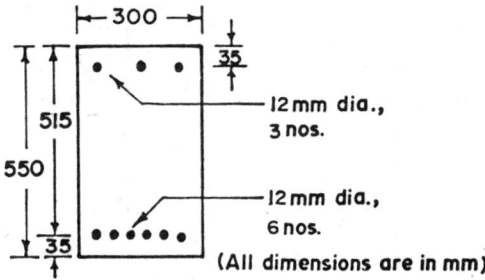

Fig. 4.3 Designed section for Example 4·1

4.4 MOMENT OF RESISTANCE OF A GIVEN BEAM SECTION

For a given beam section it is first of all necessary to find the depth of neutral axis. This is accomplished by equating the moment of the compressive area about the neutral axis to that of the tensile area about the same:

$$bnd \times \frac{nd}{2} + (1.5 \, m - 1) \, A_{sc} \, (nd - d_o)$$
$$= m \, A_{st} \, (d - nd) \qquad (4.7)$$

This is a quadratic equation in n which can be readily solved. See Fig. 4.4

Balanced value of n in terms of the permissible stresses is

$$n = \frac{1}{1 + \dfrac{\sigma_{st}}{m \, \sigma_{cbc}}}$$

On comparing the calculated value of n with the balanced value, three cases arise which are identified as follows:

(i) Calculated $n <$ than the balanced value.

For this case the stresses in the section are

$$f_{st} = \sigma_{st}$$

and $\qquad f_{cbc} < \sigma_{cbc}$

From the stress diagram

$$f_{cbc} = \frac{f_{st}}{m} \times \frac{nd}{(d - nd)}$$
$$= \frac{\sigma_{st}}{m} \times \frac{n}{1 - n} \qquad (4.8)$$

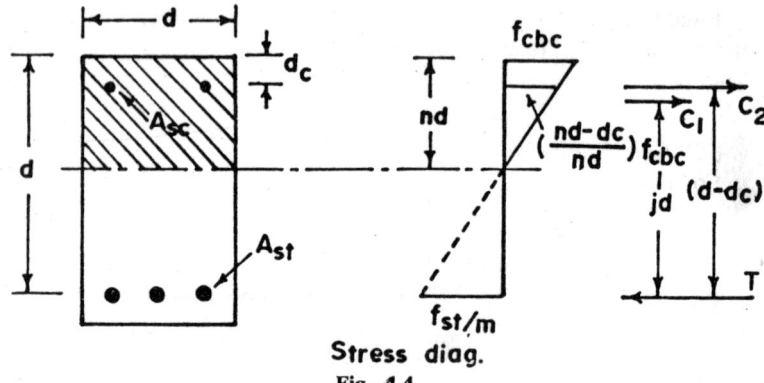

Stress diag.

Fig. 4.4

This gives the compressive stress in concrete f_{cbc}. Compressive stress in concrete at the level of compression steel is

$$= \left(\frac{nd - d_c}{nd} \right) f_{cbc}$$

Hence, the stress in compression steel is

$$= 1.5\, m \left(\frac{nd - d_c}{nd} \right) f_{cbc}$$

Compressive forces in the section can be written as

$$C_1 = \tfrac{1}{2} f_{cbc}\, bnd \qquad (4.9)$$

$$C_2 = (1.5\, m - 1)\left(\frac{nd - d_c}{nd} \right) f_{cbc}\, A_{sc} \qquad (4.10)$$

Finally, the moment of resistence is found by taking moment of the compressive forces about the tensile reinforcement. That is

$$M_r = C_1\, jd + C_2\, (d - d_c) \qquad (4.11)$$

(ii) Calculated n $=$ the balanced value.
For this case, the stresses in the section are

$$f_{st} = \sigma_{st}$$
$$f_{cbc} = \sigma_{cbc}$$

Stress in compression steel is

$$= 1.5\ m\ \left(\frac{nd - d_c}{nd} \right) \sigma_{cbc}$$

Compressive forces in the section can be written as

$$C_1 = \tfrac{1}{2}\ \sigma_{cbc}\ b\, nd \qquad (4.12)$$

$$C_2 = (1.5\ m - 1)\left(\frac{nd - d_c}{nd} \right) \sigma_{cbc}\, A_{sc} \qquad (4.13)$$

Finally, the moment of resistance is found by taking moment of the compressive forces about the centre of tensile reinforcement. That is

$$M_r = C_1 \, jd + C_2 \, (d - d_c) \tag{4.14}$$

(iii) Calculated $n >$ the balanced value.

For this case, the stresses in the section are

$$f_{st} < \sigma_{st}$$
$$f_{cbc} = \sigma_{cbc}$$

The compressive forces can be calculated from Eq. (4.12) and (4.13) and the moment of resistance from Eq. (4.14).

Example 4.2

Calculate the moment of resistance of the doubly reinforced beam section shown in Fig. 4.5 for the following cases:

(a) $d = 565$ mm, (b) $d = 500$ mm

Concrete grade: M 15 and steel grade: Fe 415.

Solution:

Permissible stresses are

$\sigma_{cbc} = 5$ N/mm^2	from Table 3.1
$\sigma_{st} = 230$ N/mm^2	from Table 3.2
$m = 18.67$	from Table 3.3

Design constants are

$$n = 0.289 \qquad \text{from Table 3.4}$$

and $j = 1 - 0.289/3 = 0.904$

(a) $d = 565$ mm.

Substituting the values in Eq. (4.7) and simplifying

$$\tfrac{1}{2} \times 300 \times 565^2 \times n^2 + (1.5 \times 18.67 - 1) \times 2 \times 113.1 \times$$
$$(565n - 35) = 18.67 \times 2 \times 314.16 \times (565 - 565n)$$

or $n^2 + 0.207n - 0.14 = 0$

or $n = 0.284$

and $j = 1 - 0.284/3$

$$= 0.905$$

Evidently, calculated n is less than the balanced value. Hence, the section is under-reinforced in tension. From Eq. (4.8)

$$f_{cbc} = \frac{230 \times 0.284}{18.67 \times (1 - 0.284)}$$
$$= 4.89 \ \text{N/mm}^2$$

From Eq. (4.9)
$$C_1 = \tfrac{1}{2} \times 4.89 \times 300 \times 0.284 \times 565 \times 10^{-3}$$
$$= 117.7 \quad kN$$

From Eq. (4.10)
$$C_2 = (1.5 \times 18.67 - 1) \; \frac{(0.284 \times 565 - 35)}{0.284 \times 565}$$
$$\times 4.89 \times 2 \times 113.1 \times 10^{-3}$$
$$= 23.7 \quad kN$$

From Eq. (4.11)
$$M_r = (117.7 \times 0.905 \times 565 + 23.7 \times (565 - 35)) \times 10^{-3}$$
$$= 72.56 \quad kN\text{-}m$$

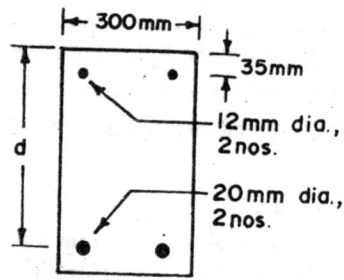

Fig. 4.5 Diagram for Example 4.2

(b) $d = 500$ mm

As before, substituting the values in Eq. (4.7) and simplifying
$$n^2 + 0.2685 \, n - 0.182 = 0$$
or $\quad n = 0.313$

and $\quad j = 1 - 0.313/3$
$$= 0.8957$$

Evidently, the calculated n is greater than the balanced value. Hence, the section is over-reinforced in tension. From Eq. (4.12)
$$C_1 = \tfrac{1}{2} \times 5 \times 300 \times 0.313 \times 500 \times 10^{-3}$$
$$= 117.375 \; kN$$

From Eq. (4.13),
$$C_2 = (1.5 \times 18.67 - 1) \times \frac{(0.313 \times 500 - 35)}{0.313 \times 500}$$
$$\times 5 \times 2 \times 113.1 \times 10^{-3}$$
$$= 23.69 \; kN$$

From Eq. (4.14)

$$M_r = (117.375 \times 0.8957 \times 500 + 23.69 \times (500 - 35))$$
$$\times 10^{-3}$$
$$= 63.58 \text{ kN-m}$$

4.5 DESIGN OF SECTION WITH $A_{sc} = A_{st}$

Beam sections where a complete reversal of bending moment is possible, are designed with equal reinforcements in tension and compression. In such sections the neutral axis factor n will always be less than the balanced value. The stresses in concrete and steel are written as:

$$f_{cbc} < \sigma_{cbc}$$
$$f_{st} = \sigma_{st}$$

The design of such a section can be done by trials as illustrated in the following example.

Example 4.3

A beam section is required to resist a reversible bending moment of 98.7 kN-m. Keeping equal reinforcements in tension and compression, work out the dimensions of the section and the area of reinforcements required. Concrete grade: M20 and steel grade: Fe 415. Keep $b \approx 0.5$ d.

Solution:

Permissible stresses are

	$\sigma_{cbc} = 7$ N/mm^2	from **Table** 3.1
	$\sigma_{st} = 230$ N/mm^2	from **Table** 3.2
and	m $= 13.33$	from **Table** 3.3

Design constants are:

n	$= 0.289$	from **Table** 3.4
j	$= 0.904$	
R	$= 0.914$	from table 3.6

Trial dimensions of the doubly reinforced section may be kept to be the same as those for the singly reinforced section. For the singly reinforced beam

$$Rbd^2 = M$$

or $0.914 \times bd^2 = 98.7 \times 10^6$

Putting $b = 0.5d$ and solving,

$d = 600$ mm

$b = 300$ mm

Tensile reinforcement

$$A_{st} = \frac{M}{\sigma_{st} \, j \, d}$$

$$= \frac{98.7 \times 10^6}{230 \times 0.904 \times 600}$$

$$= 791.2 \text{ mm}^2$$

Use 16 mm dia. bars, 4 nos. giving $A_{st} = 804.24$ mm²

Hence, $A_{sc} = 804.24$ mm², the same as tension reinforcement.

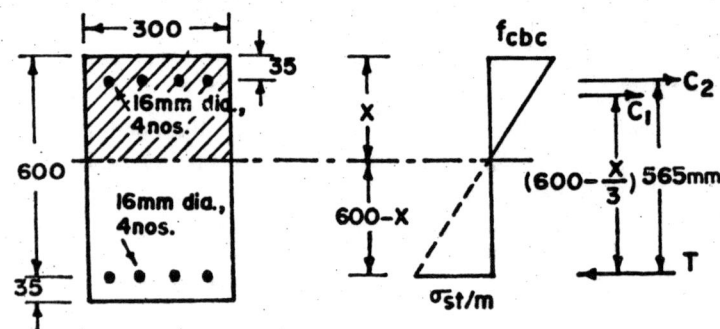

Fig. 4.6

Depth of neutral axis can now be found by taking moment of the compressive area about the neutral axis and equating it to that of the tensile area about the same. With reference to the Fig. 4.6

$$\frac{300 \, X^2}{2} + (1.5 \times 13.33 - 1) \times 804.24 \times (X - 35)$$

$$= 13.33 \times 804.24 \times (600 - X)$$

From which, $X = 145.6$ mm

From stress diagram

$$f_{cbc} = \frac{\sigma_{st} \, X}{m \, (600 - X)}$$

$$= \frac{230 \times 145.6}{13.33 \times (600 - 145.6)}$$

$$= 5.53 \text{ N/mm}^2, < 7 \text{ N/mm}^2 \text{ safe.}$$

Compressive forces are

$$C_1 = \tfrac{1}{2} f_{obc} bX$$
$$= \tfrac{1}{2} \times 5.53 \times 300 \times 145.6 \times 10^{-3}$$
$$=: 120.78 \text{ kN}$$
$$C_2 = (1.5 \text{ m} - 1) A_{sc} f_{obc} (X - d_c)/X$$
$$= (1.5 \times 13.33 - 1) \times 804.24 \times 5.53 \times$$
$$\frac{(145.6 - 35)}{145.6} \times 10^{-3}$$
$$= 64.15 \text{ kN.}$$

Moment of resistance is found by taking moment of the compressive forces about tension reinforcement and adding them

$$M_r = C_1 \left(d - \frac{X}{3} \right) + C_2 (d - d_c)$$

$$= 120.78 \times \left(600 - \frac{145.6}{3} \right) \times 10^{-3} + 64.15 \times$$
$$(600 - 35) \times 10^{-3}$$

$$= 102.85 \text{ kN·m,} > 98.7 \text{ kN-m, safe.}$$

5

T-Beam

In a reinforced concrete floor, the slab is supported over beams. In general, the slab and beam may be cast integrally and, therefore, provide a monolithic construction. Hence, a part of the slab participates with the beam in structural action and provides a flange to the beam to bear flexural compression. See Fig. 5.1.

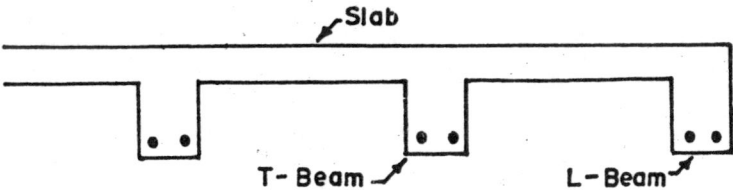

Fig. 5.1 Sectional view of a floor

In the middle part of the floor the beam would have a T shaped section, whereas, at the edge of the floor it would have an L shaped section. Accordingly, the beam may be either called a T-beam or an L-beam.

In the problem of design of a reinforced floor, first of all the c/c spacing of the beams is fixed up. Then, the slab may be designed as simply supported or continuous over the beams. The extent of the slab which may act as flange of the beam in bearing flexural compression, may be estimated from the semi-empirical rules which are given in the IS Code: 456-1978. These rules are discussed under the next para. Thus, the width b_f and the thickness D_f of the flange of a beam are known at the out set. Hence, in the design of a T-beam or an L-beam, the only unknowns which are to be determined now are the effective depth and the area of tension reinforcement.

A problem of analysis may be of reverse nature. The cross-section of the T-beam or an L-beam may be completely known and it may be required to determine its bending strength.

The width of web of a flanged beam may be fixed up on shear considerations or architectural considerations.

5.2   EFFECTIVE WIDTH OF FLANGE

The recommendations of the IS Code: 456 for the effective width of the flange are as follows:—

(a)   *Beam integral with floor slab*

    (i)   T-beam

$$b_f < \frac{l_o}{6} + b_w + 6 D_f$$

$$< b_w + \text{half the sum of the clear distances to the}$$
adjacent beams on either side.

    (ii)   L-beam

$$b_f < \frac{l_o}{12} + b_w + 3 D_t$$

$$< b_w + \text{half the clear distance to the adjacent}$$
beam.

Where

$$b_t = \text{effective width of flange of beam}$$
$$D_t = \text{thickness of flange of beam}$$
$$b_w = \text{width of web of beam}$$
$$l_o = \text{distance between the points of zero moments in}$$
a beam

(b)   *Isolated Beam*

Isolated T beams and L-beams may also be encountered in practice. IS Code: 456 recommendation for the width of flange of such beams are as follows:

    (i)   Isolated T-beam

$$b_t < \frac{l_o}{\frac{l_o}{b} + 4} + b_w$$

$$< b$$

(ii)   Isolated L-beam

$$b_f < \frac{0\,5\,l_o}{\dfrac{l_o}{b} + 4} + b_w$$

$$< b$$

where

b = actual width of flange of beam

For a simply supported beam l_o = the effective span of beam. For a simple beam overhanging a support, l_o may be taken to be the distance from the other support to the point of contraflexure. For a continuous beams IS Code 456 recommends that l_o may be taken to be 0.7 times the effective span.

5.3   DESIGN OF T-BEAM FOR BALANCED CONDITION

A balanced design is the one in which the compressive stress in the extreme fibres of concrete and the tensile stress in the reinforcement reach their permissible values at the same time.

In a design problem, since the effective width and the thickness of the flange are already known, all that remains to be determined is the effective depth of the beam and the area of tension reinforcement required to resist the given bending moment. Depending upon the depth of the neutral axis relative to the thickness of the flange, two cases arise in the design problem which are identified as follows:

(i) *Neutral axis is within the flange i.e.* $nd < D_f$. For this case the section can be treated as a singly reinforced rectangular section having width = width of flange. This is justified because the concrete below the neutral axis, being cracked in tension, is ineffective in bearing any tensile stress.

(ii) *Neutral axis lies outside the flange i.e.* $nd > D_f$. For this case the area of the concrete in compression zone is T-shaped. Hence, it becomes necessary to locate the centre of compression in addition to calculating the magnitude of compressive force.

The two cases are discussed in the following paras.

5.3.1   Neutral axis within the flange i.e. $nd < D_f$. The concrete below the neutral axis being under tensile stress is cracked and, therefore, ineffective. Hence, the section can be treated as singly reinforced rectangular section of width = b_f. For balanced design

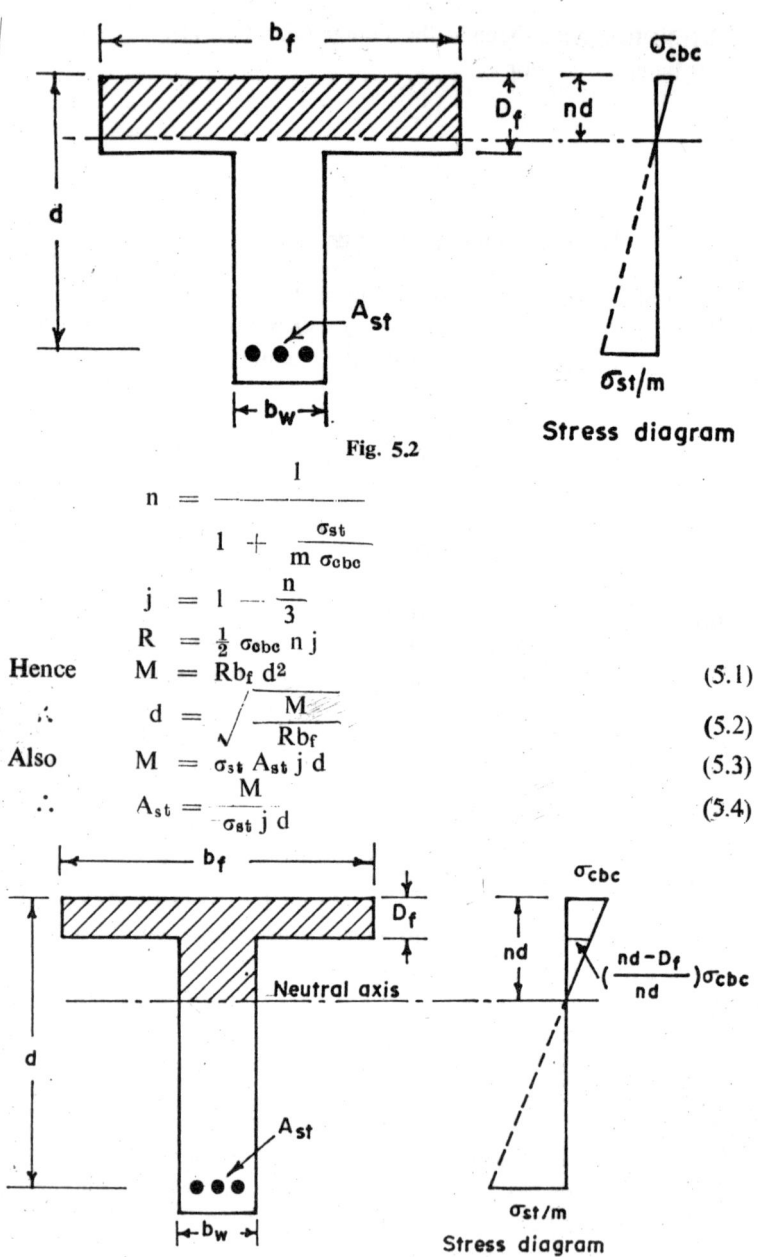

Fig. 5.2

$$n = \cfrac{1}{1 + \cfrac{\sigma_{st}}{m\,\sigma_{cbc}}}$$

$$j = 1 - \frac{n}{3}$$

$$R = \tfrac{1}{2}\,\sigma_{cbc}\,n\,j$$

Hence $M = Rb_f\,d^2$ (5.1)

∴ $d = \sqrt{\dfrac{M}{Rb_f}}$ (5.2)

Also $M = \sigma_{st}\,A_{st}\,j\,d$ (5.3)

∴ $A_{st} = \dfrac{M}{\sigma_{st}\,j\,d}$ (5.4)

Beam Section

Fig. 5.3

5.3.2 Neutral Axis Outside the Flange i.e. nd > D_f.

For balanced section

$$n = \frac{1}{1 + \dfrac{\sigma_{st}}{m\,\sigma_{cbc}}}$$

Let, $\quad r_1 = \dfrac{b_w}{b_f}$ and $r_2 = \dfrac{D_f}{nd}$

Compressive force in concrete is written as

$$C = \tfrac{1}{2}\,\sigma_{cbc}\,b_f\,nd - \tfrac{1}{2}\,\sigma_{cbc}\left(\frac{nd - D_f}{nd}\right)$$

$$\times (b_f - b_w)(nd - D_f)$$

$$= \tfrac{1}{2}\,\sigma_{cbc}\,b_f\,nd - \tfrac{1}{2}\,\sigma_{cbc}(b_f - b_w)\frac{(nd - D_f)^2}{nd}$$

$$= \tfrac{1}{2}\,\sigma_{cbc}\,b_f\,nd\left(1 - \frac{(b_f - b_w)}{b_f}\frac{(nd - D_f)^2}{(nd)^2}\right)$$

$$= \tfrac{1}{2}\,\sigma_{cbc}\,b_f\,nd\left(1 - (1 - r_1)(1 - r_2)^2\right) \qquad (5.5)$$

Moment of compressive force about the top concrete fibres is

$$M_c = \tfrac{1}{2}\,\sigma_{cbc}\,b_f\,nd\left(\frac{nd}{3}\right) - \tfrac{1}{2}\,\sigma_{cbc}\left(\frac{nd - D_f}{nd}\right)$$

$$\times (b_f - b_w)(nd - D_f) \times \left(\frac{nd - D_f}{3} + D_f\right)$$

$$= \tfrac{1}{6}\,\sigma_{cbc}\,b_f\,(nd)^2 - \tfrac{1}{6}\,\sigma_{cbc}(b_f - b_w)(nd - D_f)^2$$

$$\times \frac{(nd + 2D_f)}{nd}$$

$$= \tfrac{1}{6}\,\sigma_{cbc}\,b_f\,(nd)^2 \left\{ 1 - \frac{(b_f - b_w)}{b_f}\frac{(nd - D_f)^2}{(nd)^2} \right.$$

$$\left. \times \frac{(nd + 2D_f)}{nd} \right\}$$

$$= \tfrac{1}{6}\,\sigma_{cbc}\,b_f\,(nd)^2\left(1 - (1 - r_1)(1 - r_2)^2(1 + 2r_2)\right) \qquad (5.6)$$

The distance of the centre of compression from top concrete fibres is found by dividing Eq. (5.6) by Eq. (5.5). Thus

$$\overline{X} = \frac{M_c}{C} = \frac{nd}{3} \times \frac{1 - (1 - r_1)(1 - r_2)^2(1 + 2r_2)}{1 - (1 - r_1)(1 - r_2)^2} \qquad (5.7)$$

Hence, the arm of the internal couple is

$$a = (d - \overline{x})$$

Moment of resistance of the section is

$$M = C (d - \overline{x}) \tag{5.8}$$

also $\qquad M = T (d - \overline{x}) \tag{5.9}$

$$= A_{st} \sigma_{st} (d - \overline{x})$$

Hence, $\qquad A_{st} = \dfrac{M}{\sigma_{st} (d - \overline{x})} \tag{5.10}$

The area of reinforcement can also be found by equating tensile and compressive forces

$$T = C$$

or $\quad A_{st} \sigma_{st} = \tfrac{1}{2} \sigma_{cbc} \, b_f \, nd \, (1 - (1 - r_1) (1 - r_2)^2)$

or $\qquad A_{st} = \tfrac{1}{2} \dfrac{\sigma_{cbc}}{\sigma_{st}} \, b_f \, nd \, (1 - (1 - r_1) (1 - r_2)^2)$

or $\qquad \dfrac{A_{st}}{b_f \, d} = \dfrac{n \, \sigma_{cbc}}{2 \, \sigma_{st}} \, (1 - (1 - r_1) (1 - r_2)^2)$

$$= p \, (1 - (1 - r_1) (1 - r_2)^2) \tag{5.11}$$

where $\quad p = \dfrac{n \, \sigma_{cbc}}{2 \, \sigma_{st}}$ is the balanced proportion of reinforcement in a rectangular section.

Eq. (5.11) gives the balanced proportion of tension reinforcement in a T-beam.

5.3.3 Procedure for Balanced Design.

(*i*) *Neutral axis within the flange, i.e. nd* $\lt D_f$. The known values are b_f, D_f and the bending moment M. Firstly, the balanced value of n is calculated from the known values of σ_{st}, σ_{cbc} and m using Eq. (3.8). Effective depth of the section is then calculated from Eq. (5.2) i.e.

$$d = \sqrt{\dfrac{M}{Rb_f}}$$

Finally, the area of tension reinforcement is calculated from Eq. (5.4) i.e.

$$A_{st} = \dfrac{M}{\sigma_{st} \, j \, d}$$

(*ii*) *Neutral axis outside the flange, i.e. nd* $\gt D_f$. The known values are b_t, D_f and the bending moment M. Firstly, the balanced value of n is calculated from the known values of σ_t, σ_{cbc} and m using Eq. (3.8).

Now, as $nd > D_f$

$$d > \frac{D_f}{n}$$

Hence, choose a trial value of d equal or greater than $\frac{D_f}{n}$ and calculate the compressive force C from Eq. (5.5). Calculate $\bar{x}$ from Eq. (5.7), and the arm of the internal couple $a = (d - \bar{x})$.

Calculate the moment of resistance from Eq. (5.8) i.e.

$$M_r = C\,(d - \bar{x})$$

Check whether $M_r \geqslant$ the applied bending moment M. Otherwise repeat the procedure with a new value of trial depth d till the calculated M_r is equal to or greater than M. Finally, calculate the area of tension reinforcement

$$A_{st} = \frac{M}{\sigma_{st}\,(d - \bar{x})}$$

5.4 UNDER-REINFORCED T-BEAM SECTION

Such sections have less amount of tension reinforcement than that needed for balanced design. Hence, the tensile stress in reinforcement reaches the permissible value while the extreme fibre concrete stress remains less than the permissible value. Depth of neutral axis can be found by equating the moment of the compressive area about the neutral axis to that of the tensile area about the same.

Two cases arise, which can be identified as follows:

(i) *Neutral asix lies within the flange, i.e. nd $<$ D_t. Then,*

$$b_f\ nd\left(\frac{nd}{2}\right) = m\,A_{st}\,(d - nd) \tag{5.12}$$

Form this quadratic equation, for any given value of d the value of n can be calculated. Then the moment of resistance can be found from Eq. (5.3) i.e.

$$M_r = \sigma_{st}\,A_{st}\,jd$$

(ii) *Neutral axis lies outside the flange, i.e. nd $>$ D_f. Then*

$$b_f \, nd \left(\frac{nd}{2} \right) - (b_f - b_w) \frac{(nd - D_f)^2}{2}$$

$$= m \, A_{st} \, (d - D_f) \qquad (5.13)$$

From this quadratic equation, for any given value of d the value of n can be calculated. Then, the moment of resistance can be found from Eq. (5.9) i.e.

$$M_r = \sigma_{st} \, A_{st} \, (d - \overline{x}) \qquad (5.14)$$

Example 5.1

For a T-beam, $b_f = 2100$ mm, $D_f = 150$ mm and $b_w = 300$ mm. Calculate the effective depth and the area of tension reinforcement required if the bending moment is $= 197.5$ kN-m. Concrete grade: M15, steel grade: Fe 415.

Now,

$$\sigma_{cbc} = 5 \text{ N/mm}^2 \qquad \text{from Table 3.1}$$
$$\sigma_{st} = 230 \text{ N/mm}^2 \qquad \text{from Table 3.2}$$
$$m = 18.67 \qquad \text{from Table 3.3}$$

Balanced value of neutral axis factor is

$$n = \cfrac{1}{1 + \cfrac{\sigma_{st}}{m \, \sigma_{cbc}}} = \cfrac{1}{1 + \cfrac{230}{18.67 \times 5}}$$
$$= 0.289$$

and

$$j = 1 - \frac{0.289}{3} = 0.904$$

Assuming that the neutral axis lies within the flange, i.e. $nd < D_f$ the section can be designed like a singly reinforced beam with $b = b_f$. Now

$$R = 0.653 \quad \text{from Table 3.6}$$

Then, the effective depth from Eq. (5.2) is

$$d = \sqrt{\frac{M}{R \, b_f}} = \sqrt{\frac{197.5 \times 10^6}{0.653 \times 2100}}$$
$$= 379.5 \text{ mm, say } 380 \text{ mm.}$$

Then $\quad nd = 0.289 \times 380 = 109.8$ mm

Evidently, this is less than the thickness of flange which is 150 mm. Hence the assumption that $nd < D_f$ is justified. Area of tension reinforcement is

$$A_{st} = \frac{M}{\sigma_{st} \, jd} = \frac{197.5 \times 10^6}{230 \times 0.904 \times 380}$$
$$= 2499.7 \text{ mm}^2$$

Provide 20 m dia. bars, 8 nos. Actual area of tension reinforcement A_{st} = 2513.28 mm². See Fig. 5.4.

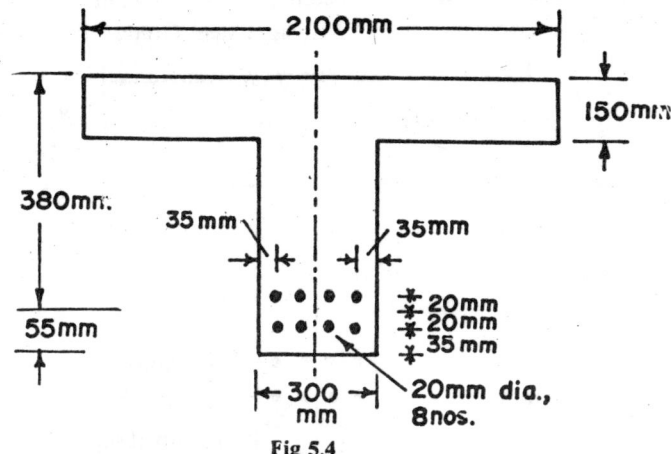

Fig 5.4

The bars are arranged in two layers of 4 bars each, as shown in Fig. 5.4. Required clear distance between bars in two vertical layers according to the IS Code: 456 is

    ≮ 15 mm

    ≮ ⅔ of the maximum aggregate size i.e. ⅔ × 20 = 13.33 mm

    ≮ dia. of bar i.e. 20 mm

Actual arrangement provides a clear vertical distance of 20 mm between bars of two layers. This is o.k. Required clear horizontal distance between bars in a layer, according to the IS Code: 456 is

    ≮ dia. of bar i.e. 20 mm

    ≮ 5 mm + maximum aggregate size, i.e. 5 + 20 = 25 mm

Actual arrangement provides a clear horizontal distance between the bars = ⅓ (300 − 2 × 35 − 3 × 20) = 56.7 mm. Required cover on the reinforcement as per IS Code: 456 is

    ≮ dia. of bar i.e. 20 mm

    ≮ 25 mm

The bars in the lower layers being centered 35 mm from the lower edge of concrete, have a clear cover on them = 25 mm. Centroid of all the 8 bars is located at 380 mm from the top concrete fibres.

Example 5.2

Calculate the moment of resistance of the T-beam section shown in Fig. 5.5. Concrete grade: M15 and steel Grade: Fe 415.

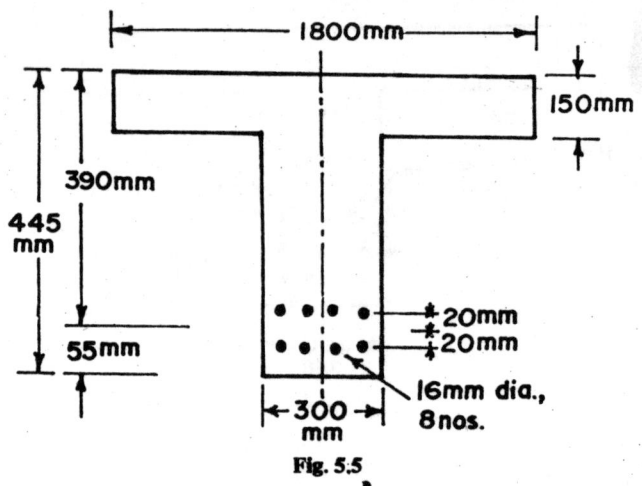

Fig. 5.5

Solution :

It is assumed that the neutral axis lies within the flange i.e. $nd < D_f$. Hence, the section can be treated as a singly reinforced rectangular section having width = 1800 mm. Permissible stresses are:

$$\sigma_{cbc} = 5 \quad \text{N/mm}^2 \quad \text{from Table 3.1}$$
$$\sigma_{st} = 230 \quad \text{N/mm}^2 \quad \text{from Table 3.2}$$
and $\quad m = 18.67 \quad\quad\quad\quad \text{from Table 3.3}$

Tension reinforcement

$$A_{st} = 8 \times 201.06 = 1608.48 \quad \text{mm}^2$$

Proportion of steel is

$$p = \frac{A_{st}}{b_f\, d} = \frac{1608.48}{1800 \times 390} = 0.00229$$

Neutral axis factor is found from Eq. (3.7), i.e.

$$n^2 + 2\, mpn - 2\, mp = 0$$

or $\quad n^2 + 2 \times 18.67 \times 0.00229\, n - 2 \times 18.67 \times 0.00229 = 0$

Hence $\quad n = 0.253$ and $j = 1 - \dfrac{0.253}{3} = 0.916$

$\therefore \quad nd = 0.253 \times 390 = 98.67$ mm, which is less than the thickness of flange i.e. 150 mm. Hence, the assumption that $nd < D_f$ is justified.

For balanced conditions,

$$n = \cfrac{1}{1 + \cfrac{\sigma_{st}}{m\, \sigma_{cbc}}} = \cfrac{1}{1 + \cfrac{230}{18.67 \times 5}}$$

$$= 0.289$$

Since the actual value of n is less than the balanced value, the section is under-reinforced and, therefore, the tensile stress in the reinforcement will reach its permissible value first. As such, the moment of resistance of the section is given by Eq. (5.3), i.e.

$$\begin{aligned}
M_r &= \sigma_{st}\, A_{st}\, jd \\
&= 230 \times 1608.48 \times 0.916 \times 390 \times 10^{-6} \\
&= 132.16 \text{ kN-m}
\end{aligned}$$

Example 5.3

For a T-beam the following data is given:

$D_f = 120$ mm, $b_f = 1650$ mm, $b_w = 240$ mm

Concrete grade: M 15, steel grade: Fe 415

Bending moment = 200 kN-m

Calculate the effective depth and the area of tension reinforcement required for a balanced design.

Solution:

Now

σ_{cbc}	$= 5$ N/mm²	from Table 3.1
σ_{st}	$= 230$ N/mm²	from Table 3.2
and m	$= 18.67$	from Table 3.3

For balanced conditions

n	$= 0.289$	from Table 3.4

and

$$j = 1 - \frac{0.289}{3} = 0.904$$

It is assumed that the neutral axis lies outside the flange i.e. $nd > D_f$. Then

$$d > \frac{D_f}{n}$$

$$> \frac{120}{0.289} \quad \text{i.e.} \quad 415 \text{ mm.}$$

Try an effective depth d = 435 mm. Then

$$nd = 0.289 \times 435 = 125.7 \text{ mm}$$

Now

$$r_1 = \frac{b_w}{b_f} = \frac{240}{1650} = 0.1455$$

$$r_2 = \frac{D_f}{nd} = \frac{120}{125.7} = 0.9547$$

Compressive force in concrete from Eq. (5.5) is

$$C = \tfrac{1}{2} \times 5 \times 1650 \times 125.7 \, (1 - (1 - 0.1455)$$
$$\times (1 - 0.9547)^2) \times 10^{-3}$$
$$= 517.58 \text{ kN}$$

Distance of centre of compression from top concrete fibres, from Eq. (5.7) is

$$\overline{X} = \frac{125.7}{3} \times \frac{1 - (1 - 0.1455)(1 - 0.9547)^2(1 + 2 \times 0.9547)}{1 - (1 - 0.1455)(1 - 0.9547)^2}$$

$$= 41.76 \text{ mm}$$

Arm of the internal couple

$$a = d - \overline{X}$$
$$= 435 - 41.76$$
$$= 393.24 \text{ mm}$$

Hence, the moment of resistance from Eq. (5.8) is

$$M_r = C(d - \overline{X})$$
$$= 517.58 \times 393.24 \times 10^{-3}$$
$$= 203.6 \text{ kN-m} > 200 \text{ kN-m} \quad \text{o.k.}$$

Area of tension reinforcement from Eq. (5.10) is

$$A_{st} = \frac{203.6 \times 10^6}{230 \times 393.24}$$

$$= 2251.09 \text{ mm}^2$$

Provide 6 bars of 22 mm dia. which give $A_{st} = 6 \times 380.13$ = 2280.78 mm². The bars are arranged in two layers of 3 bars each. The centroid all the 6 bars is located at an effective depth = 435 mm from top concrete edge as shown in the Fig. 5.6.

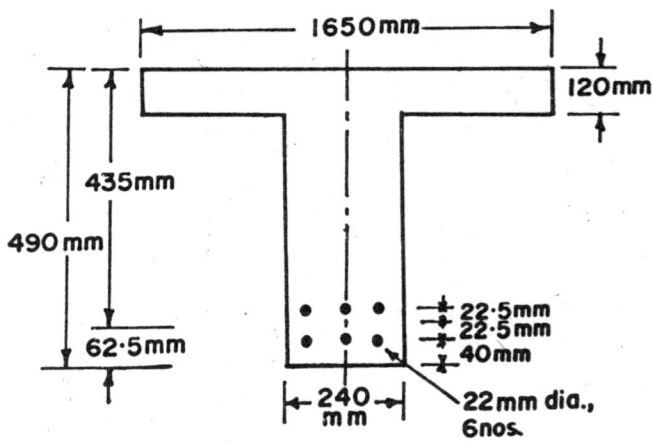

Fig 5.6

Example 5.4

Calculate the moment of resistance of the T-beam section shown in Fig. 5.7. Concrete grade: M15 and steel grade: Fe 415.

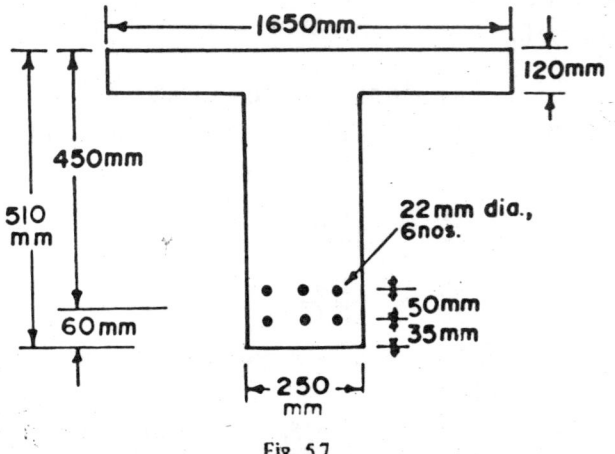

Fig. 5 7

Solution:

$$\sigma_{cbc} = 5 \text{ N/mm}^2 \qquad \text{from Table 3.1}$$
$$\sigma_{st} = 230 \text{ N/mm}^2 \qquad \text{from Table 3.2}$$

and $\quad m = 18.67 \qquad$ from Table 3.3

For balanced conditions

$$n = 0.289 \qquad \text{from Table 3.4}$$

and $\quad j = 1 - \dfrac{0.289}{3} = 0.904$

It is assumed that the neutral axis lies outside the flange i.e. $nd > D_f$. Hence from Eq. (5.13)

$$\frac{b_f (nd)^2}{2} - \frac{(b_f - b_w)(nd - D_f)^2}{2} = mA_{st} (d - nd)$$

or $\quad \dfrac{1650 \times 450^2}{2} \times n^2 - \dfrac{1400}{2} \times (450n - 120)^2$

$$= 18.67 \times 6 \times 380.13 \times 450 (1 - n)$$

or $\quad n^2 + 3.744\, n - 1.1545 = 0$

$$n = 0.2865$$

Evidently, the actual value of n is less than the balanced value. Hence, the section is under-reinforced. The tensile reinforcement will attain its permissible value while the concrete will remain understressed.

Now

$$nd = 0.2865 \times 450 = 128.93 \text{ mm}$$

Evidently, $nd > D_f$. Hence the assumption that the neutral axis lies outside the flange is justified. Then

$$r_1 = \frac{b_w}{b_f} = \frac{250}{1650} = 0.1515$$

$$r_2 = \frac{D_f}{nd} = \frac{120}{128.93} = 0.9307$$

Distance of the centre of compression from the top concrete fibres, from Eq. (5.7) is

$$X = \frac{128.3}{3} \times \frac{1 - (1 - 0.1515)(1 - 0.9307)^2 (1 + 2 \times 0.9307)}{1 - (1 - 0.1515)(1 - 0.9307)^2}$$

$$= 42.64 \text{ mm}$$

Arm of the internal couple $\quad a = (d - \overline{X}) = 450 - 42.64$
$$= 407.37 \text{ mm}$$

Hence, moment of resistance from Eq. (5.9) is

$$M_r = \sigma_{st} A_{st} (d - \overline{X})$$
$$= 230 \times 6 \times 380.16 \times 407.36 \times 10^{-6}$$
$$= 213.7 \quad kN - m$$

Example 5.5

A hall has clear dimensions in plan 6.7 m $\times$ 19.7 m. Design a T-beam floor for the hall using 4 beams, each placed along the short span, with their axes spaced at 4 m c/c. See Fig. 5.8. On the top of the slab there is a concrete tile flooring of thickness = 50 mm. Live load on the floor = 2 kN/m². Length of bearing at each end of the beams = 200 mm. Concrete grade: M 15. and steel grade: Fe 415.

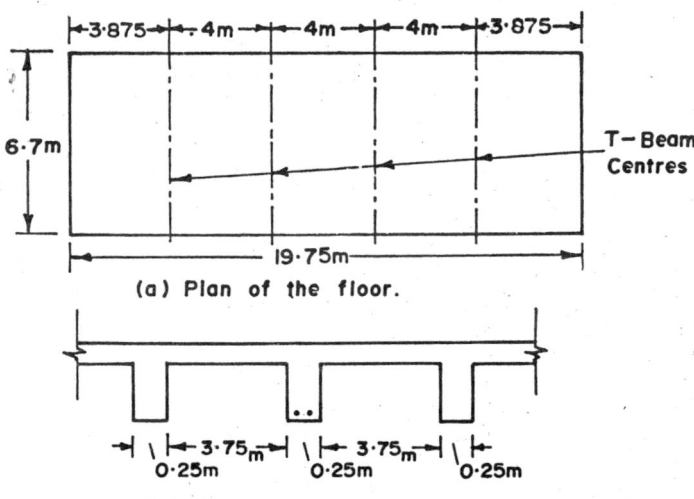

(a) Plan of the floor.

(b) Section through the floor.

Fig. 5.8

Solution:

With the arrangement of beams as shown in Fig. 5.8 the floor slab has been divided into 5 panels of 3.7 m $\times$ 6.7 m each in clear plan dimensions. The slab may be designed as one-way, spanning from beam to beam in each panel. A simple one-way slab over a clear span of 3.75 m, having the same loading as in this example, has already been designed in Example 3..7. Hence, the same design for the slab is adopted in this case also.

Overall thickness of slab = 175 mm.

Design of T-beam:

Effective span of beam

$$l = \text{c/c of the supports}$$
$$= 6.7 + 0.2$$
$$= 6.9 \text{ m}$$

and l_o = distance between points of zero bending moment in the beam

$$= 6.9 \text{ m}$$
$$D_t = 175 \text{ mm}$$

Clear distance between adjacent beams

$$= 3.7 \text{ m}$$

Effective width of flange of a beam

$$b_f < \frac{l_o}{6} + b_w + 6 \, D_t$$

i.e. $< \dfrac{6.9}{6} + 0.25 + 6 \times 0.175 = 2.45\text{m}$

Also $b_f < b_w + \frac{1}{2}$ (sum of clear distances to the adjacent beams)

i.e. $< 0.25 + \frac{1}{2} (3.75 + 3.75)$ $= 4$ m

Evidently

$$b_f = 2.45 \text{ m or } 2450 \text{ mm}.$$

A T-beam supports 4 m width of slab. Load on beam is

Weight of slab: $4 \times 0.175 \times 25$	= 17.5	kN/m
Weight of tiles: $4 \times 0.05 \times 24$	= 4.8	kN/m
Live Load: 4×2	= 8	kN/m
Self weight, say	= 1.7	kN/m
Total	= 32	kN/m

Bending moment

$$M = \frac{wl^2}{8} = \frac{32 \times 6.9^2}{8}$$
$$= 190.44 \text{ kN-m}$$

Now $\sigma_{cbc} = 5$ N/mm^2 from Table 3.1

$\sigma_{st} = 230$ N/mm^2 from Table 3.2

and $m = 18.67$ from Table 3.3

Balanced value of $n = 0.289$ from Table 3.4

$\qquad j = 0.904$

$\qquad R = 0.653$ from Table 3.6

It is assumed that the neutral axis lies within the flange i.e. $nd \leqslant D_f$.

Then, from Eq. (5.1)

$$Rb_f \, d^2 = M$$

or $0.653 \times 2450 \times d^2 = 190.44 \times 10^6$

$$d = 345 \text{ mm}$$

Then, $nd = 0.289 \times 345 = 99.7 \text{ mm} < D_f$ O.K.

Hence, the assumption that the neutral axis lies within the flange is justified.

Adopt $d = 345$ mm.

Tension reinforcement is,

$$A_{st} = \frac{M}{\sigma_{st} j d} = \frac{190.44 \times 10^6}{230 \times 0.904 \times 345}$$

$$= 2654.87 \text{ mm}^2$$

Provide 20 mm dia. bars 6 nos. and 16 mm dia. bars 4 nos., giving total $A_{st} = 2689.2$ mm². See Fig. 5.10.

Check for Shear:

$\qquad$ End reaction $= \frac{1}{2} \times 6.9 \times 32 = 110.4$ kN

Critical section occurs at an effective depth away from the face of support. See Fig. 5.9. At this section, shear force

$$V = 110.4 - 0.445 \times 32 = 96.16 \text{ kN}$$

Nominal shear stress

$$\tau_v = \frac{V}{b_w d} = \frac{96.16 \times 10^3}{250 \times 345} = 1.115 \text{ N/mm}^2$$

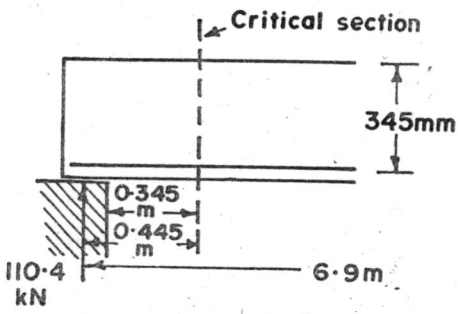

Fig. 5.9

Now,
$$\frac{100 \, A_{st}}{b_w d} = \frac{100 \times 2689.2}{250 \times 345} = 3.42$$

From Table 3.11, permissible shear stress

$$\tau_c = 0.44 \, N/mm^2$$

Evidently

$$\tau_v > \tau_c$$

Hence, shear reinforcement should be provided.
Strength of concrete in shear is

$$V_c = \tau_c \, b_w \, d = 0.44 \times 250 \times 345 \times 10^{-3}$$
$$= 37.95 \, kN$$

Required strength of shear reinforcement is

$$V_s = V - V_0 = 96.16 - 37.95$$
$$= 58.21 \, kN$$

Provide 8 mm dia. two legged stirrups of mild steel. From Eq. (3.28) the spacing of stirrups is found from

$$V_s = \frac{\sigma_{sv} \, A_{sv} \, d}{s_v}$$

i.e.
$$58.21 \times 10^3 = \frac{140 \times 100.6 \times 345}{s_v}$$

or
$$s_v = 83.5 \, mm, \text{ say } 80 \, mm \, c/c.$$

Minimum shear reinforcement is given by

$$\frac{A_{sv}}{b \, s_v} \geqslant \frac{0.4}{f_y}$$

For an 8 mm dia, stirrup of mild steel, the spacing is given by

$$\frac{100.6}{250 \times s_v} \geqslant \frac{0.4}{250}$$

or
$$s_v \leqslant 251.5 \, mm, \text{ say } 250 \, mm.$$

Shear strength of 8 mm dia. two legged stirrups of mild steel at a spacing of 250 mm c/c is
Then

$$V_s = \frac{\sigma_{sv} \, A_{sv} \, d}{s_v} = \frac{140 \times 100.3 \times 345 \times 10^{-3}}{250}$$
$$= 19.38 \, kN$$

$$V_s + V_c = 19.38 + 37.95 = 57.33 \, kN$$

Let the distance of section from the centre of support be x where the shear force = 57.33 kN. Then

$$110.4 - 32 x = 57.33$$

from which $\qquad x = 1.66$ m

Hence, the stirrup spacing may be kept = 80 mm c/c from centre of support for a distance of 1.68 m. Beyond this point, the spacing may be increased to 250 mm c/c. Details of reinforcement are given in Fig. 5.10.

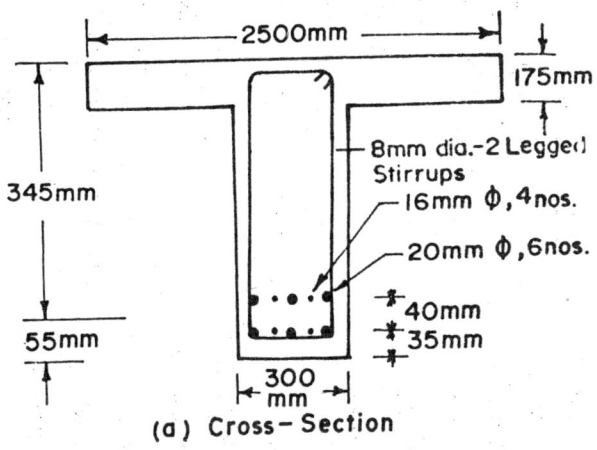

(a) Cross-Section

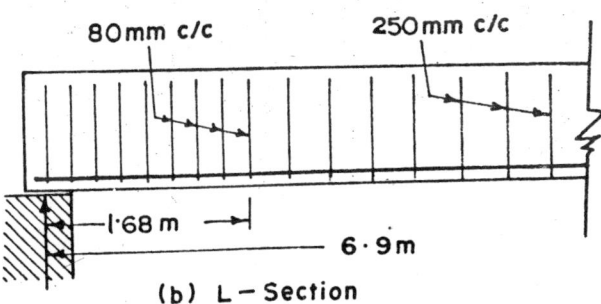

(b) L-Section

Fig. 5.10

6

Control of Deflection and Lateral Stability

6.1 LIMITS ON DEFLECTION

Reinforced concrete flexural members should possess adequate stiffness so that deflections under service loads do not exceed the permissible limits. According to the IS Code: 456 para 22.2 the permissible limits on deflection are as follows:

(i) Final deflection due to all loads including the effect of temperature, creep and shrinkage, measured below the as cast level of floors, roofs and other horizontal members should be

$$\not> \frac{\text{span}}{250}$$

(ii) Deflection after erection of partitions and the application of finishes including the effects of temperature, creep and shrinkage should be

$$\not> \frac{\text{span}}{350}$$

$$\not> 20 \text{ mm}$$

These limits are applicable to rectangular beams and slabs of uniform thickness. The effective depth and the amount of reinforcement may be varied reasonably to control the deflection of member while keeping the bending strength unchanged.

6.2 CALCULATION OF DEFLECTIONS

Deflections may be calculated by elastic theory and duly modified to take into account the effect of shrinkage and creep. Thus, the final deflection is the sum of short term deflection and the long term deflection which are discussed in the succeeding paras.

6.2.1 Short term deflection The short term deflection is calculated by elastic theory for the service loads or otherwise.
Flexural rigidity of member

$$= E_c I_c$$

where E_c = elastic modulus of concrete

$$= 5700 \sqrt{f_{ck}} \ N/mm^2$$

$$I_c = \frac{I_{cr}}{1.2 - \dfrac{M_{cr}}{M} \cdot \dfrac{a}{d} (1 - n) \dfrac{b_\omega}{b}} \qquad (6.1)$$

I_{cr} = second moment of area of cracked section

M_{cr} = cracking moment

$$= \frac{f_{cr} I_g}{y_t}$$

f_{cr} = modulus of rupture of concrete

I_g = second moment of area of gross section

y_t = distance from centroid to the extreme concrete fibre in tension

M = maximum bending moment

a = lever arm of cracked section

n = depth of neutral axis factor

b_w = width of web

b = width of compression face

However, the following limits should be satisfied:

$$I_{cr} \leqslant I_c \leqslant I_g$$

For continuous beams the values shall be modified as follows:

$$X_c = \frac{k_1 (X_1 + X_2)}{2} + (1 - k_1) X_0 \qquad (6.2)$$

where X_c = modified value of X

X = I_{cr}, I_g or M_{cr} as the case may be

X_1, X_2 = Values of X at supports

X_0 = Value of X at mid-span

k_1 = a cofficient as given in Table 6.1

Table 6.1

VALUES OF COEFFICIENT k_1

(BASED ON TABLE 21 OF IS CODE: 456 – 1878)

k_2	0.5 or less	0.6	0.7	0.8	0.9	1.0	1.1	1.2	1.3	1.4
k_1	0	0.03	0.08	0.16	0.3	0.5	0.73	0.91	0.97	1.0

where

$$k_2 = \frac{M_1 + M_2}{M_{f1} + M_{f2}}$$

$M_1, M_2 =$ support moments

$M_{f1}, M_{f2} =$ fixed end moments.

6.2.2 Long Term Deflection This may be calculated under two heads as follows:

(i) *Deflection due to shrinkage*

Deflection $d_{fs} = k_3 \, \psi \, l^2$ (6.3)

where $k_3 =$ a constant

                $= 0.53$ for cantilevers

                $= 0.125$ for simple beams

                $= 0.086$ for member continuous at one end

                $= 0.063$ for continuous member

        $\psi =$ shrinkage curvature

$$= \frac{k_4 \, \epsilon_{cs}}{D}$$

       $\epsilon_{cs} =$ ultimate shrinkage strain in concrete

        $l =$ effective span

       $k_4 =$ a coefficient as given in Table 6.2

Table 6.2

k_4	Range
$\dfrac{0.72\,(P_t - P_c)}{\sqrt{P_t}}$	$0.25 \leqslant (P_t - P_c) < 1.0$

$$\leqslant 1.0$$

$$\frac{0.65 \, (P_t - P_c)}{\sqrt{P_.}}$$

$$(P_t - P_c) \geqslant 1.0$$

$$\leqslant 1.0$$

where
$$P_t = \frac{100 \, A_{st}}{bd}$$

$$P_c = \frac{100 \, A_{sc}}{bd}$$

$$D = \text{Overall depth}$$

(ii) *Deflection due to creep*

Deflection $\quad d_{fc} = d_{cp} - d_{ip}$ $\hspace{2cm}$ (6.4)

where $\quad d_{cp} =$ Instantaneous plus creep deflection due to permanent load

$\quad d_{ip} =$ Instantaneous deflection due to permanent load.

6.3 CONTROL OF DEFLECTION BY LIMITING SPAN/DEPTH RATIO

Direct calculation of deflection is a lengthy and cumbersome process. As such, explicit calculation of deflection may be carried out only when particularly stringent deflection control is required or when the structure is abnormal.

For a flexural member made of elastic material it can be shown that the ratio $\frac{\delta}{l} \; \alpha \; \frac{1}{d}$. Hence, by setting a limit on $\frac{1}{d}$ the deflection will also be limited to a given fraction of the span. As such, as per IS Code 456 para. 22.2.1 the limit of span on deflection may be assumed to be satisfied if the members conform to the limiting ratios of span/effective depth with due moidfications described in the following paras,

6.3.1 Limits on span/effective depth ratio

Basic values for spans $\leqslant 10$ m

Cantilever	7
Simple beam	20
Continuous beam	26

Basic values for spans > 10 m

The above values may be multiplied by 10/span in meters, except for cantilever for which the deflection should be calculated.

6.3.2 Modification Factor for Tension Reinforcement The stiffness of reinforced concrete depends upon the amount of tension reinforcement. As the deflection is inversely proportional to stiffness the basic span/effective depth ratios need to be modified for the amount of tension reinforcement. For calculations, the amount of reinforcement is considered at mid-span for beams and slabs and at support for the cantilever. The modification factor as recommended in IS Code: 456 is given in Fig. 6.1.

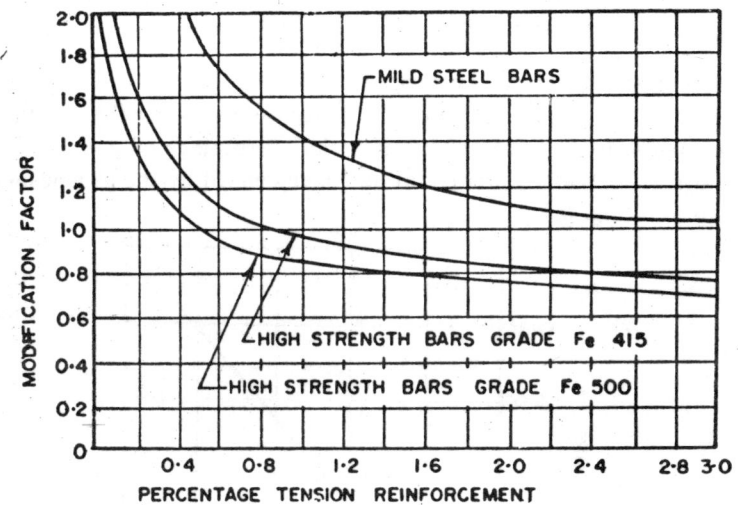

Fig. 6.1 Modification factor for tension reinforcement
(Based on Fig. 3 IS: 456—1978)

6.3.3 Modification factor for compression reinforcement Compression reinforcement partially restrains shrinkage and creep in the surrounding concrete. As such, long—term deflection is reduced.

In critical cases the best way to control deflection is by increasing the compression reinforcement without decreasing the strain in tension reinforcement. The modification factor as recommended by the IS Code: 456 is given in Fig. 6.2.

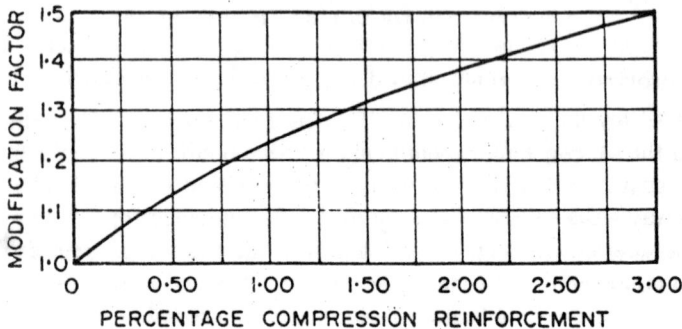

Fig. 6.2 Modification factor for compresion reinforcement
(Based on Fig. 4 IS: 456—1978)

6.3.4 Reduction factor for flanged beams In the case of flanged
beams it is proper to ignore the flanges and consider the beam as a
rectangular beam to be on the safer side. Percentage of steel is based
on $b_w d$. The modification factor as recommended by the IS Code: 456
is given in Fig. 6.3.

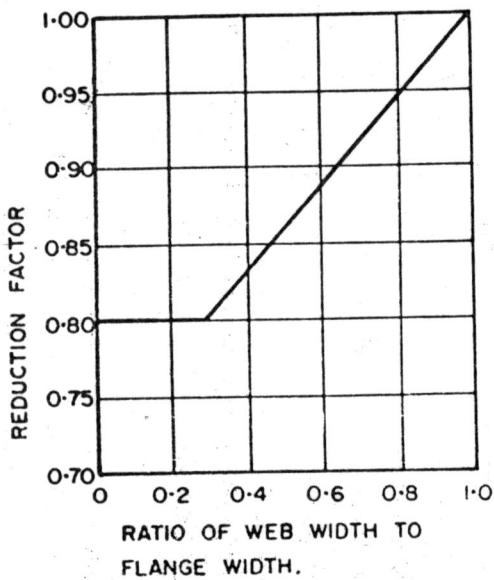

Fig. 6.3 Reduction factor for flanged beams (Based
on Fig. 5 IS: 456—1978)

6.4 CONTROL OF DEFLECTION IN SLABS

The provision for limitation of deflection of beams are applicable to slabs also.

In the case of slabs spanning in two directions, the shorter of the two spans should be used to calculate span/effective depth ratio. For two way slabs with shorter span $\leqslant$ 3.5 m, the deflection limits may be assumed to be satisfied for load $\leqslant$ 3 kN/m^2 if the following limits are maintained:

Slabs	Reinforcement	
	Mild steel	Fe 415
Simply supported	$r \leqslant 35$	$r \leqslant 28$
Continuous	$r \leqslant 40$	$r \leqslant 32$

where r = span/overall depth ratio

6.5 LATERAL STABILITY OF BEAMS

Compression flange of the beam, if not adequately supported in lateral direction, may bend and buckle sideways. As such, lateral restraint, if required, is normally provided by construction attached to the compression zone of the beam. In case the lateral restraints are spaced at certain intervals the following limitations should be observed:

For simple or continuous beams $\quad l_c \not> 60b$

$$\not> 250 \frac{b^2}{d}$$

For cantilevers $\quad\quad\quad\quad\quad\quad l \not> 25b$

$$\not> 100 \frac{b^2}{d}$$

where $\quad l_c$ = clear distance between the lateral restraints in a beam

$\quad\quad l$ = clear distance form free end to the lateral restraint in a cantilever

Example 6.1

A singly reinforced rectangular beam, designed for an imposed dead load = 15 kN/m and live load = 12 kN/m, is shown in Fig. 6.4. The section is balanced. Concrete grade: M 20 and steel grade: Fe 415. Check the beam for deflection and show whether the relevant limits as per IS Code: 456 are satisfied or not.

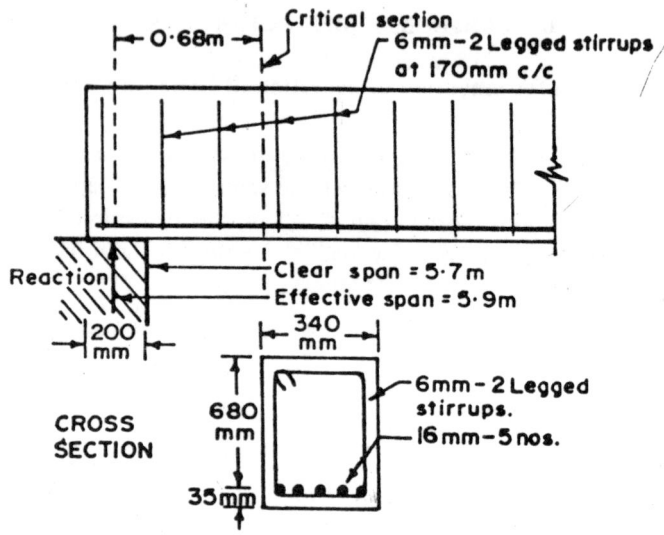

Fig. 6.4

Solution

The beam can be checked by directly computing its deflection under the loads and then comparing it with the permissible values, as well as, indirectly by comparing its actual span/effective depth ratio with the permitted value of the same.

(i) Direct check by computing deflection.

Short term deflection:

$$n = 0.289 \text{ from Table 3.4}$$
$$X = 0.289 \times 680$$
$$= 196.52 \text{ mm}$$
$$I_{cr} = \frac{bX^3}{3} + (m - 1) A_{st} (d - X)^2$$
$$= \frac{1}{3} \times 340 \times 196.52^2 + (13.33 - 1) \times 5 \times 201.06$$
$$\times (680 - 196.52)^2$$
$$= 2.902 \times 10^9 \text{ mm}^4$$

$$D = 680 + 35$$
$$= 715 \text{ mm}$$

$$I_g = \frac{bD^3}{12}$$

$$= \frac{1}{12} \times 340 \times 715^3$$

$$= 10.357 \times 10^9 \text{ mm}^4.$$

$$f_{cr} = 0.7 \sqrt{f_{ck}} \text{ modulus of rupture}$$
$$= 0.7 \sqrt{20} = 3.13 \quad \text{N/mm}^2$$
$$= D - X$$
$$= 680 - 196.52$$
$$= 483.48 \text{ mm}$$

$$M_{cr} = \frac{3.13 \times 10.357 \times 10^9}{483.48}$$

$$= 67.05 \times 10^6 \text{ N-mm.}$$

$$a = jd$$
$$= 0.904 \times 680$$
$$= 614.72 \text{ mm}$$

Self weight $\quad = 0.34 \times 0.715 \times 25$
$$= 6.08 \text{ kN/m}$$

$w =$ uniform loading on beam
$$= 12 + 15 + 6.08$$
$$= 33.08 \text{ kN/m.}$$

$$M = \frac{wl^2}{8}$$

$$= \frac{33.08 \times 5.9^2}{8}$$

$$= 143.94 \text{ kN-m} \quad \text{or} \quad 143.94 \times 10^6 \text{ N-mm}$$

Then $\quad I_c = \dfrac{2.902 \times 10^9}{1.2 - \dfrac{67.05}{143.94} \times \dfrac{614.72}{680}(1 - .289) \times 1}$

$$= 3.222 \times 10^9 \text{ mm}^4$$

Evidently, $I_{cr} \leqslant I_c \leqslant I_g$. Hence o.k.

$$E_c = 5700 \sqrt{20} = 25491 \text{ N/mm}^2$$

$\therefore$ Short term deflection at mid-span of beam

$$d_s = \frac{5}{384} \times \frac{33.08 \times 5900^4}{25491 \times 3.222 \times 10^9}$$

$$= 6.35 \text{ mm}$$

Long term deflection

$$P_t = \frac{100A_{st}}{bd}$$

$$= \frac{100 \times 5 \times 201.06}{340 \times 680}$$

$$= 0.435$$

$$P_c = \frac{100A_{sc}}{bd}$$

$$= 0$$

$\therefore$ $(P_t - P_c) = 0.435$

From Table 6.2

$$K_u = \frac{0.72 \times 0.435}{\sqrt{0.435}} = 0.475$$

ϵ_{cs} = ultimate shrinkage strain in concrete

$$= 0.0003$$

$$D = 715 \text{ mm}$$

$\therefore$ $\psi = \dfrac{K_4\epsilon_{cs}}{D} = \dfrac{0.475 \times 0.0003}{715}$

$$= 0.2 \times 10^{-6}$$

$$k_3 = 0.125$$

From Eq. (6.3) deflection due to shrinkage

$$d_{fs} = 0.125 \times 0.2 \times 10^{-6} \times 5900^2$$

$$= 0.87 \text{ mm}$$

Permanent load intensity on beam

$$= 15 + 6.08$$

$$= 21.08 \text{ kN/m}.$$

Taking creep coefficient $= 1.1$, the long term elastic modulus of concrete

$$E_{ce} = \frac{25491}{1 + 1.1}$$

$$= 12139 \ N/mm^2$$

Instantaneous deflection due to permanent load

$$= \frac{5}{384} \times \frac{21.08 \times 5900^4}{25491 \times 3.222 \times 10^9}$$

$$= 4.05 \ mm$$

Instantaneous plus creep deflection due to permanent loads

$$d_{cp} = \frac{5}{384} \times \frac{21.08 \times 5900^4}{12139 \times 3.222 \times 10^9}$$

$$= 8.5 \ mm$$

From Eq. (6.4) deflection due to creep

$$d_{fc} = d_{cp} - d_{ip}$$

$$= 8.5 - 4.05$$

$$= 4.45 \ mm$$

Final deflection due to all loads including effects of shrinkage and creep

$$\delta = d_s + d_{is} + d_{fc}$$

$$= 6.35 + 0.87 + 4.45$$

$$= 11.67 \ mm$$

Permissible deflection

$$= \frac{span}{250}$$

$$= \frac{5900}{250} = 23.6 \ mm$$

Evidently $\quad \delta < \dfrac{span}{250}$

Hence, the beam satisfies the IS Code: 456 limit on deflection.

(ii) *Indirect check from span/effective depth ratio*

Basic value of span/effective depth ratio

$$= 20$$

Percentage of tension reinforcement

$$\frac{100A_{st}}{bd} = \frac{100 \times 5 \times 201.06}{340 \times 680}$$

$$= 0.435$$

From Fig. 6.1, the modification factor

$$= 1.23$$

Hence, permissible value of span/effective depth ratio

$$= 1.23 \times 20$$

$$= 24.6$$

Actual value of span/effective depth ratio

$$= \frac{5900}{680}$$

$$= 8.68$$

This is less than the permissible value i.e. 24.6

Hence, the beam satisfies the criterion of limiting deflection.

Check for lateral stability

$$l_c = 5700 \text{ mm}$$

$$60b = 60 \times 340$$

$$= 20400$$

$$250 \frac{b^2}{d} = 250 \times \frac{340^2}{680}$$

$$= 42500$$

Evidently $\quad l_c < 60 b \qquad$ and also

$$< 250 \frac{b^2}{d}$$

Hence, the beam is safe against lateral buckling of compression zone.

Example 6.2

Roof beams overhang the supporting wall by 2.7 m as shown in Fig. 6.5. The slab which is supported by these cantilever beams, form their flanges at bottom. Thus, the cantilever beams are inverted T-beams. Check the beam for deflection and show that the relevant limits as per IS Code: 456 are satisfied.

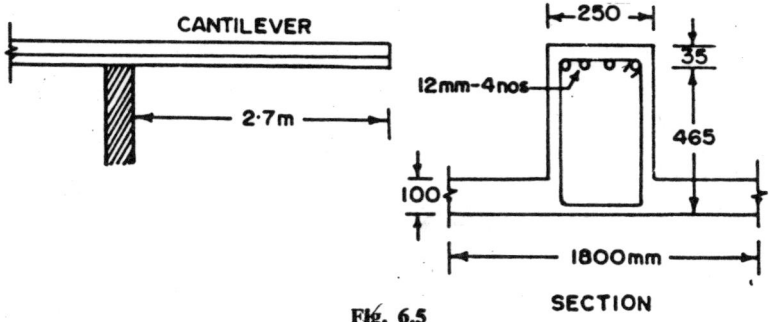

Fig. 6.5   **SECTION**

Solution

The Beam can be checked for the deflection limits by considering the limits on the span/effective depth ratio.

Basic value of the span/effective depth ratio

$$= 7.0$$

Percentage of tension reinforcement is

$$\frac{100\,A_{st}}{b_f d} = \frac{100 \times 4 \times 113.1}{1800 \times 465}$$

$$= 0.054$$

By extrapolating from Fig. 6.1, the modification factor

$$= 2.30$$

Ratio of web width to flange width

$$= \frac{250}{1800} = 0.139$$

From Fig. 6.3 the reduction factor is

$$= 0.8$$

Hence, the modified span/effective depth ratio works out to

$$= 7.0 \times 2.3 \times 0.8$$

$$= 12.88$$

Actual span/effective depth ratio of cantilever

$$= \frac{2.7}{0.465} = 5.81$$

which is less than the modified value viz. 12.88.

Hence, the beams satisfy the limit on deflection as per the IS Code: 456.

7

Design Examples

Example 7.1

DESIGN OF A CAR SHED

The Schematic diagram of a car shed is shown in Fig. 7.1. Eight columns each 300 × 300 mm in section are used to support four beams which in their turn support the roofing slab. Occassional live load = 0.75 kN/m². Design the beam and slab roofing. Concrete grade: M15 and steel grade: Fe415.

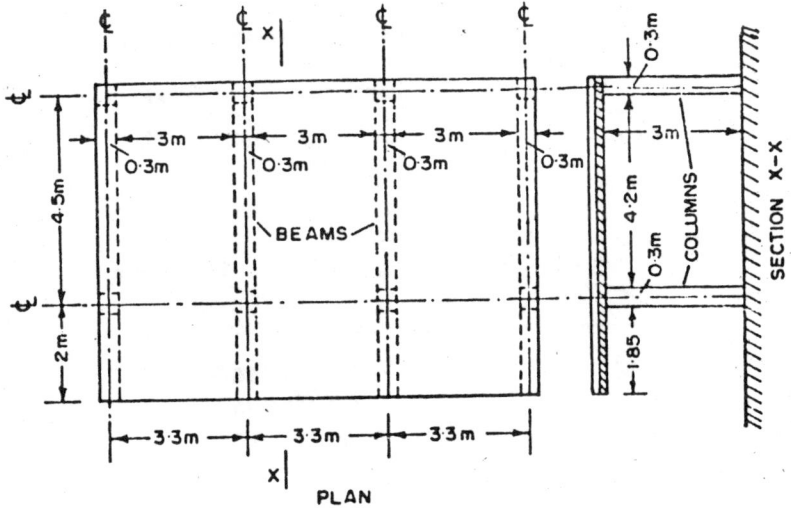

Fig. 7.1

Solution

The roofing slab is continuous over the four beams. It is to be designed for occassional live load and the self weight.

Roofing slab

Consider a strip of unit width of the slab, spanning over the four
beams. The load per m run of the strip is as follows:

Self weight of slab $0.1 \times 25 = 2.5$ kN/m

Weight of ：creed $0.025 \times 24 = 0.6$ kN/m

Occasional live load $0.75 \times 1 = 0.75$ kN/m

$$\text{Total} \quad = 3.85 \text{ kN/m}$$

The bending moment diagram for the unit width strip of slab is
shown in Fig. 7.2.

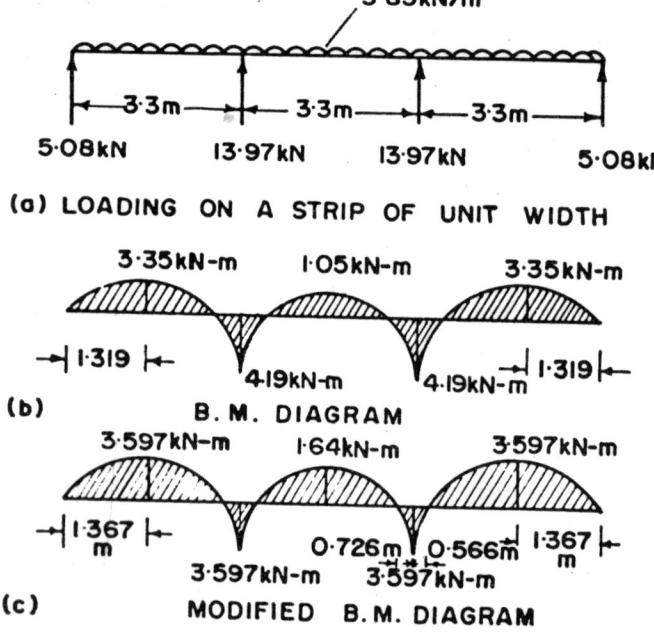

(a) LOADING ON A STRIP OF UNIT WIDTH

(b)   B. M. DIAGRAM

(c)   MODIFIED B. M. DIAGRAM

Fig. 7.2

According to the clause 43.2 of IS Code: 456 the moments over the supports may be increased or decreased upto 15%. In this case the support moments are decreased by 14% and the modified bending moment diagram is shown in Fig. 7.2 (c). Thus, the negative moment over the support is equal to the positive moment in the outer spans. Hence, effective depth

$$d = \sqrt{\frac{3.597 \times 10^6}{0.653 \times 10^3}}$$
$$= 74.2 \text{ mm.}$$

However, adopt d = 80 mm so that the overall depth may be = 80 + 20 = 100 mm at least. Therefore,

$$A_{st} = \frac{3.597 \times 10^6}{230 \times 0.904 \times 80}$$
$$= 216.2 \text{ mm}^2.$$

Provide 8 mm bars − 230 mm c/c at top over the supports and the same at bottom within the spans.

Overall depth of slab = 80 + 20
$$= 100 \text{ mm}$$

Temperature and shrinkage reinforcement is

$$= \frac{0.12}{100} \times 100 \times 1000$$
$$= 120 \text{ mm}^2/\text{m}$$

Provide 6 mm dia. bars − 235 mm c/c
Details of reinforcement are shown in Fig. 7.4.

Inner Beam

Consider one of the inner beams. The loading on the beam is as follows:

Dead load of roof $3.3 \times (2.5 + 0.6) = 10.23$ kN/m
Estimated self weight $0.2 \times 0.35 \times 25 = 1.75$ kN/m
Occasional live load 3.2×0.75 $= 2.48$ kN/m

                          Total $= 14.46$ kN/m

The loading and the bending moment diagram for the beam is shown in Fig. 7.3.

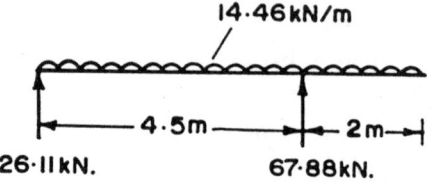

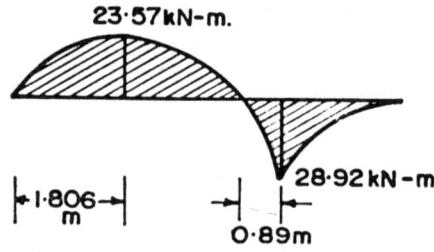

Fig. 7.3

Greater bending moment exists over the support towards the overhang. As this bending moment causes tension in the upper fibres, the beam behaves there as a rectangular beam. Let the width of the beam be 200 mm. Hence, effective depth

$$d = \sqrt{\frac{28.92 \times 10^6}{200 \times 0.653}}$$
$$= 470 \text{ mm}$$

Tension reinforcement

$$A_{st} = \frac{28.92 \times 10^6}{230 \times .904 \times 470} = 295.94 \text{ mm}^2$$

Provide 10 mm dia. bars — 4 nos.
Shear force on the inner side of the support.

$$= 67.88 - 14.46 \times 2$$
$$= 38.96 \text{ kN.}$$

The width of column support = 300 mm

Critical section occurs at an effective depth away from the face of support. Hence, shear force there is

$$V = 38.96 - (0.15 + 0.47) \times 14.46$$

$$= 30 \text{ kN}$$

$$\therefore \quad \tau_v = \frac{30 \times 10^3}{200 \times 470}$$

$$= 0.319 \text{ N/mm}^2$$

$$\frac{100 \text{ A}_{st}}{\cdot \text{ bd}} = \frac{4 \times 78.54}{200 \times 470}$$

$$= 0.334$$

From Table 3.11

$$\tau_c = 0.244 \text{ N/mm}^2$$

As $\tau_v > \tau_c$ the beam is unsafe in shear. Provide 6 mm dia. two legged stirrups.

$$V_s = 30 - 0.244 \times 200 \times 470 \times 10^{-3}$$

$$= 7.06 \text{ kN.}$$

Spacing of stirrups corresponding to this is found by equating

$$7.06 \times 10^3 = \frac{230 \times 2 \times 28.27 \times 470}{s}$$

$$s = 865.7 \text{ mm}$$

Spacing from minimum shear reinforcement requirement is found by equating

$$\frac{2 \times 28.27}{200 \times s} = \frac{0.4}{415}$$

$$\therefore \quad s = 293.3 \text{ mm}$$

Maximum permitted spacing

$$= 0.75 \times 470$$

$$= 352.5 \text{ mm}$$

Also, the spacing $\quad \nRightarrow 450 \text{ mm}$

Hence, a spacing of 290 mm which is the minimum of the three values is adopted and maintained throughout the length of the beam. In between the columns the beam behaves as a T-beam. Tension reinforcement, assuming $j = 0.925$, will be

$$A_{st} = \frac{23.57 \times 10^6}{230 \times 0.925 \times 470}$$
$$\doteq 241.2 \text{ mm}^2$$

4 bars of 10 mm dia. may be provided in the lower portion. The reinforcement bars in the overhanging portion are provided at top. Two bars may be terminated at the point of inflexion and the remaining two may be extended beyond the point of inflexion for a distance greater of the following

(i) effective depth = 470 mm

(ii) 12 × bar dia. = 120 mm

(iii) $\frac{1}{16}$ × span or $\frac{1}{16}$ × 2000 = 125 mm

i.e. 470 mm beyond the point of inflexion.

In the main span four bars are provided at bottom. Of these, two may be bent up at the point of inflexion while the remaining two may continue into the support.

As two bars are bent up at point of inflexion, only two bars remain. Hence, at this point

$$M_1 = 2 \times 78.54 \times 230 \times 470 \times 10^{-6}$$
$$= 16.98 \text{ kN-m}$$
$$\text{Shear force } V = 67.98 - 0.89 \times 14.46$$
$$= 55.11 \text{ kN}.$$
$$L_0 = 470 \text{ mm}.$$

$$\therefore \quad \frac{M_1}{V} + L_0 = \frac{16.98}{55.11} \times 10^3 + 470 = 778 \text{ mm}$$

$$l_d = \frac{10 \times 230}{4 \times 0.8 \times 1.65} = 435.6 \text{ mm}$$

Obviously $l_d \not> \dfrac{M_1}{V} + L_0$. Hence, it is o.k.

Details of reinforcement are shown in Fig. 7.4

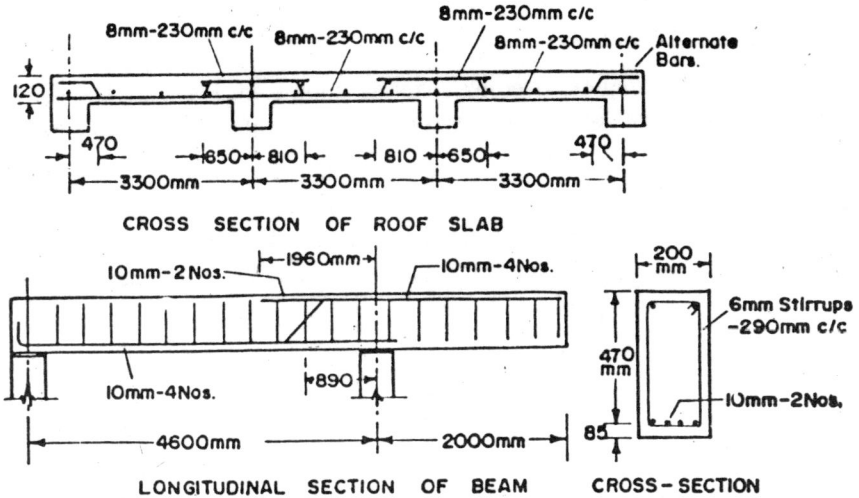

Fig. 7.4

Outer Beam

This beam which is supported in the same manner as the inner beam receives load from half the slab width from one side only which also forms its flange converting it to an L-beam. Though this beam receives lesser loading than the inner beam it is proposed to keep the same size and reinforcement for the outer beam as for the inner beam to preserve the uniformity of size and appearance.

Example 7.2
DESIGN OF AN INVERTED T-BEAM ROOF

A hall has clear dimensions 15.3 m × 6.15 m in plan. It has to be provided with an inverted T beam roofing. The roof slab is continuous over the beams. The latter are simply supported across the shorter span and are four in number. Bearing at each end = 150 mm. See Fig. 7.5. Keep width of each beam = 300 mm. Overall depth of beam should not exceed 600 mm. Live load = 1.5 kN/m². Concrete grade: M15 and Steel grade: Fe 415.

Solution

1. *Design of slab*

The slab is continuous over five spans as shown in Fig. 7.5. Design bending moments and shear forces in the slab strip will be calculated

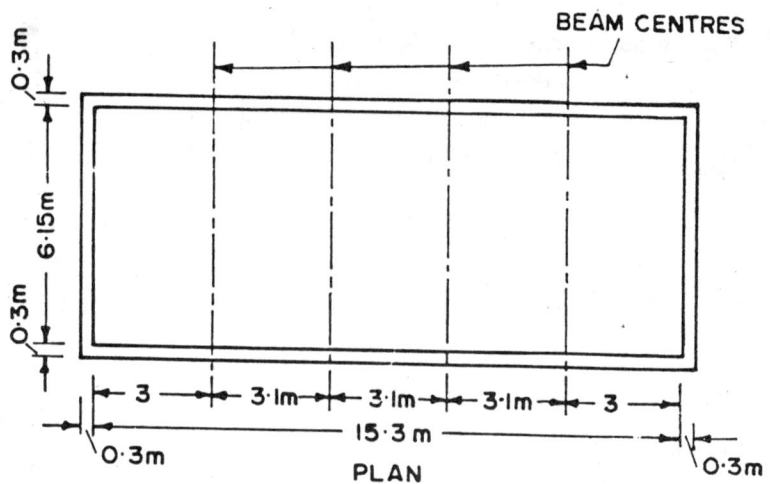

Fig. 7.5

from the coefficients given in Tables 7 and 8 respectively, of the IS Code 456. These coefficients are reproduced below in the Tables 7.1

Table 7.1

BENDING MOMENT COEFFICIENTS
(BASED ON TABLE 7 OF IS CODE: 456)

Type of load	Near middle of end span	At middle of interior span	At support next to end support	At other interior supports
Dead load	$\dfrac{wl^2}{12}$	$\dfrac{wl^2}{24}$	$-\dfrac{wl^2}{10}$	$-\dfrac{wl^2}{12}$
Live load	$\dfrac{wl^2}{10}$	$\dfrac{wl^2}{12}$	$-\dfrac{wl^2}{9}$	$-\dfrac{wl^2}{9}$

and 7.2 respectively. These are applicable to beams having 3 or more spans which do not differ by more than 15% of the longest, and which carry uniformly distributed loads.

Table 7.2

SHEAR FORCE CO-EFFICIENTS
(BASED ON TABLE 8 OF IS CODE 456)

Type of load	At end support	At support next to the endsupport		At other interior support
		Outer side	Inner side	
Dead Load	0.4 wl	0.6 wl	0.55 wl	0.5 wl
live load	0.45 wl	0.6 wl	0.6 wl	0.6 wl

Note: w = intensity of load

1 = effective span

At a support where two unequal spans meet an average of the two values for the negative moment may be taken.

For an interior span of slab
clear span = 3100 − 300

= 2800 mm

Now $\frac{1}{12} \times 2800 = 233.3$ mm

Width of support (i.e. the width of beam)

= 300 mm

which is more than $\frac{1}{12}$ of clear span.

Hence, effective span

= clear span

= 2800 mm   or   2.8 m

For the exterior span, clear span

= 2950 − 150

= 2800 mm

$\therefore$ Effective span = clear span $+ \frac{1}{2} \times$ effective depth of slab

$$= 2800 + 40$$
$$= 2840 \text{ mm} \quad \text{or} \quad 2.84 \text{ m}$$

Consider 1 m wide strip of slab.
The loading is as follows:

Dead load

Estimated self weight $0.10 \times 25 = 2.50$ kN/m
Self weight of screed $0.05 \times 24 = 1.2$ kN/m

$$\text{Total} \qquad\qquad = 3.70 \text{ kN/m}$$

Live load

$$1 \times 1.5 = 1.5 \text{ kN/m}$$

The design bending moments and shear forces calculated in the slab strip of unit width, which is continuous over 5 spans are shown in Fig. 7.6.

Redistribution of moments over the supports are not permitted for this simplified analysis based on coefficients.

Maximum bending moment in the slab strip of unit width, from Fig 7.6 (c) is

$$= 4.268 \text{ kN-m/m}$$

which exists over the first interior support.
Hence, effective depth

$$d = \sqrt{\frac{4.268 \times 10^6}{10^3 \times 0.653}}$$
$$= 80.85 \text{ mm}$$

Adopt $d = 85$ mm   then,
total thickness $= 85 + 20$
$$= 105 \text{ mm}$$

which may be kept uniform for the whole of slab.
Reinforcement calculation follows:
Over the first interior support

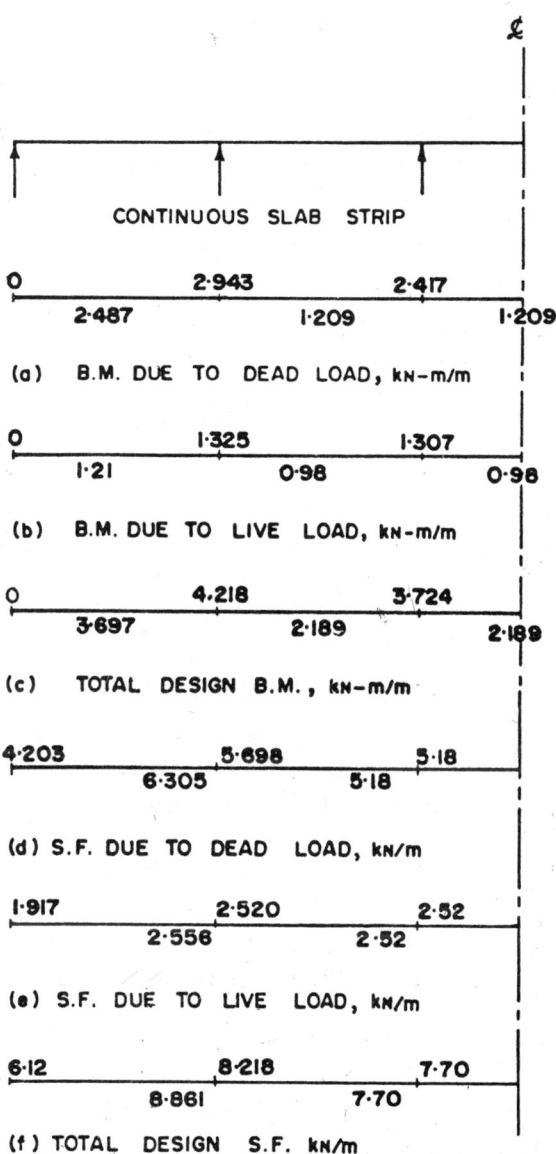

Fig. 7.6

$$A_{st} = \frac{4.268 \times 10^6}{230 \times 0.904 \times 85}$$

$$= 241.5 \text{ mm}^2/\text{m}$$

Provide 8 mm bars – 205 mm c/c

Over the other interior supports

$$A_{st} = \frac{3.724 \times 10^6}{230 \times 0.904 \times 85}$$

$$= 210.7 \text{ mm}^2/\text{m}$$

Provide 8 mm bars – 235 mm c/c

In the end span

$$A_{st} = \frac{3.697 \times 10^6}{230 \times 0.904 \times 85}$$

$$= 209.2 \text{ mm}^2/\text{m}$$

Provide 8 mm bars – 240 mm c/c.

In the interior span

$$A_{st} = \frac{2.189 \times 10^6}{230 \times 0.904 \times 85}$$

$$= 123.8 \text{ mm}^2/\text{m}$$

For 8 mm bars a spacing of 406 mm is required.

However, the maximum spacing permitted is $3 \times 85 = 255$ mm. Hence, provide 8 mm bars – 255 mm c/c

Temperature and shrinkage reinforcement is

$$= \frac{0.12}{100} \times 105 \times 1000$$

$$= 126 \text{ mm}^2/\text{m}$$

Provide 6 mm bars – 220 mm c/c throughout at right angles to the main reinforcement.

For details see Fig. 7.7

2. *Design of Inverted T-beam.*

Loading on a beam per m run is as follows:

Weight of slab $3.1 \times 0.105 \times 24 = 8.14$ kN/m

Weight of concrete screed $2.8 \times 0.05 \times 24 = 3.36$ kN/m

Self weight of beam	$0.3 \times 0.495 \times 25 = 3.71$ kN/m
Live load	$3.1 \times 1 \times 1.5 = 4.65$ kN/m

$$\text{Total load} = 19.86 \text{ kN/m}$$

Effective span of beam

$$= 6.15 + 0.15$$
$$= 6.3 \text{ m}$$

Bending moment

$$= \frac{19.86 \times 6.3^2}{8}$$
$$= 98.53 \text{ kN-m}$$

The flange of the beam is at bottom which will be in tension due to bending moment. As such, the beam will behave as a rectangular beam. As the overall depth is limited to 500 mm
The effective depth

$$= 600 - 50$$
$$= 550 \text{ mm say.}$$

Bending strength of the singly reinforced section

$$M_b = 0.653 \times 300 \times 550^2 \times 10^{-6}$$
$$= 59.3 \text{ kN-m}$$

Hence, excess bending moment

$$M = 98.53 - 59.3$$
$$= 39.23 \text{ kN-m}$$

The beam will therefore be a doubly reinforced beam

$$A_{st1} = \frac{59.3 \times 10^6}{230 \times .904 \times 550}$$
$$= 518.6 \text{ mm}^2$$
$$A_{st2} = \frac{39.23 \times 10^6}{230 \times 500}$$
$$= 341.1 \text{ mm}^2$$
$$A_{st} = 518.6 + 341.1$$
$$= 859.7 \text{ mm}^2$$

Permissible compressive stress in compression reinforcement

$$\sigma_{sc} = \frac{(0.289 \times 550 - 50)}{0.289 \times 550} \times (1.5 \times 18.67 - 1) \times 5$$

$$= 92.53 \ \text{N/mm}^2$$

Hence, $\qquad A_{s_c} = \dfrac{39.23 \times 10^6}{92.53 \times 500}$

$$= 847.9 \ \text{mm}^2$$

Provide 12 mm bars − 8 nos. at bottom and the same at top, in two layers of 2 bars each.

See details in Fig. 7.7

Critical section for shear occurs at an effective depth away from face of support. From centre of support the critical section occurs at a distance

$$= 0.075 + 0.55 \ \text{m}$$

$$= 0.625 \ \text{m}$$

Shear force at this section

$$= 19.86 \times (\tfrac{1}{2} \times 6.25 - 0.625)$$

$$= 49.65 \ \text{kN}$$

$\therefore \qquad \tau_v = \dfrac{49.65 \times 10^3}{300 \times 550}$

$$= 0.3 \ \text{N/mm}^2$$

$$\frac{100 A_{st}}{bd} = \frac{8 \times 113.01}{300 \times 550} \times 100$$

$$= 0.548$$

From Table 3.1

$$\tau_c = 0.3 \ \text{N/mm}^2$$

As $\tau_v = \tau_c$ the beam is safe in shear.

However, minimum shear reinforcement will be provided. Use 6 mm. dia. two legged stirrups. Then

$$\frac{2 \times 28.27}{300 \times s} > \frac{0.4}{415}$$

Hence, the spacing

$$s < 195.5 \ \text{mm}$$

Adopt a spacing of 195 mm c/c throughout the length of beam. For details see Fig. 7.7

Basic value of span/effective depth ratio

$$= 20$$

% of tension reinforcement

$$= 0.548$$

Modification factor from Fig. 6.1

$$= 1.14$$

% of compression reinforcement

$$= 0.548$$

Modification factor from Fig. 6.2

$$= 1.15$$

Hence, modified value of span/effective depth ratio

$$= 20 \times 1.14 \times 1.15$$
$$= 26.22$$

Actual span/effective depth ratio

$$= \frac{6.25}{0.55}$$
$$= 11.4 < 26.22 \quad \text{o.k.}$$

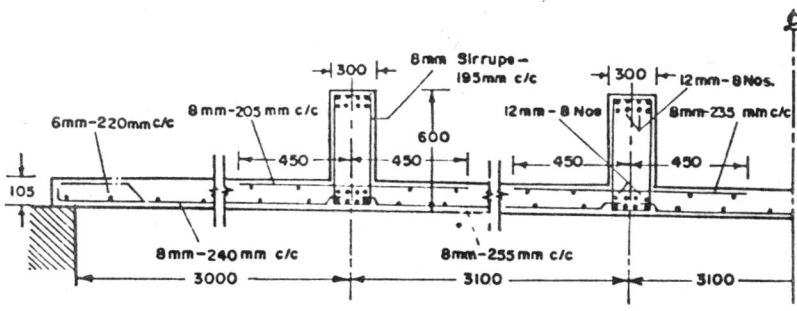

Fig. 7.7

8

Compression Members

8.1 INTRODUCTION

A compression member is a straight prismatic member designed primarily to carry axial compression which is more often accompanied by bending moment. Even with perfect axial load situation, there may arise small amount of bending moment due to construction tolerances, loading and support conditions. Hence a minimum amount of eccentricity is provided for in the design to account for these effects.

Main reinforcement for the compression member is provided parallel to the axis. Though, concrete is strong in compression and can safely bear the axial load all by itself, a minimum quantity of longitudinal reinforcement has to be provided to counter bending moments arising due to accidental lateral loading or otherwise. The longitudinal reinforcement should be placed nearer to the outer periphery of the section leaving concrete cover as required. See Fig. 8.1 (a) (b) and (c). Since the steel reinforcing bars are slender, they can buckle out under axial loading breaking off the concrete cover. To prevent this kind of failure, the longitudinal reinforcing bars should be laterally tied as shown in Fig. 8.1. (d) and (e).

If the member has a rectangular section then there will be two axes of symmetry about which buckling can take place and there will be two lateral dimensions b and D. See Fig. 8.1.(a). A compression member is classified on the basis of its slenderness

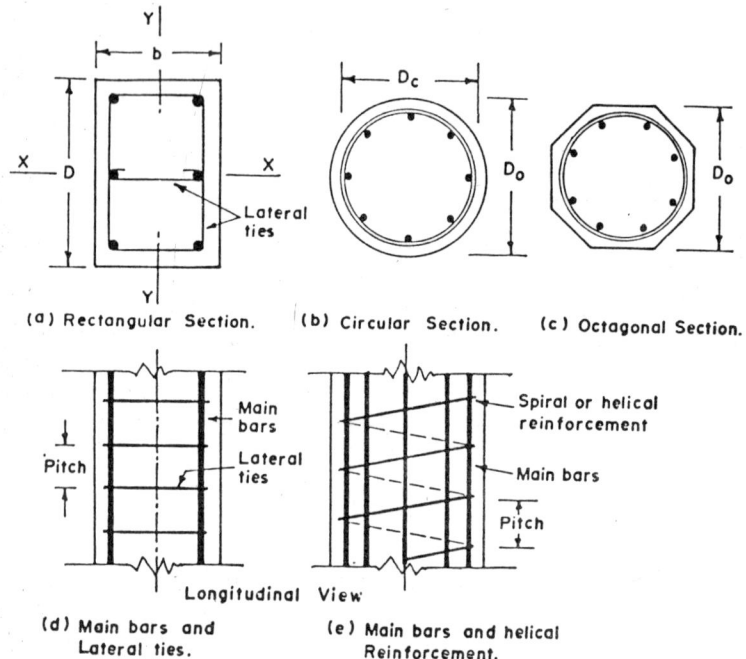

(a) Rectangular Section.   (b) Circular Section.   (c) Octagonal Section.

(d) Main bars and Lateral ties.   (e) Main bars and helical Reinforcement.

Fig. 8.1. Sections of compression members showing reinforcement

which is expressed as the ratio of the effective length to the lateral dimension. The member may buckle about X—X axis or Y—Y axis depending upon the unsupported length and the end conditions existing in the axial planes at right angles to these axes.

Let,

l_{ex} = effective length in respect of buckling about X-axis.

l_{ey} = effective length in respect of buckling about Y-axis.

D = larger lateral dimension.

b = smaller lateral dimension.

Then, the classification of compression members as per IS Code 456 is as follows:

(i) A compression member is called a pedestal if,

$$\frac{l_{ex}}{D} < 3$$

and also

$$\frac{l_{oy}}{b} < 3$$

(ii) A compression member is considered to be short if,

$$3 < \frac{l_{ox}}{D} < 12$$

and also

$$3 < \frac{l_{oy}}{b} < 12$$

(iii) A compression member is considered to be long if,

either $\dfrac{l_{ox}}{D} \geqslant 12$

or $\dfrac{l_{oy}}{b} \geqslant 12$

or both.

Circular and Octagonal cross-sections of compression members are shown in Figs. 8.1 (b) and (c) respectively. The relevant lateral dimension for circular section will be taken to be the outer diameter D_0 for tied members, and core diameter D_0 for members having helical reinforcement. For octagonal section a similar rule may be followed.

8.2 UNSUPPORTED LENGTH

The unsupported length of a compression member according to IS Code: 456 shall be taken as the clear distance between the end restraints.

Exceptions to this rule are as follows:

Column situation	*Unsupported length*
Supporting flat slab floor.	See Fig. 8.2.
Supporting Beam and Slab floor.	See Fig. 8.3.
Restrained laterally by struts.	See Fig. 8.4.
Supporting beam ends on brackets.	See Fig. 8.5.

8.3 EFFECTIVE LENGTH

The effective length of compression members with known end conditions may be taken from Table 8.1.

Table 8·1   Effective Length of Compression Members

	End Conditions		Recommended Effective Length
	Translation Permitted	Rotation Permitted	
(a) ℓ	No	No	0·65ℓ
	No	No	
(b) ℓ	No	Yes	0·8ℓ
	No	No	
(c) ℓ	No	Yes	ℓ
	No	Yes	

Contiuned . . . . .

		Yes	No	
(d)	l			$1 \cdot 2\, l$
		No	No	
		Yes	Partially	
(e)	l			$1 \cdot 5\, l$
		No	No	
		Yes	No	
(f)	l			$2 \cdot 0\, l$
		No	Yes	
		Yes	Yes	
(g)	l			$2 \cdot 0\, l$
		No	No	

Note: This is based on table 24 of IS Code: 456.

However, for columns in framed structures, the effective length may be obtained from Fig. 8.6 and 8.7. depending upon whether the member is braced or unbraced. In a braced member the sway of one end relative to the other is prevented.

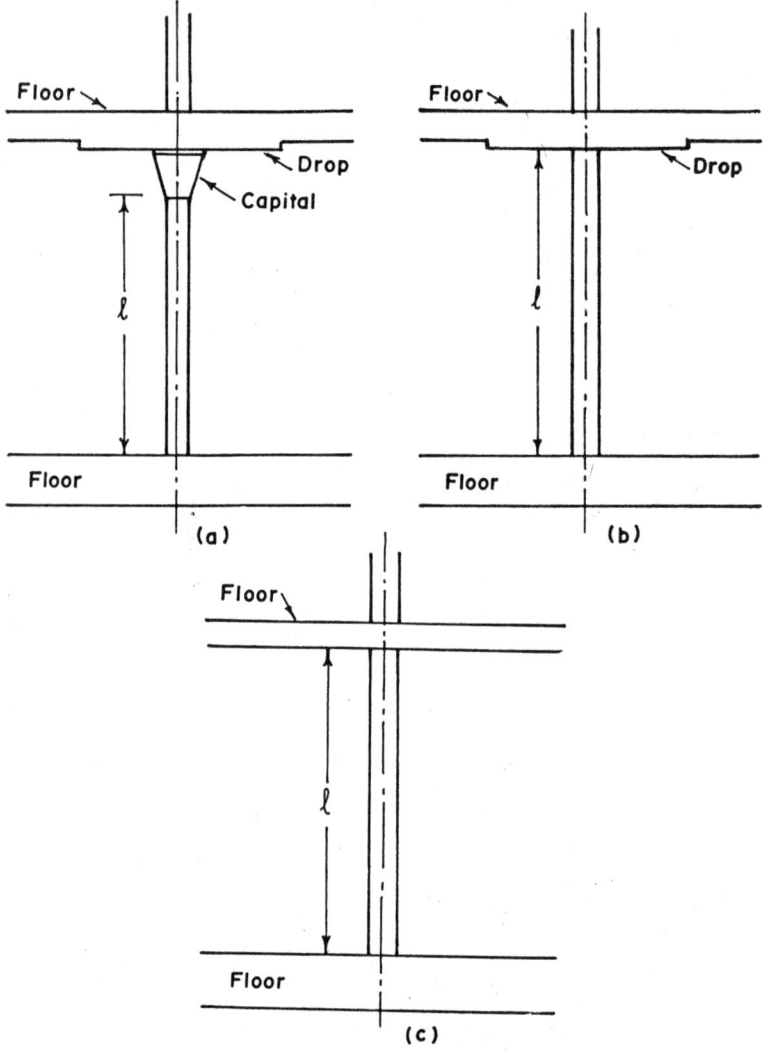

Fig. 8.2. Column supporting flat slab floor

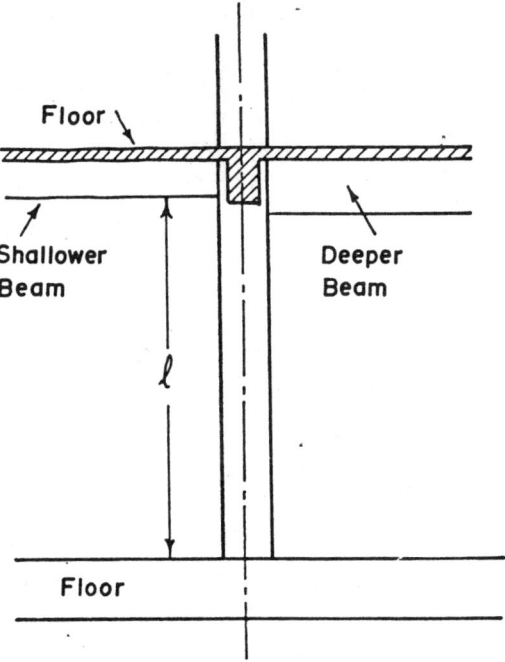

Fig. 8.3. Column supporting beam-and-slab floor

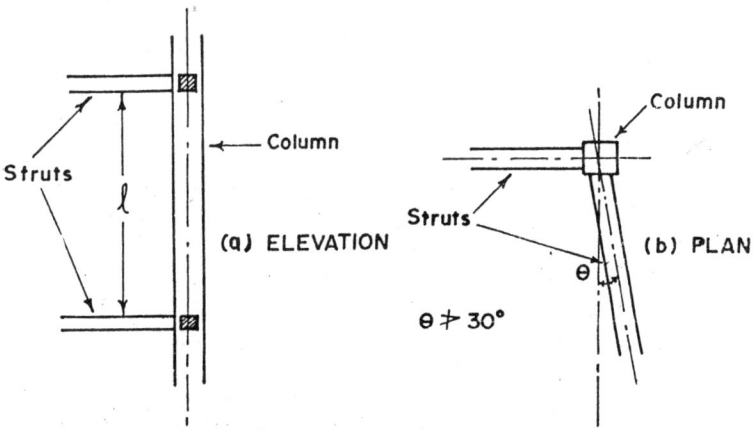

Fig. 8.4. Column restrained laterally by struts

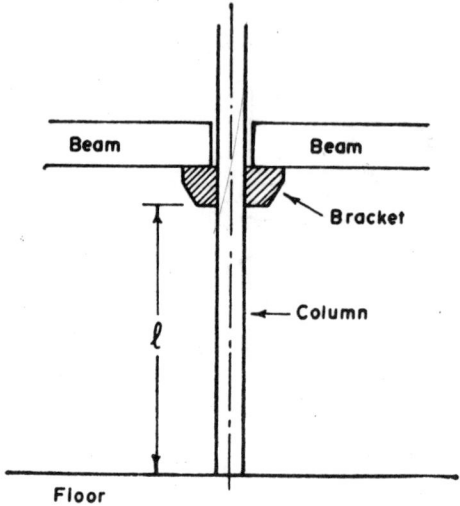

Fig. 8.5 Column supporting beam ends on brackets

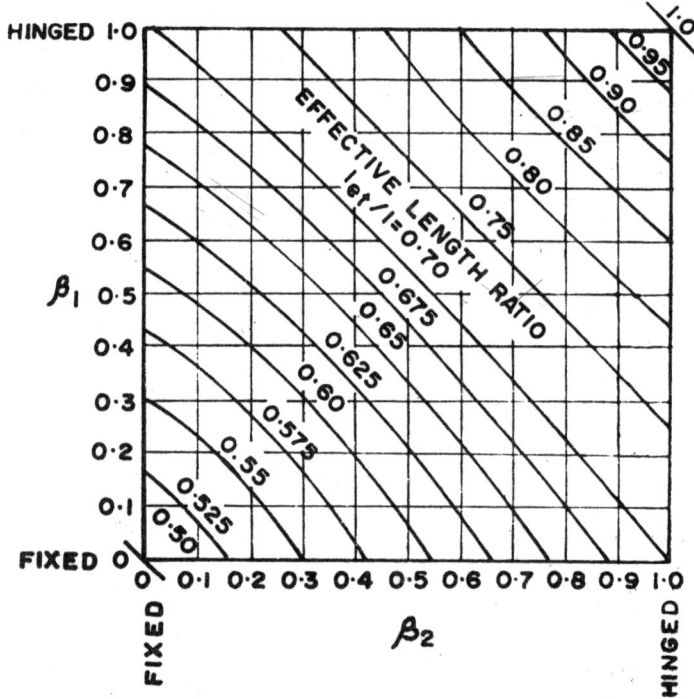

Fig. 8.6 Effective length of column in a braced frame
(Based on Fig 24 of IS Code: 456)

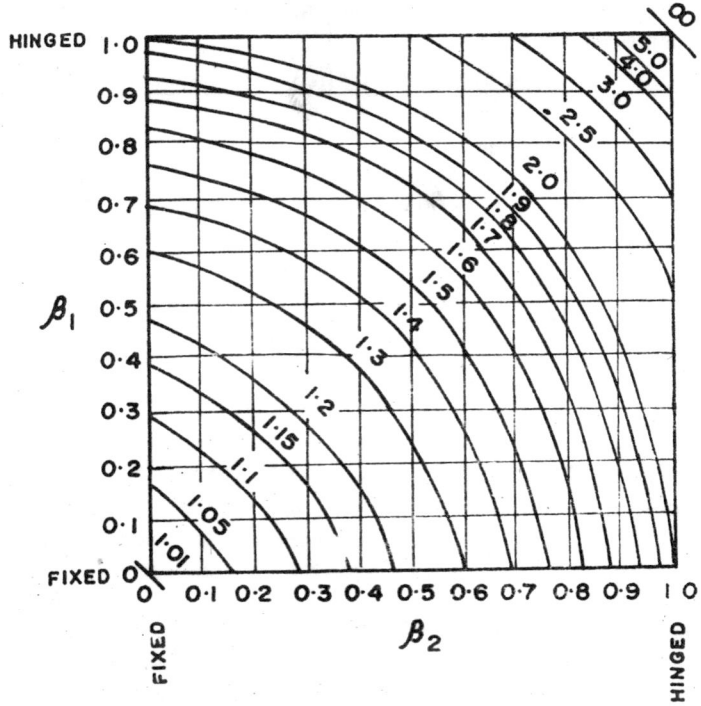

Fig. 8.7 Effective length of column in an unbraced frame
(Based on Fig. 25 of IS Code: 456)

8.4 DESIGN REQUIREMENTS AS PER IS CODE: 456

8.4.1 Slenderness limit.

(i) Between the end restraints of a compression member:
Unsupported length $1 \not> 60 \times$ least lateral dimension.

(ii) If one end of compression member is unrestrained, then:

Unsupported length $1 \not> \dfrac{100b^2}{D}$

where

b = width of section

D = depth measured in the plane under consideration.

8.4.2 Minimum eccentricity, e_{min}

All compression members shall be designed for a minimum eccentricity given by

$$e_{min} \ngtr \frac{\text{unsupported length}}{500} + \frac{\text{lateral dimension}}{30}$$

$$\ngtr 20 \text{ mm.}$$

8.4.3 Longitudinal reinforcement.

Limits of the %age of longitudinal reinforcement shall be as follows:

$$\frac{100A_s}{A_g} \geqslant 0.8\%$$

$$\leqslant 6\%$$

where

A_s = cross-sectional area of longitudinal reinforcement

A_g = gross cross-sectional area of compression member

Note: The upper limit may better be kept as 4% in order to avoid practical difficulties in placing and compaction of concrete.

If the cross-sectional area of compression member is larger than required, the minimum quantity of reinforcement shall be calculated on the concrete area required to resist the axial load, and not upon the actual area.

The minimum number of longitudinal reinforcement bars shall be as follows:

For rectangular section — 4

For circular section — 6

Diameter of longitudinal bars shall not be less than 12mm. In a helically reinforced column the number of longitudinal bars shall be at least 6 and these should be placed equidistant and in contact with the helical reinforcement. The spacing of longitudinal bars along the periphery of section shall not exceed 300 mm.

For pedestals, only nominal longitudinal reinforcement, not less than 0.15% of cross-section area shall be provided.

8.4.4. Transverse reinforcement.

The transverse reinforcement should be so arranged that every longitudinal bar is laterally supported against buckling. The transverse reinforcement may be in the form of circular rings or polygonal links with internal angle not exceeding 135°.

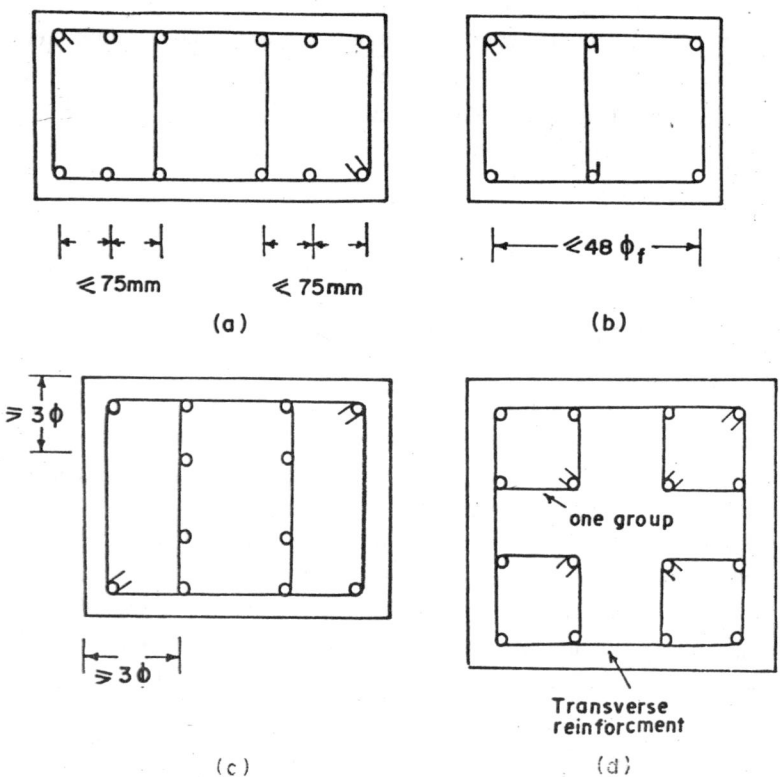

Fig. 8.8. Cross-sections of compression members showing transverse reinforcement

8.4.4.1. Arrangement of transverse reinforcement.

(i) Along any side of the section if the spacing of the longitudinal bars does not exceed 75 mm c/c, then the transverse reinforcement need go around the alternate bars only. See Fig. 8.8.(a).

(ii) Along any side of the section if the distance between the corner bars does not exceed 48 times the diameter of the tie, then the additional bars in between may be provided only with open ties in one direction. See Fig. 8.8.(b).

(iii) Longitudinal bars in the inner row may be considered to be laterally supported provided the bars in the outer row are properly tied and no bar of inner row is closer to the nearest face than three times the diameter of the bars. See Fig. 8.8.(c).

8.4.4.2 Pitch of transverse reinforcement. The pitch (of spacing c/c) shall not exceed the following limits:

(i) The least lateral dimension of the member

(ii) 16 times the diameter of the smallest longitudinal bar to be tied

(iii) 48 times the diameter of transverse reinforcement bar.

8.4.4.3 Diameter of transverse reinforcement. The diameter of transverse reinforcement shall not be less than the following:

(i) $\frac{1}{4}$ the diameter of the largest longitudinal bar

(ii) 5 mm

If the longitudinal bars are grouped and each group is properly tied, then the transverse reinforcement for the member as a whole may be provided considering each group as a single longitudinal bar, for determining the diameter and pitch of transverse reinforcement. However, the diameter of the transverse reinforcement shall not exceed 20 mm. See Fig. 8.8(d).

8.4.4.4 Helical reinforcement. The helical reinforcement shall heve a uniform pitch. Its ends shall be anchored by providing $1\frac{1}{2}$ extra turns. The pitch shall not exceed the following:

(i) 75 mm

(ii) $\frac{1}{6}$ of core diameter D_c of column.

The pitch shall not be less than the following:

(i) 25 mm

(ii) 3 times the diameter of helical bar.

The rules regarding the diameter of helical bar are the same as for the transverse reinforcement.

8.5 PERMISSIBLE STRESS

The permissible stress in direct compression in concrete (σ_{oc}) is lower than that in flexural compression. It is given in Table 3.1. The permissible stress in steel reinforcement (σ_{sc}) is also lower than that in tension. It is given in Table 3.2

8.6 PERMISSIBLE AXIAL LOAD

8.6.1 Short column with lateral ties, and pedestals. The permissible axial load is given by

$$P = \sigma_{oc} A_c + \sigma_{sc} A_{sc} \qquad (8.1)$$

where

σ_{oc} = permissible stress in concrete in direct compression

A_o = net cross sectional area of concrete

σ_{so} = permissible stress in steel reinforcement in compression

A_{sc} = cross-sectional area of longitudinal reinforcement

The minimum eccentricity discussed under para 8.4.2 is taken care of in eq. (8.1)

8.6.2 Short column with helical reinforcement. The permissible axial load for columns with helical reinforcement shall be taken to be 1.05 times that for similar member with lateral ties. This increased axial load capacity will be taken only if the ratio

$$\frac{\text{vol. of helical reinforcement}}{\text{vol. of the core}} \not< 0.36 \left(\frac{A_g}{A_{co}} - 1 \right) \frac{f_{ck}}{f_y} \qquad (8.2)$$

where

A_g = gross area of section

A_{co} = section area of core measured to outside diameter of helix

See Fig. 8.1(b).

Note: In no case shall f_y be taken more than 415 N/mm².

8.6.3 Long compression members. A reduction coefficient is applied on the axial load carrying capacity of compression member, if it is long, in order to avoid possibility of failure by buckling. It is given by

$$C_r = \left(1.25 - \frac{\text{slenderness ratio}}{48} \right) \qquad (8.3)$$

where

C_r = reduction coefficient.

For more exact calculations IS Code: 456 recommends the value as

$$C_r^{\cdot} = 1.25 - \frac{l_e}{160\, r_m} \qquad (8.4)$$

where

l_e = effective length of column

r_m = the least redius of gyration.

The capacity reduction factor may be applied only when

$$\frac{l_e}{r_m} > 40$$

8.7   COVER ON COLUMN REINFORCEMENT

The requirement as per IS Code: 456 for concrete cover (See Fig. 8.9) is as follows:

(i) *On longitudinal bar*

Concrete cover ≮ diameter of the bar

≮ 40 mm.

However, for column with section dimensions less than 200 mm, if the diameter of longitudinal bar does not exceed 12 mm, a cover of 25 mm may be used.

(ii) *On transverse reinforcement bar*

Concrete cover ≮ diameter of transverse bar

≮ 15 mm.

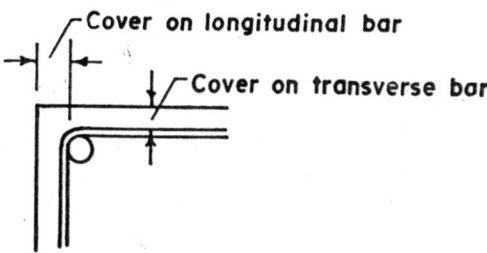

Fig. 8.9   Concrete cover on column reinforcements

An increase in cover thickness beyond the figures given above may be provided under special circumstances as follows:

(a) An increase of 15 to 50 mm when surface of concrete is exposed to the action of harmful chemicals, acidic vapour, saline atmosphere, sulphurous smoke, etc.

(b) An increase of 40 mm for reinforced concrete members immersed in sea water.

(c) An increase of 50 mm for reinforced concrete members periodically immersed in sea water or subject to sea spray.

(d) For concrete of grade M25 or above the additional cover described in (a), (b) and (c) above may be reduced to half.

However, for all such cases the total cover should not exceed 75 mm.

Example 8.1. A column is continuous through reinforced concrete beam-and-slab floors both at the top and bottom. Unsupported length of column = 5m. Total axial (Dead + Live) load = 630 kN. Concrete grade: M 20 and steel grade Fe 415. Design the column taking

(a) Square section with lateral ties

(b) Circular section with helical reinforcement.

Solution

The column may be taken to be restrained against rotation and translation both at top and bottom. It conforms to the case (a) in Table 5.1.

Hence, effective length

$$l_e = 0.65 \; l$$
$$= 0.65 \times 5$$
$$= 3.25 \; m$$

Permissible stresses are

$\sigma_{oo} = 5 \; N/mm^2$ from Table 3.1.

$\sigma_{so} = 190 \; N/mm^2$ from Table 3.2.

Initially, the column may be assumed to be short and the section dimensions worked out. Slenderness ratio may then be calculated to see whether the column is short. Otherwise, the design may be revised as for a long column.

(a) *Square section with lateral ties.*

Axial load bearing capacity of a short column is

$$P = \sigma_{oc} \, A_c + \sigma_{so} \, A_{so}$$
$$= \sigma_{co} \, (A_g - A_{so}) + \sigma_{sc} \, A_{sc}$$
$$= \sigma_{oo} \, A_g + (\sigma_{so} - \sigma_{cc}) \, A_{sc}$$
$$= \sigma_{oo} \, A_g + (\sigma_{sc} - \sigma_{oc}) \, p \, A_g$$
$$= \{\sigma_{cc} + (\sigma_{so} - \sigma_{cc}) \, p\} \, A_g \qquad (8.5)$$

where

$$p = \frac{A_{sc}}{A_g} \text{ is the proportion of longitudinal reinforcement.}$$

Let the self weight of the column be 12 kN. Then, total design axial load $P = 630 + 12 = 642$ kN.

Keeping minimum reinforcement (0.8%) in the column,

$p = 0.008$. Substituting the values in Eq. (8.5)

$$642 \times 10^3 = \{ 5 + (190 - 5) \times 0.008 \} \, A_g$$

$$A_g = 99074 \text{ mm}^2$$

$$\text{length of side} = \sqrt{99074}$$

$$= 314.76 \text{ mm}$$

Adopt a 315 × 315 mm cross-section. Slenderness ratio

$$= \frac{3.25 \times 1000}{315}$$

$$= 10.32 < 12$$

Hence, the assumption that the column is short is justified.

Longitudinal reinforcement

$$A_{sc} = 0.008 \times 315^2$$

$$= 793.8 \text{ mm}^2$$

Provide 4 nos., 16 mm dia bars, giving $A_{sc} = 804.3$ mm^2.

Transverse reinforcement in the form of lateral ties will be provided.

Diameter of tie bar

$$\not< \tfrac{1}{4} \times 16 \text{ i.e. 4 mm}$$

$$\not< 5 \text{ mm}$$

Adopt 6 mm dia. M.S. ties.

Spacing of ties

$$\not> 315 \text{ mm}$$

$$\not> 16 \times 16 \text{ i.e. 256 mm}$$

$$\not> 48 \times 6 \text{ i.e. 288 mm}$$

Adopt a spacing of 250 mm c/c throughout.

A cover of 40 mm is kept on the longitudinal bars. Hence, the cover on lateral ties will be 34 mm, which is O.K.

Fig 8.10 shows the details of the reinforcement in the column.

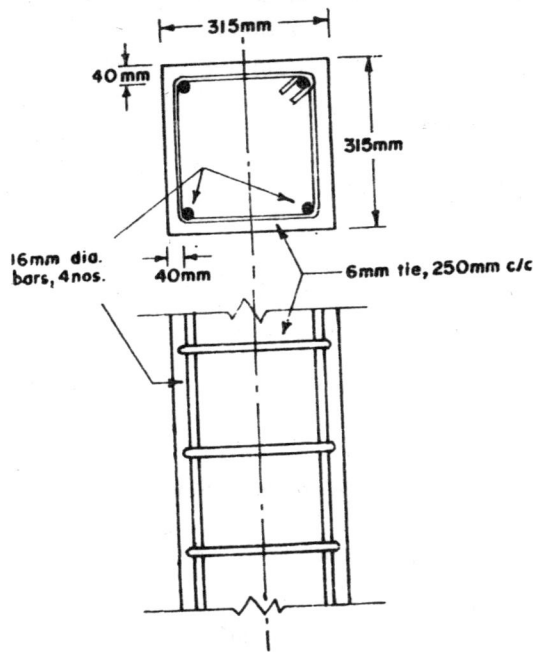

Fig. 8.10

(b) *Circular section with helical reinforcement.*

The axial load capacity of the column in this case will be 1.05 times that given by the r.h.s. of Eq. (8.5). Keeping p = 0.008 as in the previous design, and substituting the values in Eq. (8.5)

$$642 \times 10^3 = 1.05 \{ 5 + (190 - 5) \times 0.008 \} A_g$$
$$\therefore A_g = 94356.3 \text{ mm}^2$$

$$\text{Dia. of column section} = \sqrt{\frac{4 \times 94356.3}{\pi}}$$

$$= 346.6 \text{ mm}$$

Adopt a section of diameter = 350 mm

Longitudinal reinforcement

$$A_{sc} = 0.008 \times 94356.3$$
$$= 754.85 \text{ mm}^2$$

Provide 7 nos., 12 mm dia. bars giving

$A_{sc} = 791.7$ mm². Keep a 40 mm cover on the bars. Diameter of the bars forming the helical reinforcement

$\not< \frac{1}{4} \times 12$ i.e. 3 mm

$\not< 5$ mm

Try a 6 mm helical bar.

Core diameter of column

$$= 350 - 2 \times 40 + 2 \times 6$$
$$= 282 \text{ mm}$$

$A_{co} = \frac{\pi}{4} \times 282^2 = 62458.15$ mm²

The pitch of helical reinforcement

$\not> 75$ mm

$\not> \frac{1}{6} \times 282$ i.e. 47 mm

$\not< 25$ mm

$\not< 3 \times 6$ i.e. 18 mm

Hence, the pitch should lie between 25 mm and 47 mm.

Try a pitch = 40 mm.

Length of helical bar in one pitch

$$= \sqrt{\{ \pi (D_o - 6) \}^2 + \text{pitch}^2}$$
$$= \sqrt{\{ \pi (282 - 6) \}^2 + 40^2}$$
$$= 868 \text{ mm}$$

Vol. of helical reinforcement in one pitch

$$= \frac{\pi}{4} \times 6^2 \times 868$$
$$= 24542 \text{ mm}^3$$

Vol. of the core in one pitch

$$= A_{co} \times \text{pitch}$$
$$= 62458.15 \times 40$$
$$= 2498326 \text{ mm}^3$$

Hence, the L.H.S. of Eq. (5.2) is

$$= \frac{24542}{2498326}$$
$$= 0.00982$$

The R.H.S. of Eq. (5.2) is

$$= 0.36 \left(\frac{94356.3}{62458.15} - 1 \right) \frac{20}{415}$$
$$= 0.00886$$

Evidently, Eq. (8.2) is satisfied. Hence, a pitch of 40 mm is alright. Reinforcement details are shown in Fig. 8.11.

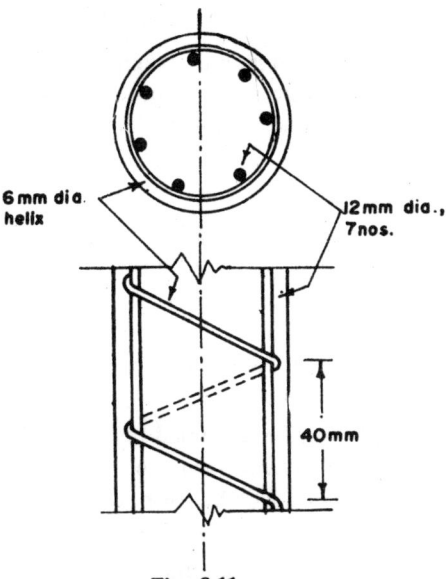

6 mm dia. helix

12mm dia., 7nos.

40mm

Fig. 8.11

Example 8.2. Design a rectangular column section keeping the ratio D/b = 2.0. It supports a beam-and-slab type R.C. floor at top and rests over an isolated footing at bottom. See Fig. 8.12.

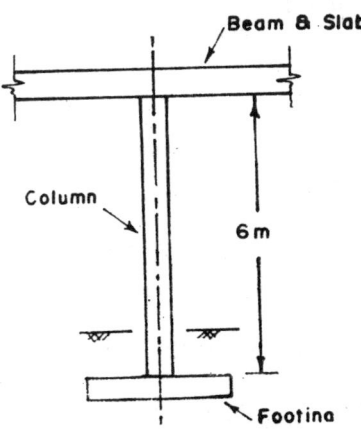

Beam & Slab floor

Column

6 m

Footing

Fig. 8.12

Unsupported length of column = 6 m as shown. Total axial (dead + live) load = 770 kN. Concrete grade: M 20 and steel grade: Fe 415.

Solution

The top end of the column may be taken to be restrained against rotation and translation, whereas, the bottom end may be taken to be free to rotate but restrained against translation. It conforms to case (b) of Table 8.1.

Hence, effective length

$$l_e = 0.8 \times \text{unsupported length}$$
$$= 0.8 \times 6$$
$$= 4.8 \text{ m}$$

Permissible stresses are

$$\sigma_{co} = 5 \text{ N/mm}^2 \text{ from Table 3.1}$$
$$\sigma_{so} = 190 \text{ N/mm}^2 \text{ from Table 3.2}$$

Initially the column may be assumed to be short and its section dimensions worked out. Then, from the value of slenderness ratio a check may be made to see whether or not the column is short. If it is found to be long, then the design may be modified.

Taking the self weight of column = 21 kN, the total design axial load

$$P = 770 + 21 = 791 \text{ kN}$$

Let the proportion of reinforcement p = 0.008

As in the previous example

$$P = \{\sigma_{co} + (\sigma_{so} - \sigma_{oo}) \, p \} \, A_g$$

or $791 \times 10^3 = \{ 5 + (190 - 5) \times .008 \} \, A_g$

∴ $A_g = 122068 \text{ mm}^2$

or $bD = 122068 \text{ mm}^2$

Putting above $b = \frac{1}{2} D$

$$D^2 = 2 \times 122068$$

∴ $D = 494 \text{ mm}$

and $b = 247 \text{ mm}$

Slenderness ratios are

$$\frac{l_{ex}}{D} = \frac{4.8 \times 10^3}{494} = 9.72 < 12$$

$$\frac{l_{ey}}{b} = \frac{4.8 \times 10^3}{247} = 19.4 > 12$$

Obviously, the column is long, and the design needs amendment. On applying the capacity reduction factor, the axial load capacity of column is written as

$$P = C_r \{ \sigma_{cc} + (\sigma_{sc} - \sigma_{cc}) \, p \} \, A_g$$

$$= \left(1.25 - \frac{4.8 \times 1000}{48\,b} \right) \{5 + (190-5) \times .008\} \, bD$$

$$= 6.48 \left(1.25 - \frac{100}{b} \right) bD$$

$$= 12.96 \left(1.25 - \frac{100}{b} \right) b^2$$

$$= 12.96 \, (1.25\,b - 100) \, b$$

Trial values of b are taken till the r.h.s $\gg 791 \times 10^3$ N.
Thus, the solution comes to be

$$b = 265 \text{ mm}$$
and $D = 2 \times 265 = 530$ mm
Longitudinal reinforcement

$$= 0.008 \times 265 \times 530$$
$$= 1123.6 \text{ mm}^2$$

Provide 12 mm dia bars, 10 nos.
Diameter of lateral ties

$$\nless \tfrac{1}{4} \times 12 \quad \text{i.e. 3 mm}$$
$$\nless 5 \text{ mm}$$

Adopt 6 mm dia lateral ties.
Spacing of ties

$$\ngtr 265 \text{ mm}$$
$$\ngtr 16 \times 12 \text{ i.e. 192 mm}$$
$$\ngtr 48 \times 6 \text{ i.e. 288 mm}$$

Adopt a spacing of 190 mm c/c.
A cover of 40 mm will be kept on the longitudinal bars. The arrangement of reinforcement is shown in Fig. 8.13. Along the shorter side of section, the c/c distance between the corner bars is

$$= 265 - 2 \times 40 - 12$$
$$= 173 \text{ mm}$$

which is less than $48 \times$ dia. of the tie. The bar in between the two corner bars is distant more than 75 mm from the corner bars on either side. Hence, the middle bars on the shorter sides are connected by an open tie as shown.

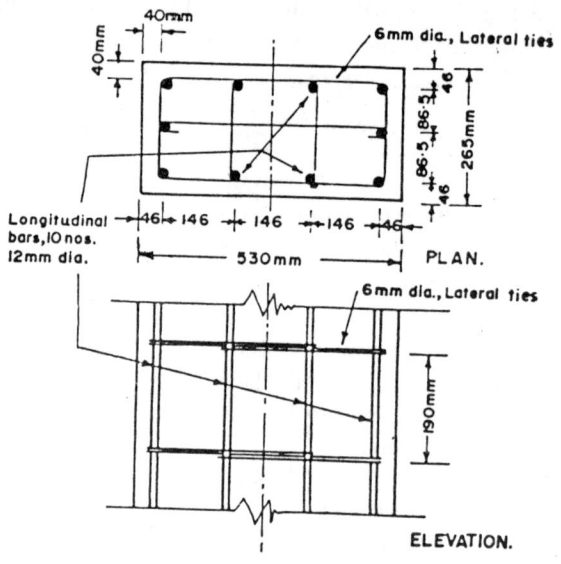

Fig 8.13

8.8 MEMBER CARRYING COMBINED AXIAL LOAD AND BENDING MOMENT

There are very few instances in practice where members are subjected to pure axial forces. Mostly, compression members are either part of framed structures or monolithic with floor. As such, bending moments are imparted to the compression members when there is no symmetry either of the structure or of loading. Thus, the members may be subjected to uniaxial or biaxial bending moments in addition to the axial forces. Design of members carrying combined axial load and bending moment may be based on two considerations described in the follwing paras.

8.8.1 Design based on uncracked section. when the bending moment is small compared with the axial force, the tensile stress caused by the bending moment is effectively countered by the compressive stress caused by the axial force. As a result, either there may be no tensile stress, or if there is some, the concrete may be able to bear it. There is no cracking of concrete and the whole section remains effective in taking up stress.

Properties of the transformed section are written as

$A_t = A_g + (1.5 \ m - 1) \ A_{sc}$ the transformed section area

Z_x = section modulus about X-axis

$= \dfrac{I_x}{D/2}$

Z_y = section modulus about Y-axis

$= \dfrac{I_y}{b/2}$

Two cases are considered in the IS Code: 456 as follows:

(a) *Axial force combined with uniaxial bending moment*

Let

P = axial force

M_x = bending moment about X-axis

Then

f_{oc} = calculated direct compressive stress

$= \dfrac{P}{A_g + (1.5 \ m - 1) \ A_{sc}}$

f_{obo} = calculated bending stress

$= \dfrac{M_x}{Z_x}$

The section will be considered safe if the following conditions are satisfied:

(i) $\dfrac{f_{oo}}{\sigma_{oc}} + \dfrac{f_{obo}}{\sigma_{obc}} \prec 1.0$

(ii) Resultant tensile stress in concrete

$\prec 0.25 \ (f_{oo} + f_{obo})$

$\prec 0.75$ (7 day modulus of rupture of concrete).

(b) *Axial force combined with biaxial bending moment.*

In this case

$f_{cc} = \dfrac{P}{A_g + (1.5 \ m - 1) \ A_{sc}}$ as in (a) above

and

$f_{ebo} = \dfrac{M_x}{Z_x} + \dfrac{M_y}{Z_y}$

The section will be considered safe if the following conditions are satisfied:

(i) $\dfrac{f_{cc}}{\sigma_{oo}} + \dfrac{f_{cbo}}{\sigma_{obc}} \prec 1.0$

(ii)   Resultant tensile stress in concrete

$$< 0.35 \ (f_{cc} + f_{obo})$$

$$< 0.75 \ (7 \ \text{day modulus of rupture of concrete})$$

8.8.2 Design based on cracked section. When the bending moment is large compared with the axial force, the resultant tensile stress in concrete may be high and the stress limits specified in para 8.8.1 may be exceeded. The area of concrete under tension is, therefore, taken to be cracked and as such neglected from consideration. The stresses in concrete and steel reinforcement are found by basing the calculations on cracked section. A cracked section is the one in which the tensile stress in concrete is ignored. See Fig. 8.14

Let

  f_{obo} = compresive stress in extreme fibres of concrete

  f_{st} = tensile stress in tensile reinforcement

  X = depth of neutral axis (no stress line) from the extreme fibres of concrete in compression

  $jd = d - \dfrac{X}{3}$

  P = axial compression force on section

  M = bending moment on section

  $e = \dfrac{M}{P}$

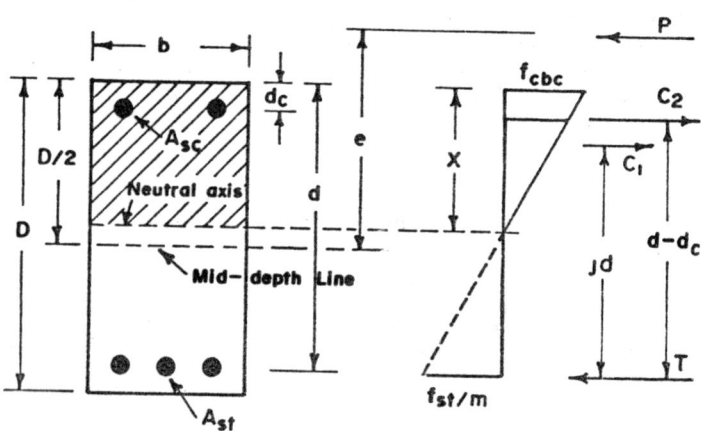

Fig. 8.1 ·

Now, stress in the reinforcement on compression side is

$$f_{st} = 1.5 \, m \left(\frac{X - d_c}{X} \right) f_{cbc} \qquad (8.5)$$

Force in concrete in compression is

$$C_1 = \tfrac{1}{2} \, b \, X \, f_{cbc} \qquad (8.6)$$

Force in the reinforcement in compression is

$$C_2 = (1.5 \, m - 1) \left(\frac{X - d_c}{X} \right) f_{cbc} \, A_{sc} \qquad (8.7)$$

Force in the reinforcement in tension is

$$T = f_{st} \, A_{st} \qquad (8.8)$$

From the stress diagram

$$f_{st} = \frac{m \, (d - X)}{X} \, f_{cbc} \qquad (8.9)$$

Considering the equilibrium of all the forces normal to the cross-section

$$P = C_1 + C_2 - T$$

$$P = \tfrac{1}{2} b \, X \, f_{cbc} + (1.5 \, m - 1) \frac{(X - d_c)}{X} \, f_{cbc} \, A_{sc} - f_{st} \, A_{st}$$

Substituting the value of f_{st} from Eq. (5.9) in the above expression and rearranging

$$P = \left\{ \tfrac{1}{2} bX + (1.5m - 1) \frac{(X - d_c)}{X} A_{sc} - m \frac{(d - X)}{X} A_{st} \right\} f_{cbc}$$

$$(8.10)$$

Taking the moment of all the forces about the tension reinforcement, equating it to zero, and rearranging

$$P \left(e + d - \frac{D}{2} \right) = C_1 \, jd + C_2 \left(d - d_c \right)$$

or $\quad P \left(e + d - \frac{D}{2} \right) = \left\{ \tfrac{1}{2} b \, X \left(d - \frac{X}{3} \right) + (1.5 \, m - 1) \right.$

$$\left. \times \frac{(X - d_c)}{X} \, (d - d_c) \, A_{sc} \right\} f_{cbc} \quad (8.11)$$

Eqs. (8.10 and 8.11) involve only two unknowns namely X and f_{cbc} which can be readily solved for any given section dimensions.

In a design problem the values of P and M are known. The procedure of design is by trials as described below:

(i)   Calculate $e = \dfrac{M}{P}$

(ii)   Take a trial section

(iii) Calculate X and f_{cbc} from Eqs. (8.10) and (8.11)

(iv) Calculate f_{st} from Eq. (8.9)

(v)   The section is safe if

        $f_{cbc} < \sigma_{cbc}$

   and $f_{st} < \sigma_{st}$

(vi) If the section is unsafe, another section may be tried.

Example 8.3. Design a column section to take up an axial force = 705 kN along with a bending moment = 90 kN — m.

Unsupported length of column = 6 m.

The column may be considered to be restrained against rotation and translation at both the ends. Concrete grade: M 20 and steel grade Fe 415.

Solution

The design will be based on uncracked section basis. Consider the trial section as shown in Fig. 8.15

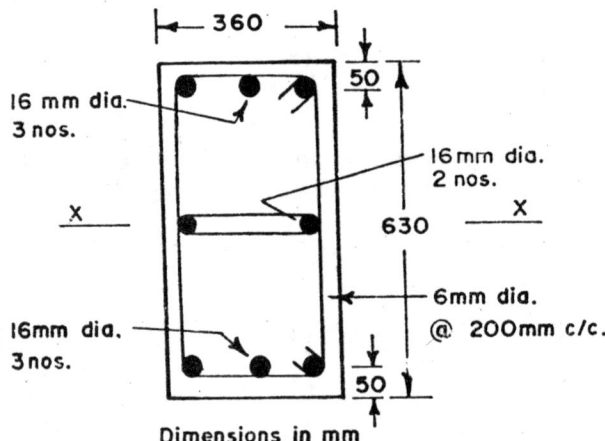

Dimensions in mm

Fig. 8.15 Trial section for Ex. (8.3)

Effective length of column = 0.65 × 6

                   = 3.9 m

Slenderness ratio $= \dfrac{l_{ey}}{b}$

$$= \dfrac{3.9 \times 1000}{360}$$

$$= 10.83 < 12$$

Evidently, the column is short.

Permissible stresses are

$\quad\sigma_{co} = 5$ N/mm² from Table 3.1.

$\quad\sigma_{cbc} = 7$ N/mm² from Table 3.1.

7 day modulus of rupture$= 2.4$ N/mm² from Table 1.3.

Modular ratio $= 13.33$ from Table 3.3.

Transformed section properties are

$$A_t = A_g + (1.5\ m - 1)\ A_{sc}$$

$$= 630 \times 360 + (1.5 \times 13.33 - 1) \times 8 \times 201.06$$

$$= 257.35 \times 10^3 \text{ mm}^2$$

$$I_x = \dfrac{1}{12} \times 360 \times 630^3 + (1.5 \times 13.33 - 1)$$

$$\times 6 \times 201.06 \times 265^2$$

$$= 9110 \times 10^6 \text{ mm}^4$$

Calculated stresses are

$$f_{co} = \dfrac{P}{A_t} = \dfrac{705 \times 10^3}{257.35 \times 10^3} = 2.74 \text{ N/mm}^2$$

$$f_{cbc} = \dfrac{M_x}{I_x} \times \dfrac{D}{2} = \dfrac{90 \times 10^6 \times 315}{9110 \times 10^6}$$

$$= 3.11 \text{ N/mm}^2$$

Hence

$$\dfrac{f_{co}}{\sigma_{co}} + \dfrac{f_{cbc}}{\sigma_{cbc}} = \dfrac{2.74}{5} + \dfrac{3.11}{7}$$

$$= 0.962 < 1.0 \quad \text{O.K.}$$

The resultant tensile stress in the section should not exceed the following limits:

(i) $0.25\ (f_{co} + f_{cbc})$

$\quad$ i.e. $= 0.25\ (2.74 + 3.11)$

$\quad\quad = 1.415$ N/mm² and

(ii) 0.75 (7 day modulus of rupture)

$\quad$ i.e. $= 0.75 \times 2.4$

$\quad\quad = 1.8$ N/mm²

Resultant tensile stress $= 3.11 - 2.74$

$$= 0.37 \text{ N/mm}^2$$

Which is less than the two limits above. Hence, the trial section is O.K.

Example 8.4. A short column is to be designed to take up an axial force $= 260$ kN. and bending moments about the two axes 1.1 kN-m and 1.3 kN-m respectively. Concrete grade: M 20 and Steel grade: Fe 415.

Solution

The design is carried out on an uncracked section basis. The section in Fig. 8.10 is taken as a trial section. Properties of transformed section are as follows:

$$A_t = 315 \times 315 + (1.5 \times 13.33 - 1) \times 4 \times 201.06$$
$$= 114502 \text{ mm}^2$$

$$I_x = \frac{1}{12} \times 315 \times 315^3 + (1.5 \times 13.33 - 1) \times 4$$
$$\times 201.06 \times 109.5^2$$

$$= 1003.6 \times 10^6 \text{ mm}^4$$
$$I_y = 1003.6 \times 10^6 \text{ mm}^4 \text{ also.}$$

The stresses are

$$f_{co} = \frac{P}{A_g + (1.5 m - 1) A_{sc}} = \frac{260 \times 10^3}{114502}$$
$$= 2.27 \text{ N/mm}^2$$

$$f_{cbc} = \frac{M_x}{Z_x} + \frac{M_y}{Z_y} = \frac{1.1 \times 10^6}{1003.6 \times 10^6} \times \frac{315}{2}$$
$$+ \frac{1.3 \times 10^6}{1003.6 \times 10^6} \times \frac{315}{2}$$

$$= 3.77 \text{ N/mm}^2$$

The permissible stresses are

$\sigma_{co} = 5 \text{ N/mm}^2$ from Table 3.1

$\sigma_{cbc} = 7 \text{ N/mm}^2$ from Table 3.1

7 day modulus of rupture $= 2.4$ N/mm^2 from table 1.3.

Hence

$$\frac{f_{cc}}{\sigma_{co}} + \frac{f_{cbc}}{\sigma_{cbc}} = \frac{2.27}{5} + \frac{3.77}{7} = 0.993 < 1 \quad \text{O.K.}$$

The resultant tensile stress in the section should not exceed the following limits:

(i) $0.35 (f_{oe} + f_{obo})$
    i.e. $= 0.35 (2.27 + 3.77)$
        $= 2.11$ N/mm^2 and
(ii) 0.75 (7 day modulus of rupture)
    i.e. $= 0.75 \times 2.4$
        $= 1.8$ N/mm^2
Resultant tensile stress
        $= 3.77 - 2.27$
        $= 1.5$ N/mm^2
which is less than the two limits above. Hence, the trial section is O.K.

Example 8.5 A compression member is required to carry an axial force of 284 kN along with a bending moment of 118 kN-m.

Design the member by taking trial sections. Concrete grade: M 20 and Steel grade: Fe 415.

Solution

As the bending moment is large compared with the axial force, the section will be designed on cracked section basis. Permissible stresses are

$\sigma_{obo} = 7$ N/mm^2 from Table 3.1
$\sigma_{st} = 230$ N/mm^2 from Table 3.1
Modular ratio $= 13.3$ from Table 3.3
The trial section is shown in Fig. 8.16

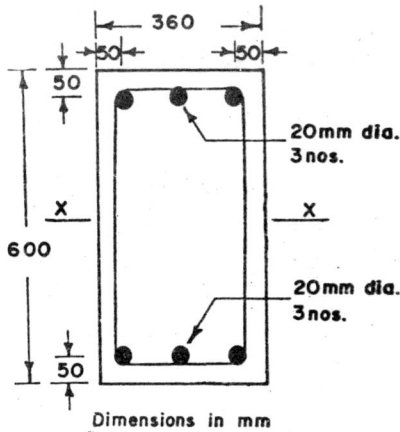

Dimensions in mm

Fig. 8.16 Trial section for Ex. (8.5)

Now

$$e = \frac{M}{P} = \frac{118 \times 10^6}{284 \times 10^3} = 415.5 \text{ mm}$$

$b = 360 \text{ mm}$

$D = 600 \text{ mm}$

$d = 550 \text{ mm}$

$A_{sc} = 3 \times 314.16 = 942.48 \text{ mm}^2$

$A_{st} = 3 \times 314.16 = 942.48 \text{ mm}^2$

Consider Eqs. (8.10) and (8.11).

Dividing Eqs. (8.11) by (8.10).

$$e + d - \frac{D}{2} =$$

$$\frac{\frac{1}{2} b X \left(d - \frac{X}{3} \right) + (1.5 m - 1) \left(\frac{X - d_c}{X} \right) (d - d_c) A_{sc}}{\frac{1}{2} b X + (1.5 m - 1) \left(\frac{X - d_c}{X} \right) A_{so} - m \left(\frac{d - X}{X} \right) A_{st}}$$

Substituting all the known values

$$665.5 =$$

$$\frac{180 X \left(550 - \frac{X}{3} \right) + 18.995 \left(\frac{X - 50}{X} \right) \times 500 \times 942.48}{180 X + 18\,995 \left(\frac{X - 50}{X} \right) \times 942.48 - 13.33 \left(\frac{550 - X}{X} \right) \times 942.48}$$

or
$$665.5 = \frac{-\frac{X^3}{3} + 550 X^2 + 49728 X - 2.486 \times 10^6}{X^2 + 169.26 X - 43361}$$

(8.12)

The solution of this equation gives

$$X = 240 \text{ mm}$$

Substituting all the values in Eq. (8.10) and evaluating

$$f_{cbc} = 6.9 \text{ N/mm}^2 < 7 \text{ N/mm}^2, \text{ safe.}$$

Hence, from Eq. (8.9)

$$f_{st} = 13.33 \times \frac{(550 - 240)}{240} \times 6.9$$

$$= 118.8 \text{ N/mm}^2 < 230 \text{ N/mm}^2, \text{ safe.}$$

The trial section is O.K.

9

Design of Staircases

9.1 INTRODUCTION

Staircases are the structural forms utilized by people to move from one level to another. In this chapter the staircases having ractangular plan forms only are discussed. There are different types of stairs depending upon their structural forms. If the steps are non-structural, they may be built over an inclined slab. The latter may be supported at its ends by two beams at right angles as shown in Fig. 9.1., or it may be integral at its ends with the landings spanning in the same direction as shown in Fig. 9.2. If the steps are themselves structral elements, they may be supported over a stringer beam as shown in Flg. 9.3., or cantilevered out from a supporting wall or an edge as shown in Fig. 9.4. A staircase having more than one flight is shown in Fig. 9.5., in which the non-structural steps are built over inclined slabs which at their ends are integral whith the landings spanning at right angles.

9.2 EFFECTIVE SPAN

For the staircases having non-structural steps built over inclined slabs, the effective span shall be taken as the horizontal distance as follows:

(a) If the inclined slab is supported at its two ends by beams at right angles (see Fig. 9.1), then

  l = c/c of beams.

(b) If the inclined slab is supported at its ends with landings spanning in the same direction (see Fig. 9.2), then

  l = c/c of supports.

(c) If the inclined slab is integral at its ends with the landings spanning at right angles (see Fig. 9.5), then the effective span is as given in Table 9.1.

Table 9.1

(BASED ON FIG. 16 OF IS CODE 456 : 1978)

Effective span m	X m	Y m
G + X + Y	< 1	< 1
G + X + 1	< 1	⩾ 1
G + 1 + Y	⩾ 1	< 1
G + 1 + 1	⩾ 1	⩾ 1

If the steps are structural members supported over a stringer beam, the effective span shall be the effective span of the beam itself,

For the steps cantilevered from the support, the effective span shall be the clear projection beyond the face of support.

9.3 LIVE LOAD

The live load is idealized as uniformly distributed over the entire span of staircase. It is specified in IS Code 875 and is mentioned below in table 9.2, for ready reference.

Table 9.2

(BASED ON TABLE 1 OF IS CODE 875-1964)

Locations	Live load KN/m^2
Stairs and landings in dwelling houses tenements, hospital wards, hostels and dormitories.	
(i) Not liable to overcrowding	3.0
(ii) Liable to overcrowding	5.0

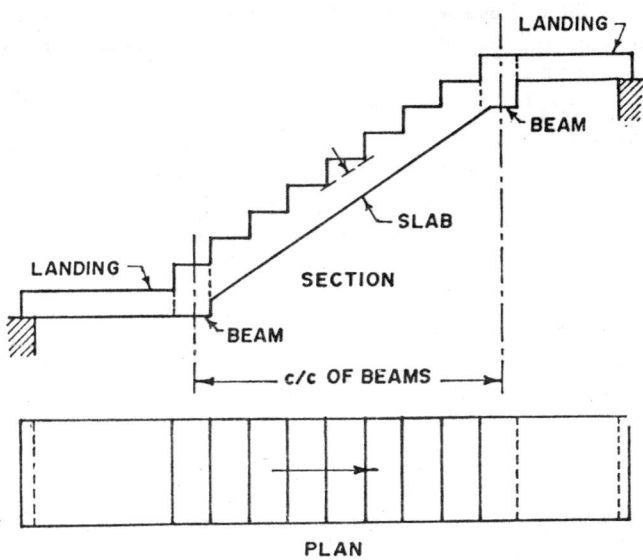

Fig. 9.1

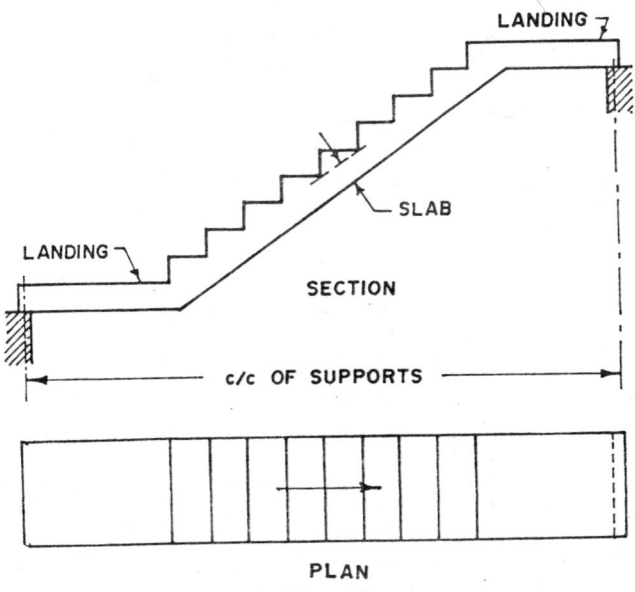

Fig. 9.2

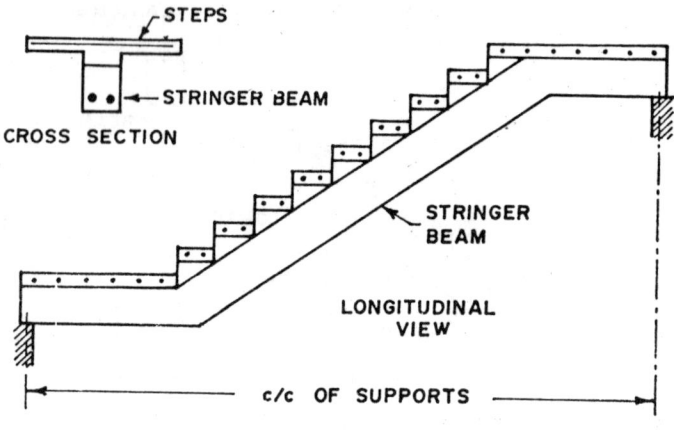

Fig. 9.3

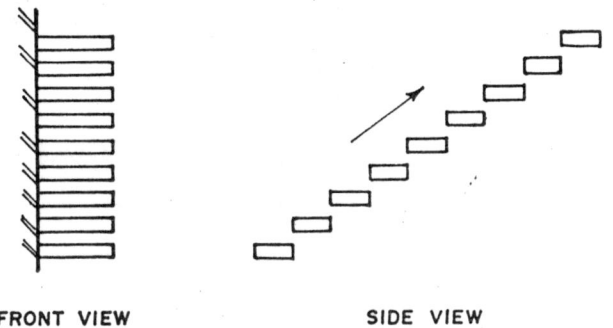

FRONT VIEW   SIDE VIEW

Fig. 9.4

9.4 DISTRIBUTION OF LOAD

In stairs with open wells the spans cross at right angles to each other on common landing as shown in Fig. 9.6. The load on the common landing may be assumed to be equally divided in the two directions.

If the flights or landings have their side edges embedded into walls for a length not less than 110 mm, then a 150 mm, strip may be deducted from the loaded area and the effective width taken to be the projected width plus 75 mm. See Fig 9.7

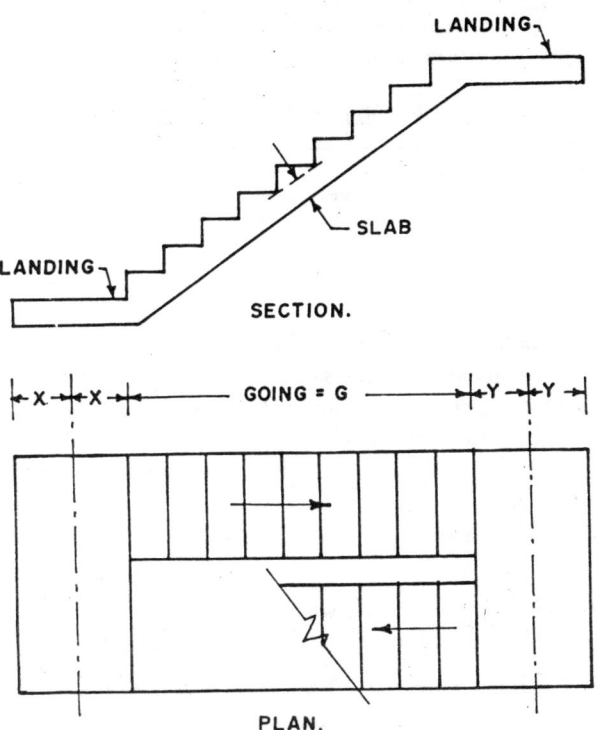

SECTION.

PLAN.

Fig. 9. 5

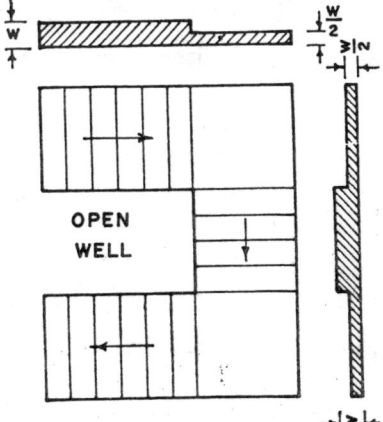

OPEN WELL

Fig. 9.6

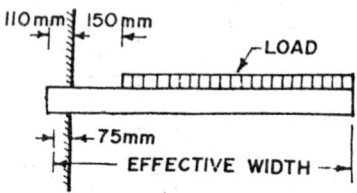

Fig. 9.7

Example 9.1

A staircase of single flight has the inclined slab integral at its ends with landings spanning in the same direction, as shown in Fig. 9.2. Total rise from one landing to another = 1.8 m. Going of the stairs = 3 m. Length of landing at each end = 1.2 m. Bearing at each end = 150 mm. Design the staircase keeping its width = 1.2 m. Concrete grade: M 15 and Steel grade: Fe 415. Live load = 3 kN/m².

Solution

Let there be 10 steps, each with a rise = 180 mm and tread = 300 mm. Effective span is the horizontal distance c/c of support i.e.

$$= \tfrac{1}{2} \times 0.15 + 1.2 + 3 + 1.2 + \tfrac{1}{2} \times 0.15$$
$$= 5.55 \text{ m. See Fig 9.9}$$

The inclination of the inclind portion is given by

$$\text{Tan } \theta = \frac{1.8}{3} = 0.6$$

Loading on staircase per horizontal meter run:

(i) On landing

Weight of slab (assuming thickness = 280 mm)

	= 0.28 × 1.2 × 25	= 8.4 kN/m.
Live load	= 3 × 1.2	= 3.6 kN/m.
	Total	= 12.0 kN/m.

(ii) On inclined slab

Weight of slab	= 0.28 × 1.2 × 25/Cos θ	= 9.8 kN/m.
Weight of steps	= $\dfrac{0.18 \times 0.3 \times 1.2 \times 25}{2 \times 0.3}$	= 2.7 k/Nm.
Live load	= 3 × 1.2	= 3.6 kN/m.
	Total	= ^{1}6.1 kN/m.

The loaded span and the corresponding bending moment diagram is shown in Fig. 9.8. Points B and C correspond to the kinks in the slab.

Permissible stress are :

$$\sigma_{cbc} = 5 \text{N/mm}^2, \ \sigma_{st} = 230 \text{ N/mm}^2$$

Modular ratio $= 18.67$

Design constant are :

$n = 0.289$ from Table 3.4,

$j = 0.904$

$R = 0.653$ from Table 3.6

Effective depth of slab for the maximum bending moment is

$$d = \sqrt{\frac{M}{Rb}} = \sqrt{\frac{58.66 \times 10^6}{0.653 \times 1200}}$$

$$= 273.6 \text{ mm.}$$

Adopted $= 275$ mm.

Tension reinforcement

$$A_{st} = \frac{M}{\sigma_{st}jd} = \frac{58.66 \times 10^6}{230 \times 0.904 \times 275}$$

$$= 1025.9 \text{ mm}^2.$$

9 nos. 12 mm dia. bars may be provided.

At the upper kink, bending moment$=40.65$ kN-m.

Tension reinforcement required at that point is

$$A_{st} = \frac{40.65 \times 10^6}{230 \times 0.904 \times 275}$$

$$= 710.9 \text{ mm}^2.$$

7 nos. 12 mm dia. bars may be provided.

Details of reinforcement are shown in Fig. 9.9.

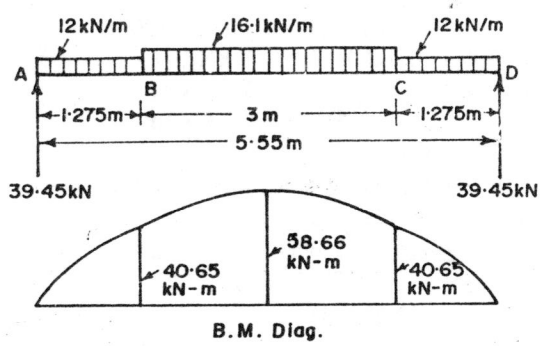

B.M. Diag.

Fig. 9.8

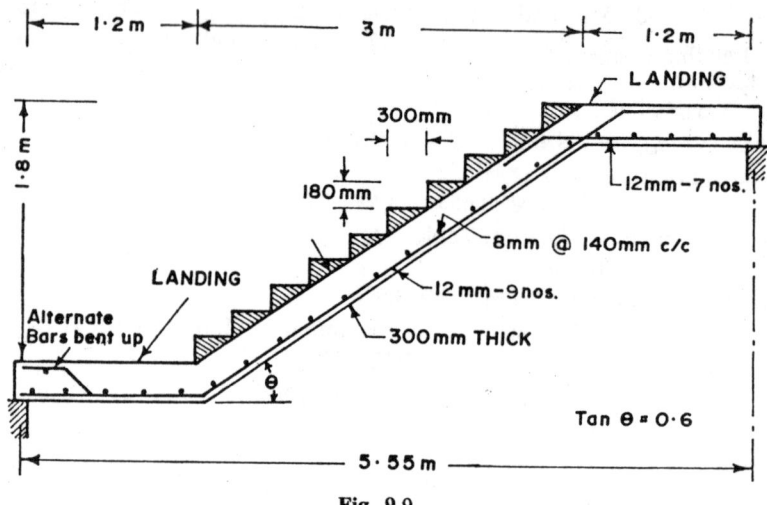

Fig. 9.9

Evample 9.2.

Redesign the staircase of Ex. (9.1) if the inclined slab be supported at its ends over beams running in transverse direction. The landings are parallel to the stairs as shown in Fig. 9.1.

Solution:

Effective span for the inclined slab is
= c/c of beams horizontally
= 2.7. See Fig. 9.10.

Loading per horizontal meter run is

(i) on landing.

Weight of slab (assuming thickness = 100 mm)
= 0.1 × 1.2 × 25 = 3 kN/m

Live load = 1.2 × 3   = 3.6 kN/m

            Total   = 6.6 kN/m

(ii) On inclined slab.

Weight of slab (assuming thickness = 150 mm)
= 0.15 × 1.2 × 25/Cosθ = 5.25 kN/m

Weight of steps = $\dfrac{0.18 \times 0.3 \times 1.2 \times 25}{2 \times 0.3}$ = 2.7 kN/m

Live load   = 1.2 × 3   = 3.6 kN/m

                  Total   = 11.55 kN/m

Design of inclined slab:

Bending moment $= \dfrac{11.55 \times 2.7^2}{8}$

$= 10.52$ kN/m

Effective depth $d = \sqrt{\dfrac{10.52 \times 10^6}{0.653 \times 1200}}$

$= 115.9$ mm.

Adopt effective depth $= 120$ mm.

Area of reinforcement

$$A_{st} = \dfrac{10.52 \times 10^6}{230 \times 0.904 \times 120} = 421.6 \text{ mm}^2$$

4 nos. bars of 12 mm dia. may be provided. An overall depth of 145 mm may be adopted.

Temp. and shrinkage reinforcement

$$= \dfrac{0.12 \times 145 \times 1000}{100} = 174 \text{ mm}^2/\text{m}$$

6 mm dia. bars at 160 mm c/c may be provided.

Design of landing:

Effective span $= 1.3$ m.

Bending moment $= \dfrac{6.6 \times 1.3^2}{8}$ 1.39 kN-m

Effective depth $d = \sqrt{\dfrac{1.39 \times 10^6}{0.653 \times 1200}} \qquad = 42.1$ mm

Adopt $d = 65$ mm

Steel reinforcement

$$A_{st} = \dfrac{1.39 \times 10^6}{230 \times 0.904 \times 65} = 102.9 \text{ mm}^2$$

4 nos. bars of 8 mm dia. may be provided.

Temp. and shrinkage reinforcement

$$= \dfrac{0.12 \times 100 \times 1000}{100} = 120 \text{ mm}^2/\text{m}$$

6 mm dia. bars @ 230 mm c/c may be provided.

Details of reinforcement are shown in Fig. 9.10.

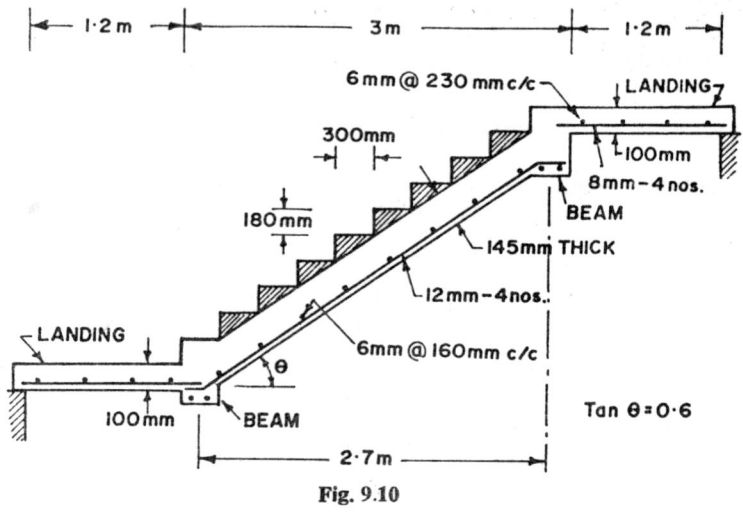

Fig. 9.10

Example 9.3

Redesign the staircase of Ex. (9.1) if the inclined slab at its ends be integral with the landings spanning at right angles as shown in Fig. 9.5. Thickness of landing = 150 mm. Rest of the data is same.

Solution

Effective span is the horizontal distance
$$= \tfrac{1}{2} \times 1.2 + 3 + \tfrac{1}{2} \times 1.2$$
$$= 4.2 \text{ m. See Fig 9.12.}$$

Loading per horizontal meter run is

(i) On landing.

Self weight $= \tfrac{1}{2} \times 0.15 \times 1.2 \times 25 = 2.25$ kN/m

Live load $= 1.2 \times 3$ $= 3.6$ kN/m

Total $= 5.85$ kN/m

(ii) On inclined slab.

Weight of slab (assuming thickness = 200 mm)

$$= 0.2 \times 1.2 \times 25/\text{Cos } \theta = 7 \text{ kN/m}$$

Weight of steps $= \dfrac{0.18 \times 0.3 \times 1.2 \times 25}{2 \times 0.3} = 2.7$ kN/m

Live load $= 1.2 \times 3$ $= 3.6$ kN/m

Total $= 13.3$ kN/m

The loaded span and the corresponding bending moment diagram are shown in Fig. 9.11.

Effective depth of the inclined slab is

$$d = \sqrt{\frac{28.09 \times 10^6}{1200 \times 0.653}} = 189.3 \text{ mm}$$

Adopt d = 190 mm
Tension reinforcement

$$A_{st} = \frac{28.90 \times 10^6}{230 \times 0.904 \times 190} = 711 \text{ mm}^2$$

7 nos. bars of 12 mm dia. may be provided.
Adopt overall thickness = 215 mm
Temp. and shrinkage reinforcement

$$= \frac{0.12 \times 125 \times 1000}{100} = 150 \text{ mm}^2/\text{m}$$

6 mm dia. bars at 180 mm c/c may be provided.
Details of reinforcement are shown in Fig. 9.12.

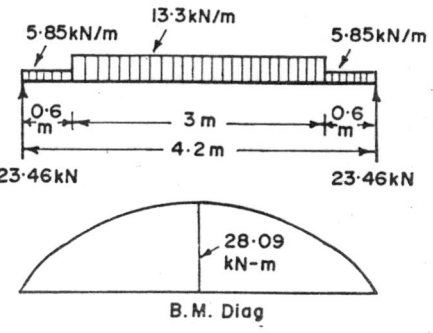

Fig. 9.11

Example 9.4

Redesign the single flight of the staircase of Ex. (9.1) if the steps be themselves structural elements fixed on the top of a single stringer beam, as shown in Fig. 9.3. Width of stringer = 200 mm. Rest of the data is same.

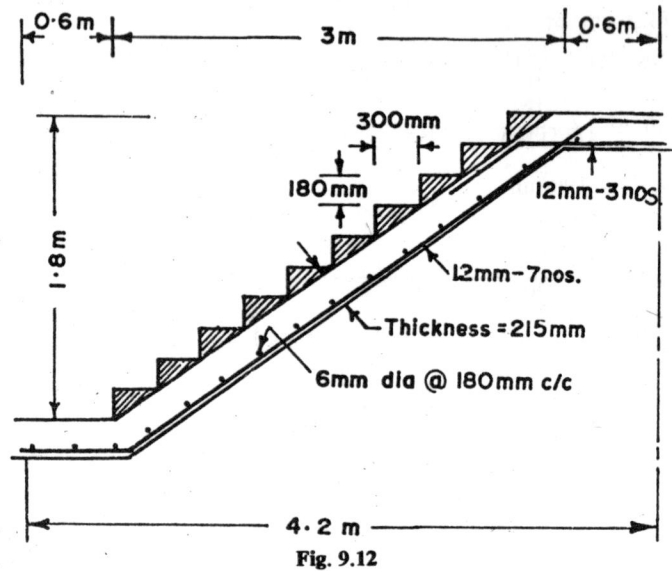

Fig. 9.12

Solution

The step overhangs the sides of the stringer. It will be designed as cantilever. Overhang

$$= \tfrac{1}{2}(1.2 - 0.2) = 0.5 \text{ m on each side.}$$

Design of steps:

Loading per meter run of the overhang is as follows:

Self weight (assuming thickness = 100 mm)

$$= 0.1 \times 0.3 \times 25 = 0.75 \text{ kN/m}$$

Live load $= 0.3 \times 3 \qquad = 0.9 \text{ kN/m}$

$$\text{Total} \quad = 1.65 \text{ kN/m}$$

Bending moment $\Rightarrow \dfrac{1.65 \times 0.5^2}{2}$

$$= 0.206 \text{ kN-m}$$

Effective depth

$$d = \sqrt{\frac{0.206 \times 10^6}{0.653 \times 300}}$$

$$= 34.4 \text{ mm}$$

Adopt d $= 75$ mm, so that overall depth may be 100 mm

Reinforcement

$$A_{st} = \frac{0.206 \times 10^6}{203 \times 0.904 \times 75} = 13.2 \text{ mm}^2$$

Temp. and shrinkage reinforcement

$$= \frac{0.12 \times 100 \times 300}{100} = 36 \text{ mm}^2$$

Provide 6 mm dia. bars 2 nos. in each step. These bars may be embedded in the beam as shown in Fig. 9.13.

Design of beam:

Loading per horizontal meter run is as follows:

(i) On landing.

Weight of slab $= 0.1 \times 1 \times 1.2 \times 25 = 3$ kN/m

Weight of beam (assuming depth $= 500$ mm)

$$= 0.2 \times 0.5 \times 25 = 2.5 \text{ kN/m}$$

Live load $= 1 \times 1.2 \times 3 \qquad = 3.6 \text{ kN/m}$

Total $= 9.1$ kN/m

(ii) On inclined slab.

Weight of steps $= \dfrac{0.1 \times 0.3 \times 1.2 \times 25}{0.3} = 3$ kN/m

Weight of beam $= 0.2 \times 0.5 \times 25/\text{Cos } \theta = 2.92$ kN/m

Live load $= 1 \times 1.2 \times 3 = 3.6$ kN/m

Total $= 9.52$ kN/m

The loading span and the corresponding bending moment diagram are shown in Fig. 9.13.

Maximum bending moment $= 36.31$ kN m

Effective depth

$$d = \sqrt{\frac{36.31 \times 10^6}{0.653 \times 200}} = 527.3 \text{ mm}$$

Adopt $d = 530$ mm., and overall depth $= 565$ mm

Tension reinforcement

$$= \frac{36.31 \times 10}{230 \times .904 \times 530} = 329.5 \text{ mm}^2$$

3 nos. bars of 12 mm dia. may be provided.

Bending moment at upper kink $= 23.66 \text{ mm}^2$

Corresponding tension reinforcement
$$A_{st} = \frac{23.66 \times 10^6}{230 \times .904 \times 530} = 214.7 \text{ mm}^2$$

2 nos. bars of 12 mm dia. may be provided.

Distance of critical section for shear force from centre of support
$$= 0.53 + \tfrac{1}{2} \times 0.15$$
$$= 0.605 \text{ m}$$

At this section, shear force
$$V = 25.85 - 0.605 \times 9.1$$
$$= 20.375 \text{ kN}$$
$$\tau_v = \frac{20.375 \times 10^3}{200 \times 530} = 0.19 \text{ N/mm}^2$$
$$\frac{100 A_{st}}{bd} = \frac{100 \times 3 \times 113.1}{200 \times 530} = 0.32$$

From Table 3.11
$$\tau_c = 0.24 \text{ N/mm}^2$$

Evidently, $\tau_v < \tau_c$. Nominal shear reinforcement may be provided. It is given by Eq. (3.31)

Try 6 mm dia. 2 legged stirrups. Then
$$\frac{2 \times 28.27}{200 \times S_v} \geqslant \frac{0.4}{415}$$
$$\therefore \; S_v \leqslant 293.3 \text{ mm c/c}$$

Adopt a spacing of 290 mm c/c throughout.

Details of reinforcement are shown in Fig. 9.14.

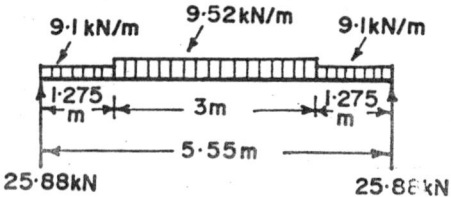

25·88kN   25·88 kN

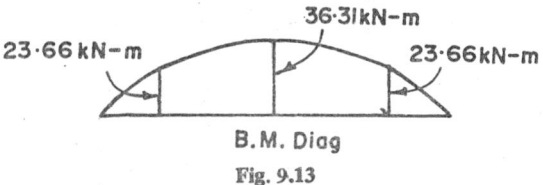

36·31kN-m

23·66 kN-m   23·66kN-m

B.M. Diag

Fig. 9.13

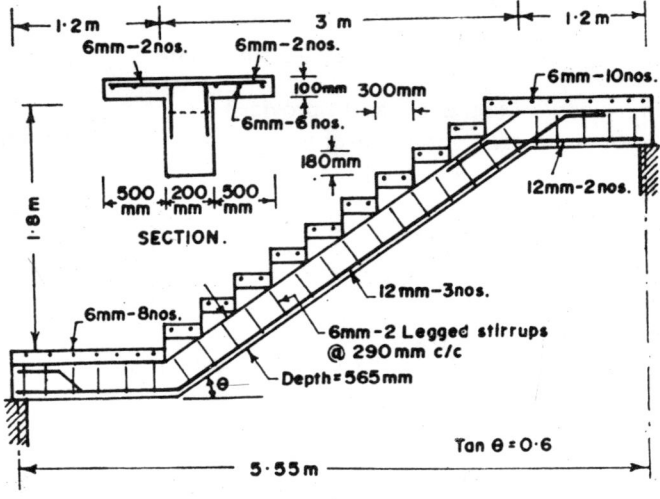

Fig. 9.14

10

Two Way Slabs

10.1 INTRODUCTION

A rectangular slab supported along all the four edges spans in two directions and supports the load by two-way bending action. See Fig.10.1. Bending moments occur along both the short and the long span. If the length to width ratio of slab is large (say $\geqslant 2$) then the slab has a pre-dominant one-way action. It has a large bending moment along the shorter span so much so that the bending moment along the longer span is comparatively insignificant. Such slabs have already been discussed under para 3.10. For slabs with length to width ratio small (i.e. less than 2) the bending moments along the longer span also have appreciable magnitude compared with those along the shorter span. Thus, bending moments need to be calculated along both the spans and the reinforcement provided for both. Hence, the slab is known as a two-way slab.

A two-way slab is statically indetrminate and as such its elastic analysis is quite cumbersome. Therefore, recourse has been made to empirical procedures of analysis. The latter have been developed broadly for two types of supporting conditions for the slab. In one type, the corners of the simply supported rectangular slab are free to lift up when the slab bends under transverse loading. Such a slab develops bending moments in the directions of the two spans while the corner twisting moments are negligible. In the other type the corners of the transversely loaded slab are considered to be held down. In such a slab there will develop appreciable amount of twisting moments in the regions near the corners while bending moments will develop in the remaining middle region. Empirical methods of analysis are available for both the types of supporting conditions. The methods of analysis which have been adopted by the IS Code: 456 (Appendix C) are discussed in the succeeding para.

10.2 EMPIRICAL ANALYSIS AS PER IS CODE: 456

10.2.1. Simply supported rectangular slab with corners free to lift. The slab is assumed to take up the transverse loading by bending action along the two spans. The effect of twisting is assumed to be negligible. Consider the simply supported slab shown in Fig.10.1 which is uniformely loaded.

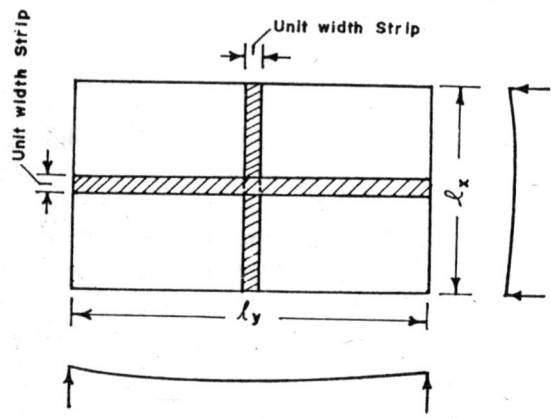

Fig. 10.1. Slab with corners free to lift

The intensity of load w is shared between strips running along the two directions i.e.

$$w = w_x + w_y$$

where w_x = intensity of load shared along shorter direction

w_y = intensity of load shared along longer direction

In the Grashoff—Rankine theory the two intensities are found by equating the deflections of two central strips of unit width i.e.

$$\frac{5}{384} \frac{w_x l_x^4}{E I} = \frac{5}{384} \frac{w_y l_y^4}{E I}$$

From which

$$w_x = \frac{w r^4}{1 + r^4}$$

and $\qquad w_y = \dfrac{w}{1 + r^4}$

where $\qquad r = \dfrac{l_y}{l_x}$

Bending moments per unit width in the two directions are:

$$M_x = \frac{w_x\, l_x^2}{8}$$

$$= a_x\, w\, l_x^2 \qquad\qquad (10.1)$$

$$M_y = \frac{w_y\, l_y^2}{8}$$

$$= a_y\, w\, l_x^2 \qquad\qquad (10.2)$$

where $\qquad a_x = \dfrac{1}{8}\dfrac{r^4}{(1 + r^4)}$

$$a_y = \frac{1}{8}\frac{r^2}{(1 + r^4)}$$

This procedure of calculating the bending moments has been adopted by the IS Code: 456 where the ready calculated values of a_x and a_y are given in Table 23 under Appendix C. These values are given in Table10.1 below.

Table 10.1.

BENDING MOMENT COEFFICIENTS

(BASED ON TABLE 23 OF IS CODE: 456-1978)

l_y/l_x	1.0	1.1	1.2	1.3	1.4	1.5	1.75	2.0	2.5	3.0
a_x	0.062	0.074	0.084	0.093	0.099	0.104	0.113	0.118	0.122	0.124
a_y	0.062	0.061	0.059	0.055	0.051	0.046	0.037	0.029	0.020	0.014

Reinforcement detail as per IS Code: 456-1978)

At least 50% of the tension reinforcement provided at mid-span should extend into the supports. The remaining 50% should extend upto within $0.1\ l_x$ or $0.1\ l_y$ of the support as the case may be. See Fig.10.2.

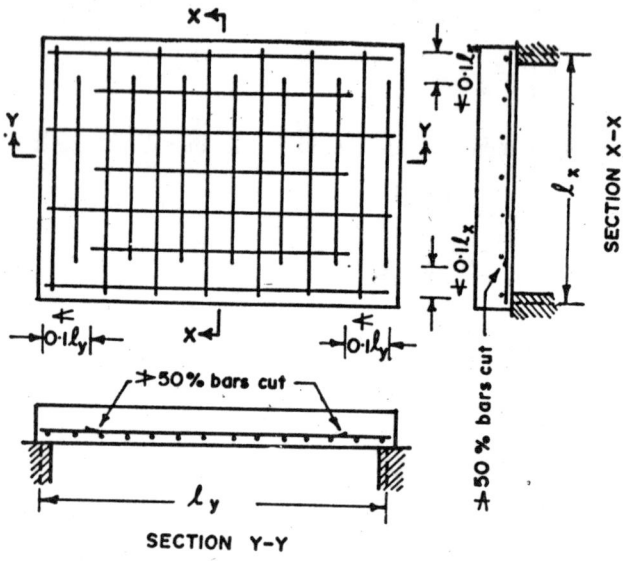

Fig. 10.2. Reinforcement for simply supported slab with corners free to lift

10.2.2. Rectangular slab panel with corners prevented from lifting. In a transversely loaded simply supported rectangular slab, if the corners are held down, twisting moments are created in vertical planes parallel to the edges in the corner regions. However, if the slab is confinous over an edge, the twisting moments vanish in the vertical plane at right angles to the edge. If the slab is continous over the two adjacent edges, there will be no twisting moments in the vertical planes at right angles to the edges. Hence, the corner is free from twisting moments.

The bending moment coefficients for the working stress design as given in the IS Code are the same as for the moments obtained by analysis based on yield line method. The bending moments per unit width in the slab are given by

$$M_x = a_x \, w \, l_x^2 \tag{10.3}$$

$$M_y = a_y \, w \, l_x^2 \tag{10.4}$$

where, a_x and a_y are the coefficients which are given in Table 10.2 for various support conditions.

TABLE 10.2

BENDING MOMENT COEFFICIENTS FOR RECTANGULAR PANELS SUPPORTED ON FOUR SIDES WITH PROVISION FOR TORSION AT CORNERS (BASED ON TABLE 22 IS CODE: 456-1978)

Case No.	Type of panel and moments considered	Short span coefficients α_x (Values of l_y/l_x)								Long span coefficients α_y for all values of l_y/l_x
		1.0	1.1	1.2	1.3	1.4	1.5	1.75	2.0	
(1)	(2)	(3)	(4)	(5)	(6)	(7)	(8)	(9)	(10)	(11)
1. *Interior panels:*										
	Negative moment at continuous edge	0.032	0.037	0.043	0.047	0.051	0.053	0.060	0.065	0.032
	Positive moment at mid-span	0.024	0.028	0.032	0.036	0.039	0.041	0.045	0.049	0.024
2. *One short edge discontinuous:*										
	Negative moment at continuous edge	0.037	0.043	0.048	0.051	0.055	0.057	0.064	0.068	0.037
	Positive moment at mid-span	0.028	0.032	0.036	0.039	0.041	0.044	0.048	0.052	0.028
3. *One long edge discontinuous:*										
	Negative moment at continuous edge	0.037	0.044	0.052	0.057	0.063	0.067	0.077	0.085	0.037
	Positive moment at mid span	0.028	0.033	0.039	0.044	0.047	0.051	0.059	0.065	0.028

Continued

4. *Two adjacent edges discontinuous:*									
Negative moment at continuous edge	0.047	0.053	0.060	0.065	0.071	0.075	0.084	0.091	0.047
Positive moment at mid-span	0.035	0.040	0.045	0.049	0.053	0.056	0.063	0.069	0.035
5. *Two short edges discontinuous:*									
Negative moment at continuous edge	0.045	0.049	0.052	0.056	0.059	0.060	0.065	0.069	—
Positive moment at mid-span	0.035	0.037	0.040	0.043	0.044	0.045	0.049	0.052	0.035
6. *Two long edges discontinuous:*									
Negative moment at continuous edge	—	—	—	—	—	—	—	—	0.045
Positive moment at mid-span	0.035	0.043	0.051	0.057	0.063	0.068	0.080	0.088	0.035
7. *Three edges discontinuous (One long edge continuous):*									
Negative moment at continuous edge	0.057	0.064	0.071	0.076	0.080	0.084	0.091	0.097	—
Positive moment at mid-span	0.043	0.048	0.053	0.057	0.060	0.064	0.069	0.073	0.043
8. *Three edges discontinuous (One short edge continuous):*									
Negative moment at continuous edge	—	—	—	—	—	—	—	—	0.057
Positive moment at mid-span	0.043	0.051	0.059	0.065	0.071	0.076	0.087	0.096	0.043
9. *Four edges discontinuous:*									
Positive moment at mid-span	0.056	0.064	0.072	0.079	0.085	0.089	0.100	0.107	0.056

Reinforcement detailing as per IS Code: 456.

For the purpose of providing reinforcement the slab is considered to be divided into middle strips and edge strips. See Fig.10.3. The middle strip is 3/4 of the width and each edge strip 1/8 of the width.

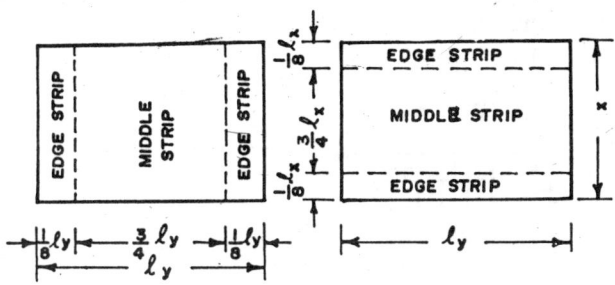

Fig. 10.3.

Reinforcements for the bending moments M_x and M_y are provided only within the middle strips. The edge strips contain reinforcement for temperature and shrinkage only. For a slab panel with all the four edges discontinous, the positive reinforcement in the middle strip shall extend upto a point within 0.15 of span from the edge. See Fig.10.4. To resist twisting moments at the corners, torsion reinforcement in the form of square mesh near top as well

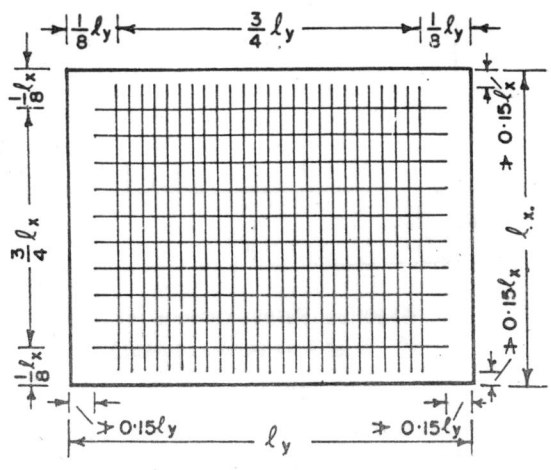

Fig. 10.4.

as bottom surface is provided. See Fig.10.5 (a). Each of the four layers of the corner reinforcement shall have upto 3/4 of the positive reinforcement intensity provided along the short span. Negative reinforcement near the top surface shall have an intensity of upto 1/2 of that of the positive reinforcement at the mid-span and shall extend from the discontinuous edge for a distance = 0.1 of span.

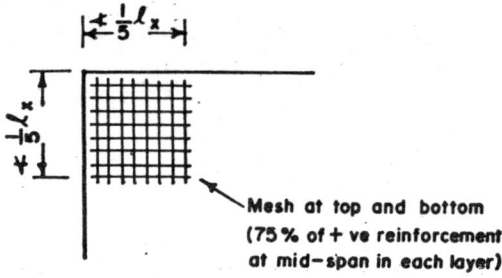

Mesh at top and bottom
(75 % of + ve reinforcement
at mid-span in each layer)

(a) Torsion Reinforcement at corner.

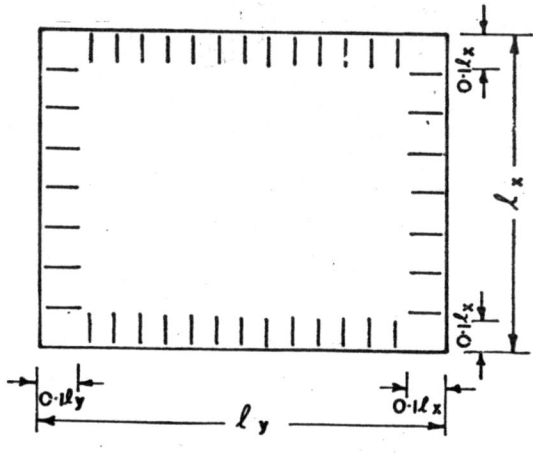

**(b) Negative Reinforcement at top
(50 % of + ve reinforcement)**

Fig. 10.5.

For a slab panel with all the four edges continuous the positive reinforcement in the midddle strip shall extend upto a point within 0.25 of span from the edge. See Fig.10.6 (a). No torsion reinforcement is needed at the corners. At least 50% of the negative reinforcement near the top surface shall extend from the continuous edge for a distance upto 0.3 of span, and the remaining upto 0.15 of span. See Fig.10.6 (b). If one of the edges is continuous and the other is discontinuous at a corner then the torsion reinforcement shall be reduced to half of what is required when both the edges are discontinuous.

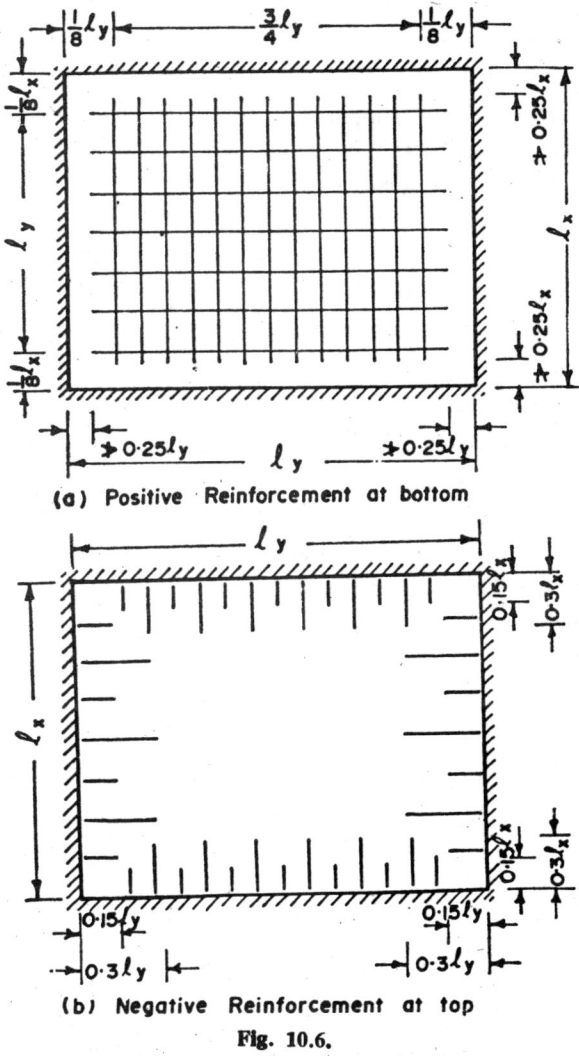

(a) Positive Reinforcement at bottom

(b) Negative Reinforcement at top

Fig. 10.6.

No torsion reinforcement need be provided if both the edges meeting at a corner are continuous.

10.3 SHEAR STRESS IN SLAB

The shear stress in two way slabs is very small and generally lies within safe limits. As such, revision of the amount of the main reinforcement may not be required. However, for small span slabs carrying very heavy loads, check for shear may be done.

Example 10.1

A slab is simply supported over a clear opening in plan 6m × 4.5 m. Its corners are free to lift. Live load = 3 kN/m². Design a two way slab. Concrete grade: M 15 and steel grade: Fe 415.

Solution:

Permissible stress are:

$$\sigma_{cbc} = 5 \quad N/mm^2 \text{ from Table 3.1.}$$
$$\sigma_{st} = 230 \ N/mm^2 \text{ from Table 3.2.}$$
and $\quad m = 18.7 \qquad$ from Table 3.3.
$$n = 0.289 \qquad \text{from Table 3.4.}$$
$$j = 0.904$$
$$R = 0.653 \qquad \text{from Table 3.6.}$$

For estimating the self weight and the effective spans of slab assume D = 175 mm and d = 150 mm.

Self weight = 0.175 × 25 = 4.375 kN/m²

Live load $\qquad\qquad$ = 3 $\quad$ kN/m²

Total $\qquad\qquad\qquad$ = 7.375 kN/m²

Effective span will be

$$l_x = 4.5 + 0.15 = 4.65 \text{ m}$$
$$l_y = 6 \ + 0.15 = 6.15 \text{ m}$$

Ratio $\quad \dfrac{l_y}{l_x} = \dfrac{6.15}{4.65} = 1.32$

From Table 7.1, by interpolation

$$\alpha_x = 0.0942$$
$$\alpha_y = 0.0542$$

Bending moments per unit width at mid-span are:

$$M_x = a_x \, w \, l_x^2 = 0.0942 \times 7.375 \times 4.65^2$$
$$= 15.022 \text{ kN-m/m}$$
$$M_y = a_y \, w \, l_x^2 = 8.643 \text{ kN·m/m}$$

Effective depth for the larger moment

$$d = \sqrt{\frac{15.022 \times 10^6}{1000 \times 0.653}} = 151.7 \text{ mm}$$

Adopt $\quad d = 155$ mm

Tension reinforcement:

Along short span $A_{st} = \dfrac{15.022 \times 10^6}{230 \times .904 \times 155}$

$$= 466.1 \text{ mm}^2/\text{m}$$

Provide 10 mm dia. bars — 165 mm c/c.

Along long span

$$A_{st} = \frac{8.643 \times 10^6}{230 \times .904 \times 145}$$
$$= 286.7 \text{ mm}^2/\text{m}$$

Provide 10 mm dia. bars — 270 mm c/c.

Check for shear:

Load distribution along short span

$$w_x = \frac{w \, r^4}{1 + r^4} = \frac{7.375 \times 1.32^4}{1 + 1.32^4}$$
$$= 5.548 \text{ kN/m}^2$$

Critical section occurs at a distance d from the face of support. At this section

$$V = (\tfrac{1}{2} \times 4.5 - 0.155) \times 5.548$$
$$= 11.62 \text{ kN/m}$$

Nominal shear stress

$$\tau_v = \frac{V}{bd} = \frac{11.62 \times 10^3}{1000 \times 155} = 0.075 \text{ N/mm}^2$$

Reinforcement per m width of slab

$$A_{st} = \frac{78.5 \times 1000}{165} = 475.8 \text{ mm}^2/\text{m}$$

$$\frac{100 \, A_{st}}{bd} = \frac{100 \times 475.8}{1000 \times 155} = 0.307$$

From Table 3.11, permissible shear strees

$\tau_0 = 0.236$ N/mm^2

Evidently, τ_0 is far greater than τ_v. Hence, the slab is safe in shear. Alternate bars of the main reinforcement may be bent at a point 0.1 of span from the support. An overall depth of 175 mm may be used.

Details of reinforcement are shown in Fig.10.7.

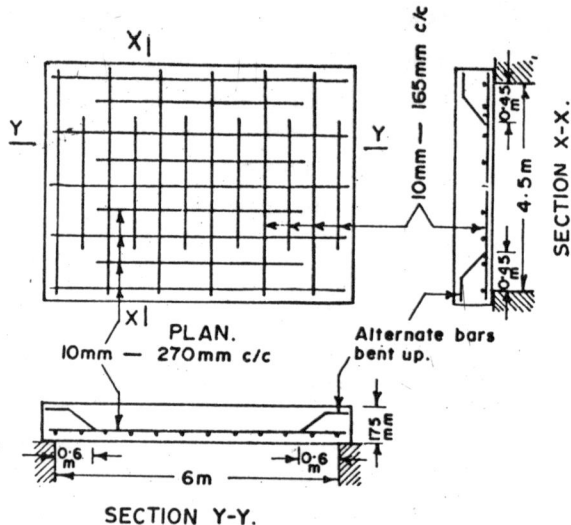

PLAN.
10mm — 270mm c/c

Alternate bars bent up.

SECTION Y-Y.

Fig. 10.7.

Example 10.2.

Redesign the slab in Ex.10.1 if the corners are not free to lift.

Solution:

For es'imating self weight and the effective spans, assume D = 160 mm and d = 140 mm.

Self weight = 0.16 × 25 = 4 kN/m^2

Live load   = 3 kN/m^2

Total   w = 7 kN/m^2

Effective spans will be

$l_x = 4.5 + 0.14 = 4.64$ m

$l_y = 6 \ + 0.14 = 6.14$ m

Ratio $\dfrac{l_y}{l_x} = \dfrac{6.14}{4.64} = 1.32$

From Table10.2, by interpolation

$$\alpha_x = 0.0802$$

and $\alpha_y = 0.056$

Bending moments per unit width at mid-span are:

$$M_x = \alpha_x \, wl_x^2 = 0.0802 \times 7 \times 4.64^2$$
$$= 12.087 \text{ kN-m/m}$$
$$M_y = \alpha_y \, w \, l_x^2 = 0.056 \times 7 \times 4.64^2$$
$$= 8.44 \text{ kN-m/m.}$$

Effective depth for larger moment

$$d = \sqrt{\frac{12.087 \times 10^6}{1000 \times 0.653}}$$
$$= 136 \text{ mm}$$

Adopt $d = 140$ mm

Tension reinforcement:

Along short span, $\quad A_{st} = \dfrac{12.087 \times 10^6}{230 \times .904 \times 140}$
$$= 415.2 \text{ mm}^2/\text{m}$$

Provide 10 mm dia. bars — 185 mm c/c.

Width of middle strip $= 0.75 \times 6.14 = 4.6$ m

Along long span, $\quad A_{st} = \dfrac{8.44 \times 10^6}{230 \times .904 \times 130}$
$$= 312.3 \text{ mm}^2/\text{m}$$

Provide 10 mm dia. bars — 250 mm c/c. Width of middle strip $= 0.75 \times 4.64 = 3.48$ m., say 3.5 m

Check for shear is not needed.

Torsion reinforcement at corners:

One-fifth of short span

$$= \frac{4.64}{5}$$
$$= 0.93 \text{ m}$$

Adopt 0.95 m × 0.95 m as plan dimensions for the corner mesh.

Intensity of reinforcement in one layer

$$= 0.75 \times \text{intensity along short span.}$$
$$= 0.75 \times 415.2$$
$$= 311.4 \text{ mm}^2$$

Provide 10 mm dia. bars — 250 mm c/c. in each layer. Corner mesh will be provided near top as well as bottom surfaces in each of the four corners.

Adopt an overal depth of slab = 160 mm.

Temperature and shrinkage reinforcements is

$$= \frac{0.12}{100} = \times 160 \times 1000$$

$$= 192 \text{ mm}^2/\text{m}$$

Provide 8 mm dia. bars — 250 mm c/c. along the edge strip
Details of reinforcement are shown in Fig.10.8.

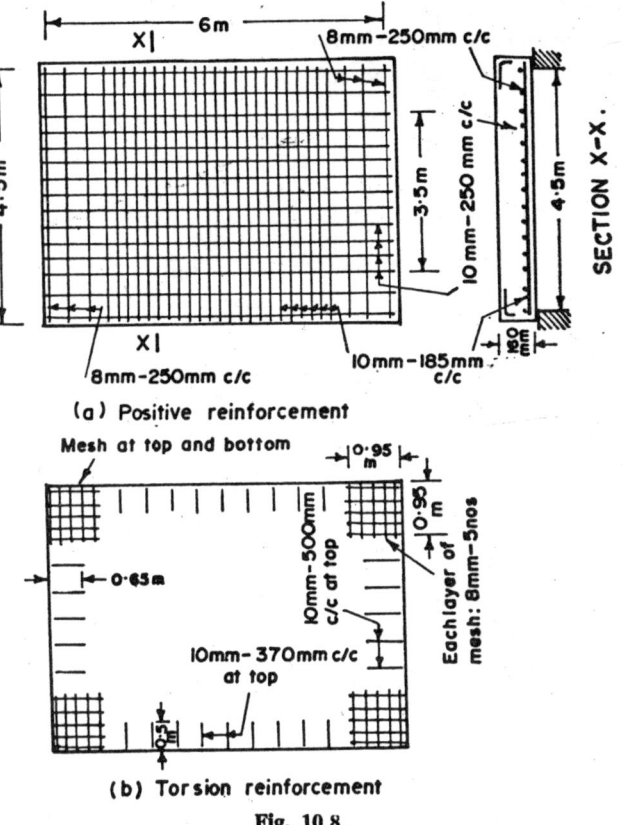

(a) Positive reinforcement

(b) Torsion reinforcement

Fig. 10.8.

Example 10.3

Redesign the slab in Ex.10.1, if one long and one short edges are
continuous and the remaining two discontinuous. Corner is not free
to lift.

Solution:

For estimating the self weight and the effective span, assume D = 150 mm and d = 130 mm.

Self weight $= 0.15 \times 25 = 3.75$ kN/m²

Live load $\qquad\qquad = 3 \quad$ kN/m²

Total $\quad w = 6.75$ kN/m²

Effective spans will be

$$l_x = 4.5 + 0.13 = 4.63 \text{ m}$$
$$l_y = 6 \ + 0.13 = 6.13 \text{ m}$$

Ratio $\dfrac{l_y}{l_x} = \dfrac{6.13}{4.63} = 1.32$

From Table 7.2

(i) For negative moments

$\alpha_x = 0.0662$

$\alpha_y = 0.047$

(ii) For positive moments

$\alpha_x = 0.0498$

$\alpha_y = 0.035$

Bending moments per unit width are:

Negative moments

$M_x = \alpha_x \ w \ l_x{}^2 = 0.0662 \times 6.75 \times 4.63^2$
$\qquad = 9.579$ kN-m/m

$M_y = \alpha_y \ w \ l_x{}^2 = 0.047 \times 6.75 \times 4.63^2$
$\qquad = 6.801$ kN-m/m

Positive moments

$M_x = \alpha_x \ w \ l_x{}^2 = 0.0498 \times 6.75 \times 4.63^2$
$\qquad = 7.206$ kN-m/m

$M_y = \alpha_y \ w \ l_x{}^2 = 0.035 \times 6.75 \times 4.63^2$
$\qquad = 5.064$ kN-m/m

Effective depth for the largest moment

$$d = \sqrt{\dfrac{9.579 \times 10^6}{1000 \times 0\ 653}}$$

$\qquad = 121.1$ mm

Adopt d = 125 mm

Width of middle strip along short span

$\qquad = 0.75 \times 6.13 \quad = 4.6$ m

Width of middle strip along long span

    $0.75 \times 4.63 = 3.47$ m say 3.5 m.

Tension reinforcement:

For negative bending moment along short span

$$A_{st} = \frac{9.579 \times 10^6}{230 \times .904 \times 125}$$

$$= 368.6 \text{ mm}^2/\text{m}$$

For 10 mm dia. bars, spacing $= \dfrac{78.54}{368.6} \times 1000 = 213$ mm c/c,

say 210 mm c/c.

Alternate bars may extend for a distance $= 0.3 \times 4.63 = 1.39$ m, say 1.40 m, while the remaining bars may extend for a distance $= 0.15 \times 4.63$ m $= 0.695$ m, say 0.7 m.

For negative bending moment along long span

$$A_{st} = \frac{6.801 \times 10^6}{230 \times .904 \times 115}$$

$$= 284.4 \text{ mm}^2/\text{m}$$

For 10 mm dia. bars, spacing $= \dfrac{78.54}{284.4} \times 1000 = 276.2$ mm, say 275 mm c/c.

Alternate bars may extend for a distance $= 0.3 \times 6.13 = 1.839$ m, say 1.9 m while the remaining bars may extend for a distance $= 0.15 \times 6.13 = 0.92$ m, say 0.95 m.

For positive bending moment along short span

$$A_{st} = \frac{7.206 \times 10^6}{230 \times .904 \times 125}$$

$$= 277.3 \text{ mm}^2/\text{m}$$

For 10 mm dia. bars, spacing $= \dfrac{78.54}{211.8} \times 1000 = 283.2$ mm, say 280 mm c/c.

For positive bending moment along long span

$$A_{st} = \frac{5\,064 \times 10^6}{230 \times .904 \times 115}$$

$$= 211.8 \text{ mm}^2/\text{m}$$

For 10 mm dia. bars spacing $= \dfrac{78.54}{211.8} \times 1000 = 370.8$ mm.

Maximum permitted spacing = 3 × 115 = 345 mm c/c.

Hence, adopt a spacing of 345 mm c/c.

Overall depth may be kept = 145 mm.

Temperature and shrinkage reinforcement

$$= \frac{0.12 \times 145 \times 1000}{100}$$

$$= 174 \text{ mm}^2/\text{m}$$

Corner reinforcement:

In the corner where two discontinuous edges meet, full corner reinforcement should be provided. There will be a mesh near top as well as near bottom surface.

Reinforcement intensity in one layer of mesh

= 0.75 × reinforcement intensity along short span

= 0.75 × 277.3

= 208mm²/m

For 8 mm dia. bars, spacing = $\frac{50.3}{208}$ × 1000 =∴ 241.8 mm

Plan dimension for corner reinforcement

= $\frac{1}{5}$ × 4 63

= 0.926 m, Say 0.95 m

8 mm dia. bars 5 — nos. may be used in each layer giving a spacing of = 237.5 mm c/c.

In the corner where one continuous and one discontinuous edges meet, corner reinforcement may be reduced to half. As such, use 6 mm bars, 5 nos. in each layer giving a spacing of 237.5 mm c/c.

In the corner where continuous edges meet, corner reinforcement is not required.

In the edge strip, temperature and shrinkage reinforcement i.e. 8 mm bars at 280 mm c/c may be provided.

Negative reinforcement at discontinuous edge:

6 mm dia. — 200 mm c/c near top surface extending along short span for a distance = 0.1 × 4.63

= 0.463 m, say 0.5 m, may be provided.

6 mm dia. — 250 mm c/c near top surface extending along long span for a distance = 0.1 × 6.13

= 0.613 m, say 0.625 m may be provided.

Details of reinforcement are shown in Fig. 10.9.

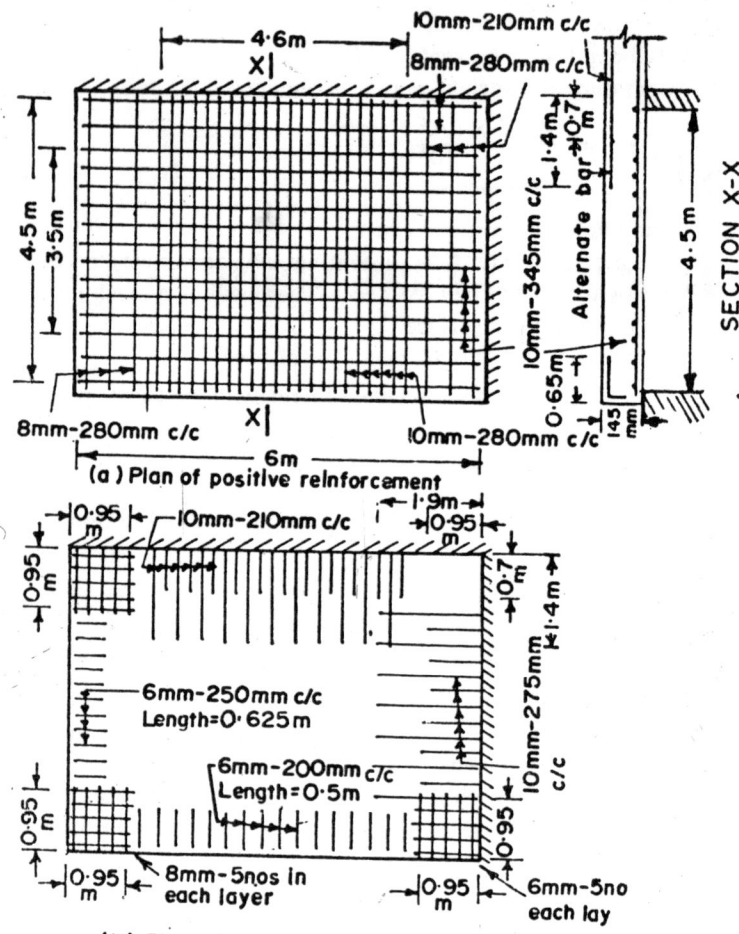

(a) Plan of positive reinforcement

(b) Plan of negative reinforcement
and corner reinforcement

10.9.

11

Design of Flat Slabs

11.1. INTRODUCTION

Flat slabs are the slabs which directly rest over columns. Beams are not provided generally. As such, flat slab construction results in more head room compared with the beam and slab construction for a given storey height. The amount of form work is also comparitively less. Flat slab rests over a number of columns. Often, the tops of the columns are flared. Such flared portions of columns are known as column heads or capitals. These are intended primarily to increase the resistance of slab against punching shear. A portion of the slab thickened in the vicinity of the column is known as a drop. It is provided to reduce the shear stress as well as the amount of negative reinforcement in the region of slab surrounding the column head. If the slab does not have a drop, and rests over columns which do not have capitals, it is known as a flat plate. Otherwise it is known as a flat slab. See Fig. 8.2. (c)

Consider the schemetic plan of the flat slab shown in Fig.11.1. A part of the slab bounded on the four sides by the center lines of columns is known as a panel. Transverse loading of slab causes two-way bending action in a panel. A portion of slab of uniform width may be visualized to be acting as a beam supported over columns in one direction. This slab band is called a column strip. The remaining portion of slab, bounded by column strips on the two sides, is called a middle strip. Along a given direction, the middle strip spans over transversely running column strips.

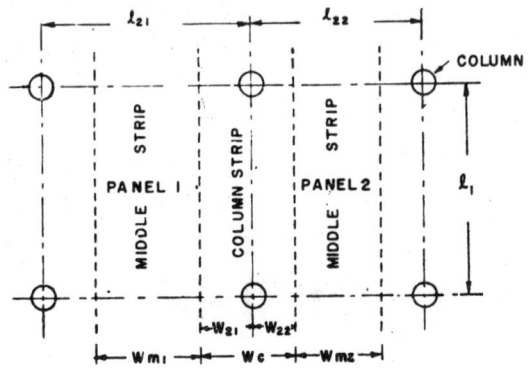

Fig. 11.1.

11.2 WIDTH OF COLUMN STRIP

With reference to Fig.11.1., the width of column strip is written as

$$w_c = w_{21} + w_{22}$$

$$\ngtr 0.5 \; l_1 \text{ in any case.}$$

where

l_1 = span along which moments are being considered, measured c/c of columns.

$$w_{21} = 0.25 \, l_{21}$$

$$w_{22} = 0.25 \, l_{22}$$

l_{21} and l_{22} are the spans transverse to l_1 and measured to the centres of the column to the left and to the right, respectively.

when $l_{21} = l_{22} = l_2$

then $w_o = 0.5 \, l_2$

$$\ngtr 0.5 \; l_1 \text{ in any case.}$$

11.3. WIDTH OF MIDDLE STRIP

The middle strip is bounded on each side by column strips. Its width is the width of slab panel available between the column strips.

11.4. PROPORTION AS PER I.S. CODE 456

11.4.1. Drops. Drops shall be rectangular in plan as shown in Fig.11.2. The dimensions shall be as follows

$$D_1 \nless \frac{l_1}{3}$$

$$D_2 \nless \frac{l_2}{3}$$

In exterior panels, width of drop at right angles to the discontinuous edge measured from the centre line of the columns shall be equal to one-half of the width of drop for interior panels.

11.4.2. Column heads. The portion of column head which lies within the largest circular cone or pyramid having a vertex angle of 90° and which lies within the outlines of the column and the column head, shall be considered for the design.

For slab without drop and column without column head, see Fig.11.3 (a). For slab without drop and column with column head, see Fig. 11.3 (b). For slab with drop and column with column head see Fig.11.3 (c).

11.4.3. Thickness of flat slab. The thickness of slab with or without drop is goverened by the following rules:

(i) Effective depth $\nless \dfrac{\text{Longer span}}{26 \, M_t}$

where, M_t is the modification factor depending upon the area and the type of steel for tension reinforcement in the slab. See Fig. 6.2.

(ii) Overall thickness $\nless$ 125 mm

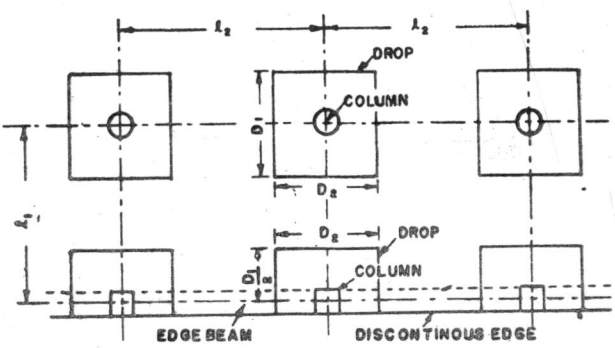

Fig. 11.2

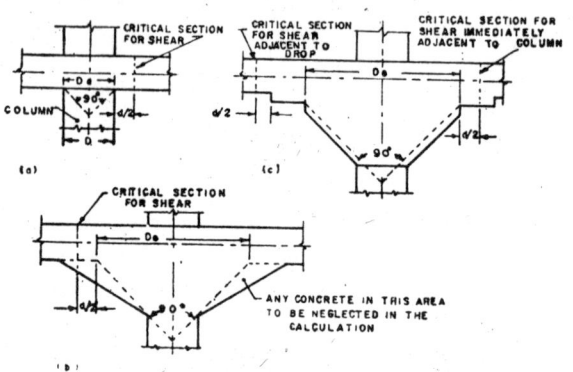

Fig. 11.3

11.5. PANELS WITH MARGINAL BEAMS AND WALLS

If the slab is supported by a marginal beam (See Fig.11.2) having a depth greater than 1.5 times the thickness of slab, or by a wall, then

(i)   the load supported by the beam or the wall shall be the sum of the load directly supported by it plus one quarter of the load on the slab, uniformly distributed.

(ii)   the bending moment in the half column strip running along the beam or wall shall be one-quarter of the bending moment in the first interior column strip.

11.6   DIRECT DESIGN METHOD

For this method to be applicable, the following conditions should be fulfilled:

11.6.1.   Limitations.

(i)   There shall be at least three continuous spans in each direction.

(ii)   Panels should be rectangular with the ratio of the longer to the shorter span not exceeding 2.0.

(iii)   Columns may be off set upto 10% of the span in that direction.

(iv) Succesive span lengths in any direction shall not differ by more than one-third of the longer span. End span shall not be greater than the interior span.

(v) Live load shall not be more than 3 times the dead load.

11.6.2. Total design moment. Total design moment for a span shall be determined for a slab width included between the centre lines of the two adjacent panels. It is given by

$$M_o = \frac{W \, l_n}{8} \tag{11.1}$$

$$= \frac{w \, l_2 \, l_n^2}{8} \tag{11.2}$$

where M_o = total design moment

W = total design load

$= w \, l_2 \, l_n$

w = uniformely distributed design load

l_n = clear span face to face of columns, capitals or walls

$\not< 0.65 \, l_1$ in any case.

For this purpose, circular supports shall be treated as square supports having the same area.

l_1 = span, c/c of columns in the direction of M_o

l_2 = length of span transverse to l_1

$= \frac{(l_{21} + l_{22})}{2}$ when $l_{21} \neq l_{22}$

See Fig. 11.1

Note: When a span parallel to an edge is being considered l_2 shall be the distance from the edge to the centre line of the panel.

11.6.3. Negative and positive design moments. Circular supports are treated as square supports having the same area. The negative design moment shall be located at the face of the rectangular supports.

(i) For interior span.

Negative design moment $= 0.65 \, M_o$

Positive design moment $= 0.35 \, M_o$

(ii) For exterior span.

Negative design moment:

Over interior support $= \left(0.75 - \dfrac{0.10}{1 + \dfrac{1}{\alpha_0}} \right) M_o$

Over exterior support $= \left(\dfrac{0.65}{1 + \dfrac{1}{\alpha_c}} \right) M_o$

Positive design moment $= \left(0.63 - \dfrac{0.28}{1 + \dfrac{.1}{\alpha_c}} \right) M_o$

where $\alpha_c = \dfrac{\Sigma K_c}{K_s}$

$\Sigma K_c =$ sum of the flexural stiffness of the exterior columns meeting at the joint.

$K_s =$ flexural stiffness of slab.

These design moments may be modified upto 10%. However, their sum in any direction should not be less than the calculated value of M_o as per 11.6.2.

11.7 EQUIVALENT FRAME METHOD

11.7.1 Analysis for moments. The bending moments and shear forces may be determined by analysing the flat slab as a continuous frame. A multistoreyed building may have a number of flat slab floors through which the supporting columns may be continuous. Such a structure may be considered to be made up of vertical frames in two orthogonal directions. Each frame consists of a row of columns as the vertical members, and bands of floor slab bounded laterally by the centre lines of the panel on each side as horizontal members. For frames adjacent and parallel to an edge the external boundry shall be the edge itself.

Each of the equivalent frames in the two orthogonal directions may be analysed for the given loading by the usual elastic analysis. Approximate methods of analysis based on subframing may be employed. Relative stiffnesses of column and slab members of each frame are based on gross cross-section of concrete. If the slab has drops, then, its stiffness should be suitably modified. Effect of flared column head on the stiffness of the coloumn may be ignored.

11.7.2. Nagative and positive design moments. Two cases arise depending upon the magnitude of live load/dead load ratio.

11.7.2.1. Live load $<$ 0.75 Dead load.

Maximum moments are assumed to occur at all sections when full design live load is on the entire slab system.

11.7.2.2. Live load > 0.75 Dead load.

The following situations of live load may be considered:

(i) For maximum positive moment in the panel near mid-span three quarters of the design live load is placed on that panel and on alternate panels. See Fig.11.4 (b).

(ii) For maximum negative moment in the panel at a support three-quarter of the full design live load is placed on adjacent panels only. See Fig.11.4 (c).

(iii) The design positive and negative moments shall in no case be less than those calculated with full design live load placed on all the panels. See Fig.11.4 (a).

11.7.3. Location for negative design moment. For this purpose, circular or polygonal shaped supports shall be treated as square supports having the same area. The location of the critical section for negative moment shall be as follows:

(i) At internal support:

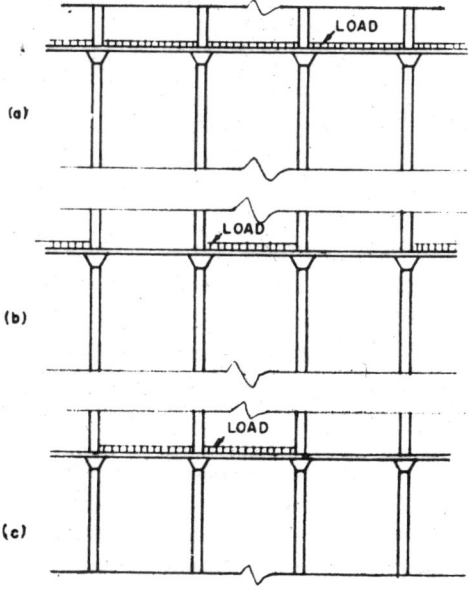

Fig. 11.4.

At the face of support, but not greater than $0.175 \, l_1$ from centre of coloumn.

(ii) At exterior support:
From the face of support, not greater than one-half the projection of bracket or capital.

11.7.4. Modification of maximum moments. The negative and positive design moments determined by equivalent frame method may be reduced in such proportion that their numerical sum is not less than M_0 specified in 11.6.2.

11.8. BENDING MOMEMTS FOR DESIGN OF SLAB

11.8.1. Column strip.
(a) *Negative bending moment.* At an interior support the column strip shall be designed to resist 75% of the panel negative moment there.

At an exterior support the column strip shall be designed to resist whole of the panel negative moment there. However, if the length of the exterior support is equal to or more than three-quarters of the span in transeverse direction, the negative moment there shall be considered to be uniformly distributed across the width of panel.

(b) *Positive bending moment.* For each span the column strip shall be designed to resist 60% of the panel positive moment.

11.8.2. Middle strip. The portion of negative or positive panel moment not resisted by column strip shall be assigned to the adjacent middle strip.

Each middle strip shall be designed to resist the sum of the bending moments assigned to its two half portions.

In an exterior panel the midle strip parallel to an edge supported by a wall shall be designed to resist twice the bending moment assigned to half the middle strip corresponding to the first row of interior columns.

11.9. BENDING MOMENTS IN COLUMNS

As the columns are built integrally with the slab system, they should be designed to resis the moments arising from the loads on the slab. The moments may be estimated by the following rules.

11.9.1. The columns above and below the Slab shall be designed to resist the moment M in direct proportion to their stiffness as follows:

(i) *At an external support*:

$$M = \frac{0\,65}{1 + \dfrac{1}{\alpha_c}} \quad \frac{(w_d + w_l)\, l_2\, l_n^2}{8}$$

(ii) *At internal support*:

$$M = \frac{0.08\,(w_d + 0.5\,w_l)\, l_2\, l_n^2 - w_d'\, l_2'\, l_n'^2}{1 + \dfrac{1}{\alpha_c}}$$

where

w_d, w_l = intensities of design dead and live load respectively

l_2, l_n = as defined in 8.6.2

$\alpha_c = \dfrac{\Sigma\, K_c}{\Sigma\, K_s}$

$\Sigma\, K_c$ = sum of the flexural stiffnesses of the columns above and below the slab

$\Sigma\, K_s$ = sum of the flexural stiffness of the slab panels meeting at the joints

w_d', l_2' and l_n' refer to the shorter span.

11.9.2 Modifications for pattern loading. When $w_l > 0.5\,w_d$ the following modifications shall be applied as per I S Code: 456:

(i) $\Sigma\, K_c$ shall be such that $\alpha_c \geqslant \alpha_{omin}$ defined in Table 11.1.

(ii) If $\Sigma\, K_c$ is such that $\alpha_c < \alpha_{omin}$ defined in Table 8.1, then the positive design moments for the panel shall be multiplied by the coefficient β_s
where

$$\beta_s = 1 + \left(\frac{2 - \dfrac{w_d}{w_l}}{4 + \dfrac{w_d}{w_l}} \right) \left(1 - \frac{\alpha_c}{\alpha_{omin}} \right)$$

Table 11.1
MINIMUM PERMISSIBLE VALUES OF α_c
(BASED ON TABLE 11 OF IS CODE: 456-1978)

w_1/w_d	l_2/l_1	α_{cmin}
0.5	0.5 to 2.0	0
1.0	0.5	0.6
	0.8 to 1.0	0.7
	1.25	0.8
	2.0	1.2
2.0	0.5	1.3
	0.8	1.5
	1.0	1.6
	1.25	1.9
	2.0	4.9
3.0	0.5	1.8
	0.8	2.0
	1.0	2.3
	1.25	2.8
	2.0	13.0

11.9.3. Transfer of bending moment from slab to column.
Transfer of moment from slab to column takes place on account of unbalanced loading of panels on either side, lateral loads such as wind, earthquake or otherwise. Flexural stresses shall be investigated in the slab for an effective width given by

Effective width $= D + 3 t$

where

D = diameter of column or column head

t = thickness of slab or drop as the case may be.

Flexural stresses in the slab should be investigated for a moment given by

$M_s = \alpha$ times the moment.

where

$$a = \frac{1}{1 + \frac{2}{3}\sqrt{\frac{a_1}{a_2}}}$$

$a_1 = c_1 + d$

$a_2 = c_2 + d$

c_1 = dimension of column or capital in the direction in which moment acts

c_2 = dimension of column or capital in the transverse direction

d = Effective depth of slab or drop.

Additional reinforcement may be provided in the slab over the coloumn head, if required, to resist this moment.

11.10 CONSIDERATION OF SHEAR STRESSES

11.10.1. Critical section. The critical section for shear shall be taken at right angles to the slab at a distance $d/2$ away from the peripliery of the column or capital, where d is the effective depth of section. The shapes in plan are shown in Fig.11.5. For columns located near the free edge of a slab, the plans of critical section are shown in Fig.11.6. Openings in slab shall not encroach upon column or column head. For the openings located at a distance less than 10 times slab thickness from the centroid of support reaction and for openings located within the column strips, the periphery of the critical section around the column or capital shall be reduced by the part included within the radial lines drawn from the opening to the centroid of reaction. See Fig.11.7.

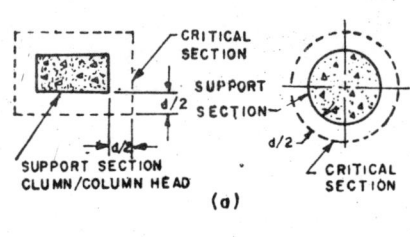

(a)

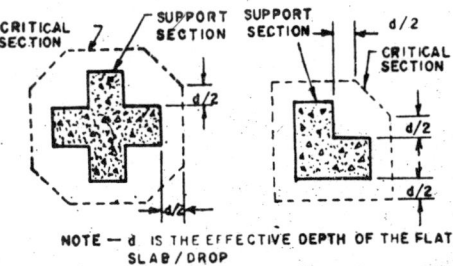

NOTE — d IS THE EFFECTIVE DEPTH OF THE FLAT SLAB / DROP

Fig. 11.5

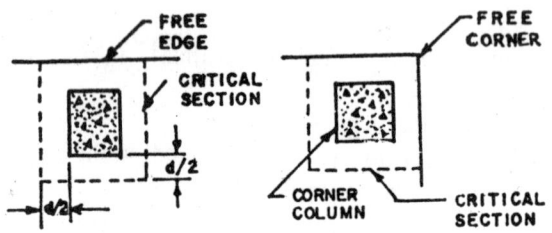

Fig. 11.6

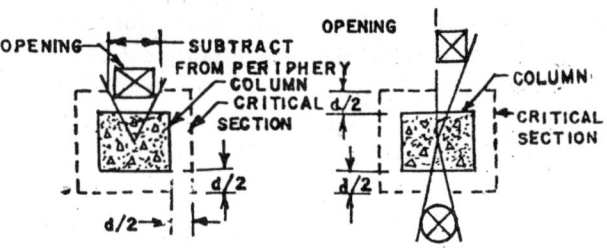

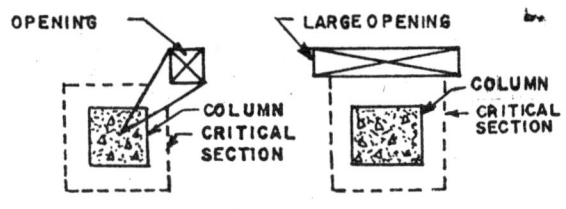

Fig. 11.7

11.10.2   Evaluation of shear stress. Shear stress in the slab shall be the sum of the values calculated as follows:

(i)   Nominal shear stress given by

$$\tau_v = \frac{V}{b_o\, d}$$

where

V = design shear force

b_o = periphery of critical section

d = effective depth.

(ii)   Shear stress on account of transfer of bending moment between slab and column head. Shear stresses shall be taken to be linearly varying about the centroid of the critical section. See Fig. 11.8. A fraction of the moment shall be considered for this purpose, which is given by.

$(1 - \alpha)$ times the moment.

where α is defined in 11.9.3.

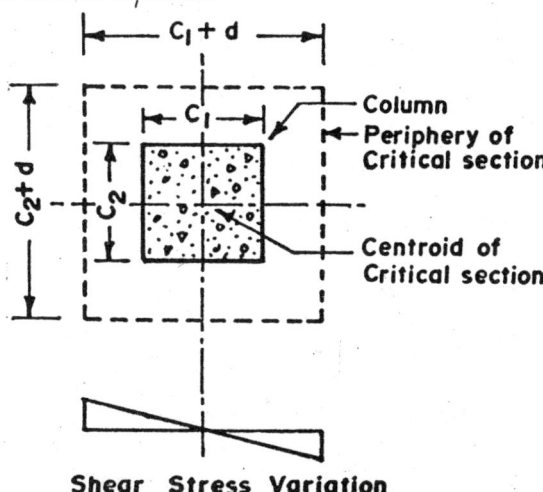

Shear Stress Variation

Fig. 11.8

11.10.3 Permissible shear stress. When the slab has no shear reinforcement, the calculated shear stress shall not exceed a value

$$k_s \tau_c$$

where

$$k_s = 0.5 + \beta_c$$

but $\not> 1.0$ in any case

β_c = ratio of short side to long side of column or capital.

$\tau_c = 0.16 \sqrt{f_{ck}}$ in working stress design.

11.10.4 Safety in shear. The following rules are recommended by IS Code: 456.

(i) If calculated shear stress $> k_s \tau_c$

but $< 1.5 \tau_c$

shear reinforcement should be provided. Concrete shall be assumed to take-up shear stress up to $0.5 \tau_c$. Remaining shear stress shall be assigned to shear reinforcement.

(ii) If calculated shear stress $> 1.50 \tau_c$

the flat slab should be redesigned taking greater thickness.

11.11 TENSION REINFORCEMENT IN SLAB

Once the effective depth of the slab has been fixed up, reinforcement is calculated in the usual way. However, when drops are provided, the thickness of drop for determining the area of reinforcement shall be the lesser of

(i) Actual thickness

(ii) slab thickness + 0.25 of the distance between edge of drop and edge of capital.

11.11.1. Spacing. The spacing of tension reinforcement shall not exceed 2 times the slab thickness.

11.11.2. Minimum length of bars.

(i) Lengths of reinforcement bars shall not be less than those specified in Fig. 11.9.

(ii) For unequal adjacent spans, the extension of negative reinforcement beyond each face of the column shall be based on the longer span.

11.11.3. Anchorage of reinforcement. Slab reinforcement perpendicular to a discontinuous edge shall be anchored beyond the internal face of spandrel beam, wall or column as follows:

(i) For positive reinforcement: $\not< 150$ mm

(ii) For negative reinforcement: not less than the development length.

If the slab overhangs a support, the reinforcement bars shall be anchored within the slab.

11.11.4. Provision of openings. Openings may be provided and their sides strengthened as follows:

(ii) Within the middle half of the span in each direction, openings of any size may be provided. However, full reinforcement as required for panel without opening may be maintained.

(ii) Within the area commom to two column strips openning of size not more than one-eigth of the width of column strip in each direction may be provided. Reinforcement interrupted by the opening shall be added on the sides of the openings.

(iii) Within the area common to a column strip and a middle strip, the size of opening shall be restricted so that not more than one-quarter of the reinforcement in either strip shall be interrupted. The interrupted reinforcement shall be added on the sides of the opening.

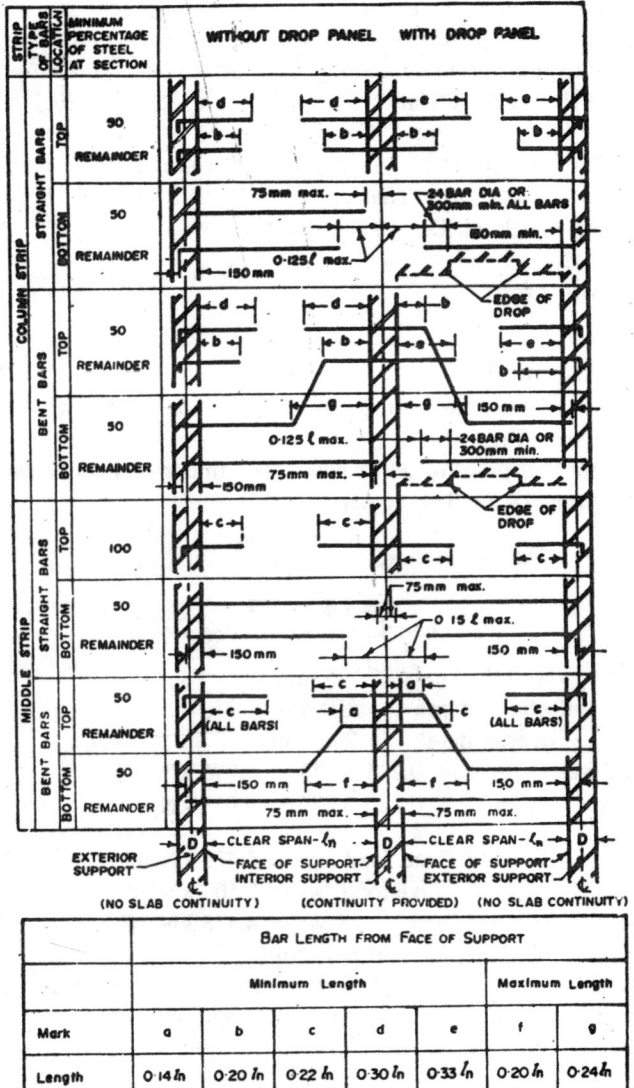

	BAR LENGTH FROM FACE OF SUPPORT						
	Minimum Length					Maximum Length	
Mark	a	b	c	d	e	f	g
Length	$0.14\,l_n$	$0.20\,l_n$	$0.22\,l_n$	$0.30\,l_n$	$0.33\,l_n$	$0.20\,l_n$	$0.24\,l_n$

• Bent bars at exterior supports may be used if a general analysis is made.

Fig. 11.9

Example 11.1

Design an ؛interior panel of a flat slab which is supported over columns, provided at a spacing of 5.4 m c/c along one direction and 4.5 m c/c along the orthogonal direction. Live load on slab is 2.5 kN/m². The slab has a 25 mm thick layer of concrete tiles on the top. Concrete grade : M 15 and steel grade: Fe 415. The supporting columns are circular in section with diameter = 360 mm. The columns have a column head with diameter = 0.96 m. Design the slab,

(i) without drop

(ii) with drop.

Solution:

(i) Slab without drop
Widths of the column and the middle strips in the two directions are shown in Fig.11.10.
For the purpose of self weight, the thickness may be estimated as per para 11.4.3. Taking $M_t = 1.0$

$$\text{effective depth} \nless \frac{\text{longer span}}{26}$$

$$\text{i.e. } \nless \frac{5.4 \times 10^3}{26} \text{ or } 207.7 \text{ mm}$$

Also, overall thickness $\nless$ 125 mm
Try effective depth = 210 mm
and overall depth = 235 mm
Self weight of slab, $0.235 \times 25 = 5.875 \text{ kN/m}^2$
weight of tiles $\quad 0.025 \times 24 = 0.6 \quad \text{kN/m}^2$

$$\therefore \text{ Dead load } w_d = 6.475 \text{ kN/m}^2$$
$$\text{Live load } w_1 = 2.50 \text{ kN/m}^2$$
Total distributed load $\quad w = 8.975 \text{ kN/m}^2$
Direct design method will be adopted.
Diameter of column head = 960 mm.
Length of side of a square of the same area

$$= 960 \sqrt{\frac{\pi}{4}}$$

$$= 850.8 \text{ mm}$$

Total design moment will now be calculated as per para 11.6.2.

Along the longer span:

$l_n = 5.4 - 0.85 = 4.55$ m,which is not less than 0.65×5.4 i.e. 3.51 m. Hence o.k.

$l_2 = 4.5$ m.

Then

$$M_o = \frac{w l_2 \, l_n{}^2}{8}$$

$$= \frac{8.975 \times 4.5 \times 4.55^2}{8}$$

$$= 104.52 \text{ kN-m}.$$

Along the shorter span

$l_n = 4.5 - 0.85 = 3.65$ m,which is not less than 0.65×4.5 i.e. 2.925 m. Hence o.k.

$l_2 = 5.4$ m.

Then

$$M_o = \frac{w \, l_2 \, l_n{}^2}{8}$$

$$= \frac{8.975 \times 5.4 \times 3.65^2}{8}$$

$$= 80.71 \text{ kN-m}$$

Calculated panel design moments as per para 11.6.3 are as follows:

	Long span	Short span
Positive design moment $(0.35 \, M_o)$	36.68	28.25
Negative design moment $(0.65 \, M_o)$	-67.94	-52.46

Calculated bending moments in column and middle strips as per para 8.8 are as follows:

	Long span kN-m	Short span kN-m
column strip:		
Negative bending moment (75%)	-50.96	$-39.35\cdot$
Positive bending moment (60%)	21.95	16.95
Middle strip:		
Negative bending moment	-16.98	-13.11
Positive bending moment	14.63	11.90

Design of slab:

$\sigma_{cbc} = 5$ N/mm², $\sigma_{st} = 230$ N/mm², $m = 18.67$

$n = 0.289$, $j = .904$ and $R = 0.653$

Effective depth is based on maximum negative bending moment which occurs in the column strip along the longer span. Hence

$$d = \sqrt{\frac{M}{b\,R}} = \frac{50\,96 \times 10^6}{2.25 \times 10^3 \times 0.653}$$

$$= 176.7 \text{ mm, say } 180 \text{ mm}$$

Overall depth $= 180 + 25 = 205$ mm

Check for shear:

Periphery of critical section is at $\frac{1}{2} \times 180 = 90$ mm away from the column face. Hence

Length of periphery $b_0 = \pi\,(960 + 90 + 90)$

$$= 3581.4 \text{ mm}$$

Design shear force $V = 8.975\,(5.4 \times 4.5 - \dfrac{\pi}{4} \times 1.14^2)$

$$= 208.94 \text{ kN}$$

Nominal shear stress

$$\tau_v = \frac{V}{b_0\,d} = \frac{208.94 \times 10^3}{3581.4 \times 180}$$

$$= 0.324 \text{ N/mm}^2$$

Permissible shear stress is found according to para 11.10.3.

$$k_s = 0.5 + \beta_c$$

$$= 0.5 + 1$$

$$= 1.5 > 1.0$$

Hence, take $k_s = 1.0$

$$\tau_c = 0.16 \sqrt{f_{ck}}$$

$$= 0.16 \sqrt{15} = 0.62 \text{ N/mm}^2$$

Permissible shear stress

$$= k_s\,\tau_c$$

$$= 1.0 \times 0.62$$

$$= 0.62 \text{ N/mm}^2$$

As $\tau_v < \tau_o$ the slab is safe in shear.

Tension reinforcement

$$= \frac{50.96 \times 10^6}{230 \times .904 \times 180} = 1361.6 \text{ mm}^2$$

% reinforcement $= \dfrac{1361.6 \times 100}{2500 \times 180} = 0.3$ %

From Fig. 16.2 $M_f = 1.40$

$\therefore$ Minimum effective depth $= \dfrac{5.40 \times 10^3}{26 \times 1.4} = 148.4$ mm.

Actual effective depth is more than this. Hence o.k.

Reinforcement in column strip:

Maximum permitted bar spacing $= 2 \times 205 = 410$ mm.

(a) In long direction.

Temperature and shrinkage reinforcement

$$= \frac{0.12}{100} \times 205 \times 2250 = 553.5 \text{ mm}^2.$$

Negative reinforcement $= \dfrac{50.96 \times 10^6}{230 \times .904 \times 180} = 1861.6$ mm^2
(d $= 180$ mm)

Provide 12 mm dia. bars — 12 nos.

Positive reinforcement $= \dfrac{21.95 \times 10^6}{230 \times .904 \times 180} = 586.5$ mm^2
(d $= 180$ mm)

Provide 10 mm dia. bars — 8 nos.

(b) In short direction.

Temperature and shrikage reinforcement

$$= \frac{0.12}{100} \times 205 \times 2250 = 553.5 \text{ mm}^2$$

Negative reinforcement $= \dfrac{39.35 \times 10^6}{230 \times .904 \times 168} = 1126.5$ mm^2
(d $= 168$ mm)

Provide 12mm dia. bars, 10 nos.

Positive reinforcement $= \dfrac{16.95 \times 16^6}{230 \times .904 \times 170} = 479.54$ mm^2
(d $= 170$ mm)

This is less than 553.5 mm^2. Hence, adopt 553.5 mm^2.
Provide 10 mm dia. bars, 7 nos.

Reinforcement in middle strtp:
(a) In long direction.

Temperature and shrinkage reinforcement $= \dfrac{0.12}{100} \times 205 \times 2250$

$$= 553.5 \text{ mm}^2.$$

Negative reinforcement $= \dfrac{16.98 \times 10^6}{230 \times .904 \times 180} = 453.7$ mm^2.
(d $= 180$ mm)

This is less than 553.5 mm^2. Hence, adopt 553.5 mm^2.
Provide 10 mm dia. bars — 7 nos.

Positive reinforcement $= \dfrac{14.63 \times 10^6}{230 \times .904 \times 160} = 439.8$ mm^2.
(d = 160 mm)

This is less than 553.5 mm^2. Hence, adopt 553.5 mm^2.
Provide 10 dia. bars — 7 nos.

(b) In short direction.

Temperature and shrinkage reinforcement $= \dfrac{0.12}{100} \times 205 \times 3150$

$$= 774.9 \text{ mm}^2.$$

Negative reinforcement $= \dfrac{13.11 \times 10^6}{230 \times .904 \times 169} = 373$ mm^2.
(d = 169 mm)

This is less than 774.9 mm^2. Hence, adopt 774.9 mm^2.
Provide 10 mm dia. bars — 9 nos.

Positive reinforcement $= \dfrac{11.90 \times 10^6}{230 \times .904 \times 170} = 336.7$ mm^2.
(d = 170 mm)

This is less than 774.9 mm^2. Hence, adopt 774.9 mm^2.
Provide 10 mm dia. bars —10 nos.

Projection of negative reinforcement into the span (See Fig.11.9):

Column strip.

In long direction = half width of support + 0.3 l_n
  $= \frac{1}{2} \times 0.85 + 0.3 \times 4.55$
  = 1.79 m, say 1.8 m.

In short direction $= \frac{1}{2} \times 0.85 + 0.3 \times 3.65$
  = 1.52 m, say 1.525 m.

Middle strip.

In long direction = half width of support + 0.22 l_n
  $= \frac{1}{2} \times 0.85 + 0.22 \times 4.55$
  = 1.426 m, say 1.45 m.
In short direction $= \frac{1}{2} \times 0.85 + 0.22 \times 3.65$
  = 1.228 m, say 123 m.

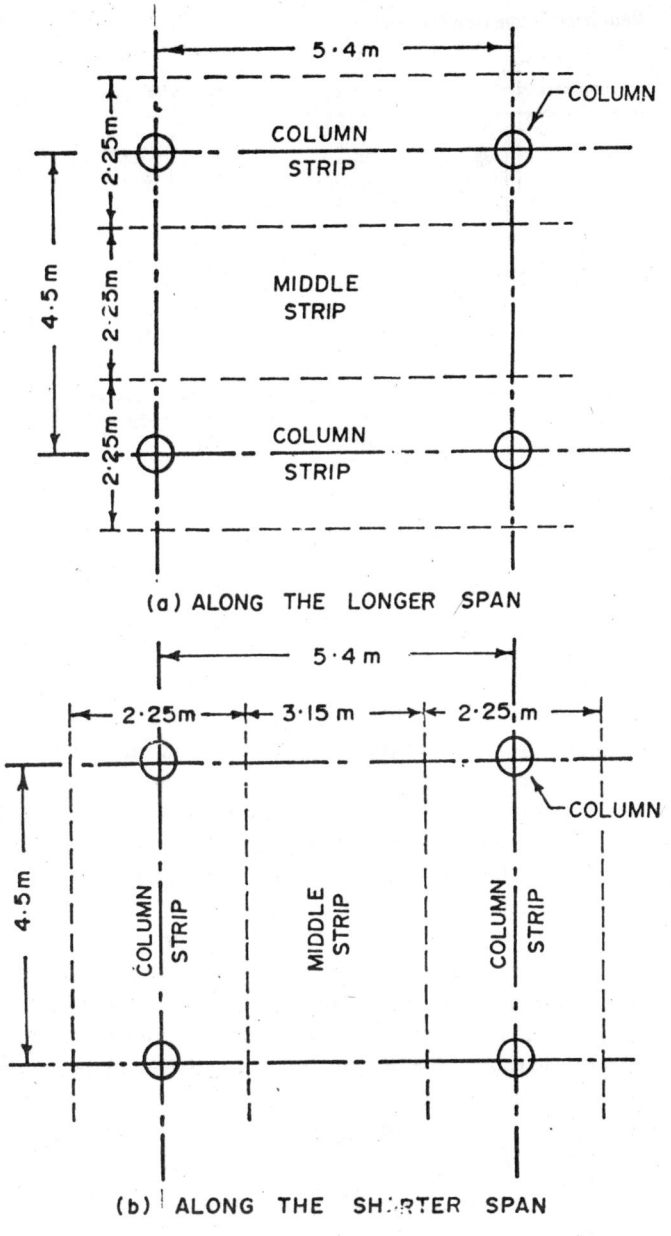

(a) ALONG THE LONGER SPAN

(b) ALONG THE SHORTER SPAN

Fig. 11.10.

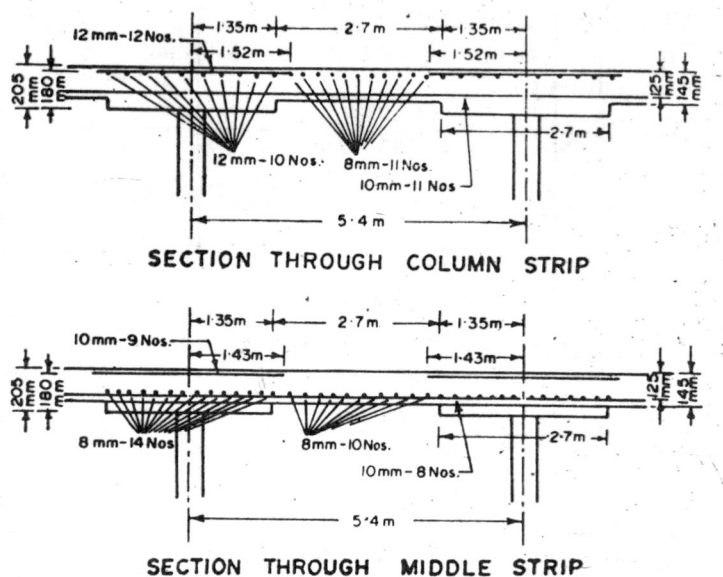

SECTION THROUGH COLUMN STRIP

SECTION THROUGH MIDDLE STRIP

Fig. 11.11.

(ii) Slab with drop.

Drop dimensions in plan are taken to be equal to the width of the column strip in each direction. See Fig. 11.12. As such, the design bending moments in the column strip and middle strip in the two directions will be the same as in the previous example. Thickness of the drop will be the same as that for the slab in the previous example. Negative reinforcement in the drop in the two directions will be the same as that for the column strip in two directions calculated in the previous example. The only calculations needed are for the thickness of slab required for positive bending moment in column strip in the mid-span portion and the corresponding reinforcements. Taking the column strip positive bending moment to be the same as in the previous example, the effective depth is found for the maximum positive bending moment along the longer span:

$$\text{effective depth} = \sqrt{\frac{21.95 \times 10^6}{2.25 \times 10^3 \times 0.653}}$$

$$= 122.2 \text{ mm, say } 125 \text{ mm}$$

Overall depth $= 125 + 20 = 145$ mm.

Reinforcement in column strip.

Maximum permitted spacing = 2 × 145 = 290 mm

(a) In long direction.

Temperature and shrinkage reinforcement = $\dfrac{0.12}{100}$ × 145 × 2250

$$= 391.5 \text{ mm}^2$$

Positive reinforcement = $\dfrac{21.95 \times 10^6}{230 \times .904 \times 125}$
(d = 125 mm)

$$= 844.6 \text{ mm}^2$$

Provide 10 mm dia. bars-11 nos.

Spacing = $\dfrac{2250}{11}$ = 204.5 mm c/c.

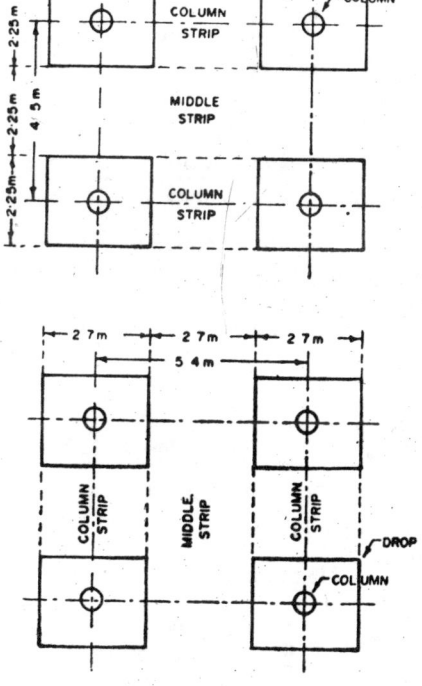

Fig. 11.12

(b) In short direction

Temperature and shrinkage reinforcement $= \dfrac{0.12}{100} \times 145 \times 2700$

$$= 469.8 \text{ mm}^2.$$

Positive reinforcement $= \dfrac{16.95 \times 10^6}{230 \times .904 \times 116}$
(d = 116 mm)

$$= 702.8 \text{ mm}^2$$

Provide 8 mm dia. bars-14 nos.

Spacing $= \dfrac{2700}{14} = 192.9$ mm. O.K.

Reinforcement in middle strip.

(a) In long direction.

Negative reinforcement $= \dfrac{16.98 \times 10^6}{230 \times .904 \times 125}$
(d — 125 mm)

$$= 653.3 \text{ mm}^2.$$

Provide 10 mm dia. bars-9 nos.

Spacing $= \dfrac{2250}{9} = 250$ mm. c/c.

Positive reinforcement $= \dfrac{14.63 \times 10^6}{230 \times .904 \times 125}$
(d — 125 mm)

$$= 562.9 \text{ mm}^2.$$

Provide 10 mm dia. bars-8 nos.

Spacing $= \dfrac{2250}{8} = 281.3$ mm. c/c.

(b) In short direction.

Negative reinforcement $= \dfrac{13.11 \times 10^6}{230 \times .904 \times 116}$
(d = 116 mm)

$$= 543.6 \text{ mm}^2.$$

Provide 8 mm dia. bars-11 nos.

Spacing $= \dfrac{2700}{11} = 245.5$ mm c/c.

Positive reinforcement $= \dfrac{11.90 \times 10^6}{230 \times .904 \times 116}$
(d = 116 mm)

$$= 493.4 \text{ mm}^2$$

Provide 8 mm dia. bars-10 nos.

$$\text{Spacing} = \frac{2700}{10} = 270 \text{ mm c/c.}$$

Check for shear:

Shear around the column head need not be checked as it has already been done in the last example.

Shear needs to be checked along a periphery of critical section distant $\frac{110}{2} = 55$ mm away from the periphery of the drop.

Periphery of critical section $= 2 (2.7 + 2 \times .05 + 2.25$
$$+ 2 \times .05) = 10.3 \text{ m.}$$

Shear force causing shear $= 8.975 (5.4 \times 4.5 - 2.8 \times 2.35)$
$$= 159 \text{ kN.}$$

Nominal shear stress $\tau_v = \dfrac{159 \times 10^3}{10.3 \times 1000 \times 100}$
$$= 0.14 \text{ N/mm}^2.$$

Permissible shear stress:

$$\beta_c = \frac{2.8}{2.35} = 1.19$$

$$k_s = 0.5 + 1.19 = 1.69 > 1.0$$

Take $\quad k_s = 1.0$

$$\tau_c = 0.16 \sqrt{f_{ok}} = 0.16 \sqrt{15} = 0.62 \text{ N/mm}^2$$

Permissible shear stress $= k_s \, \tau_c$
$$= 0.62 \text{ N/mm}^2$$

Nominal shear stress is less than the permissible shear stress. Hence, the slab is safe in shear.

Projection of negative reinforcent into the span.

Column strip:

In long direction $=$ half width of support $+ 0.24 \, l_a$
$$= \tfrac{1}{2} \times 0.85 \times 0.24 \times 4.55$$
$$= 1.517 \text{ m, say } 1.52 \text{ m.}$$

In short direction $= \tfrac{1}{2} \times 0.85 + 0.24 \times 3.65$
$$= 1.30 \text{ m.}$$

Middle strip :

In long direction $=$ half width of support $+ 0.22 \, l_a$
$$= \tfrac{1}{2} \times 0.85 + 0.22 \times 4.55$$
$$= 1.426 \text{ m, say } 1.43 \text{ m.}$$

In short direction $= \tfrac{1}{2} \times 0.85 + 0.22 \times 3.65$
$$= 1.22.8 \text{ m, say } 1.23 \text{ m.}$$

Details of reinforcement are shown in Fig.11.13

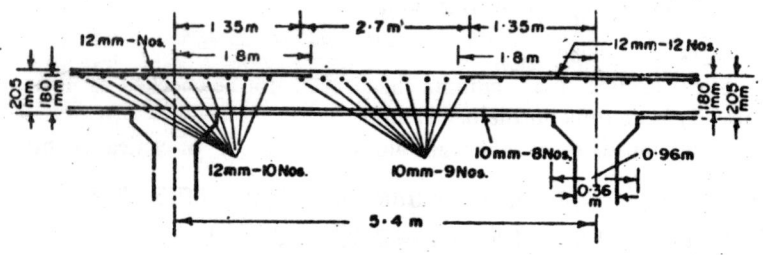

SECTION THROUGH COLUMN STRIP

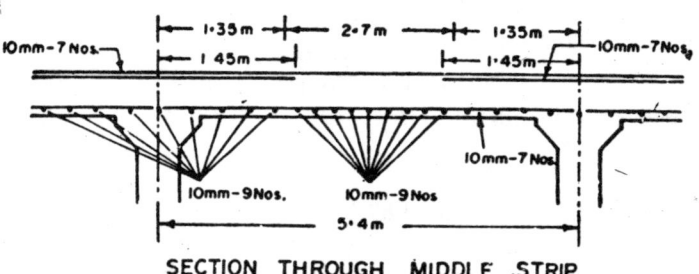

SECTION THROUGH MIDDLE STRIP

Fig. 11.13

Example 11.2

The plan and the section of an exterior panel of a flat slab are shown in Fig. 11.14. Slab thickness $= 200$ mm. It has a 25 mm thick plain concrete tile floor on top. Live load $= 2.5$ kN/m². Clear height between two consecutive floor slabs $= 4.2$ m. Depth of marginal beam $= 420$ mm. Calculate the following design values:

(a) Design load on the marginal beam.

(b) Bending moment in the half column strip adjacent and parallel to the marginal beam.

(c) Negative and positive design moments along the span at right angles to the edge.

(d) The moments taken up by the columns above and below the floor, at the exterior and the interior supports.

Solution:

Dead load on slab:

Due to concrete tiles	$= 0.025 \times 25 = 0.625$ kN/m².	
Self weight of slab	$= 0.2 \times 25 = 5.0$ kN/m².	

$$w_d = 5.625 \text{ kN/m}^2.$$

Live load
$$w_l = 2.5 \text{ kN/m}^2.$$

Total $w = 8.125$ kN/m².

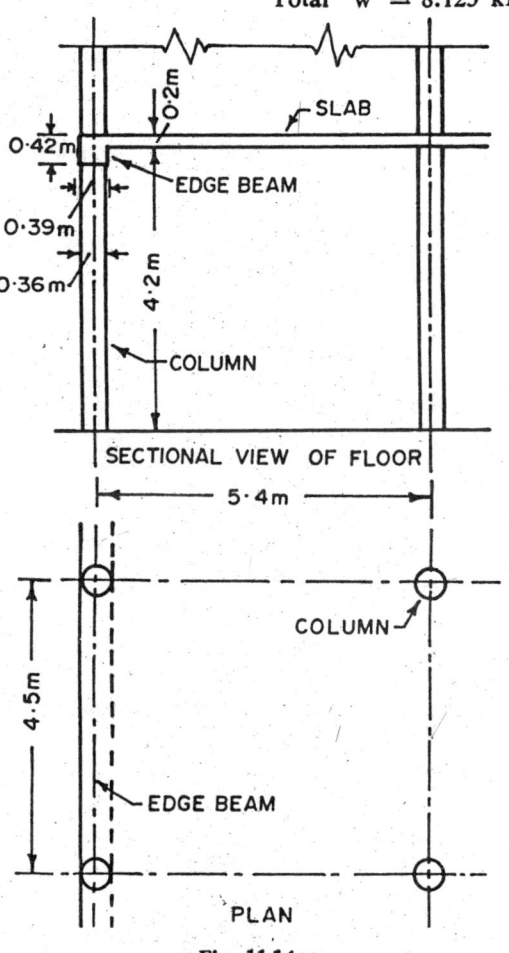

SECTIONAL VIEW OF FLOOR

PLAN

Fig. 11.14

(a) Self weight of marginal beam

$$= 0.42 \times 0.39 \times 25$$
$$= 4.095 \ \text{kN/m}$$

Load contributed by slab panel

$$= \frac{\text{quarter of the slab load}}{4.5}$$
$$= \frac{0.25 \times 5.4 \times 4.5 \times 8.125}{4.5}$$
$$= 10.969 \ \text{kN/m}.$$

∴ Total distributed load on the marginal beam

$$= 4.095 + 10.969$$
$$= 15.064 \ \text{kN/m}.$$

(b) Bending moment in the half column strip adjacent and parallel to the edge beam = one quarter of that for the first interior column strip.

Now, the length of the side of a square having the same area as that of the circular column

$$= 360 \ \sqrt{\frac{\pi}{4}}$$
$$= 319 \ \text{mm}.$$

∴ $l_n = 4.5 - 0.319$
$$= 4.181 \ \text{m}$$
$l_2 = 5.4 \ \text{m}$

Then

$$M_o = \frac{w \, l_2 \, l_n^2}{8} = \frac{8.125 \times 5.4 \times 4.181^2}{8}$$
$$= 95.87 \ \text{kN-m}.$$

Negative design moment $= 0.65 \times 95.87 = 62.32$ kN-m.
Positive design moment $= 0.35 \times 95.87 = 33.55$ kN-m.

Hence, bending moments in the half column strip along the edge will be as follows:

Negative bending moment $= \frac{1}{4} \times 0.75 \times 62.32$
$$= 11\,69 \ \text{kN-m}.$$

Positive bending moment $= \frac{1}{4} \times 0.60 \times 33.55$
$$= 5.03 \ \text{kN-m}.$$

(c) For the span at right angles to the edge beam.

$l_n = 5.4 - 0.319 = 5.081 \ \text{m}$
$l_2 = 4.5 \ \text{m}$

Then

$$M_o = \frac{w\, l_2\, l_n^2}{8} = \frac{8.125 \times 4.5 \times 5.081^2}{8}$$

$$= 118 \text{ kN-m.}$$

For a coloumn section

$$I_o = \frac{\pi}{64} \times 0.36^4$$

$$= 8.245 \times 10^{-4} \text{ m}^4.$$

For one column

$$K_a = \frac{I_c}{l_o} = \frac{8.245 \times 10^{-4}}{4.2}$$

$$= 1.963 \times 10^{-4} \text{ m}^3.$$

Considering the upper and lower column

$$\Sigma K_o = 2 \times 1.963 \times 10^{-4}$$

$$= 3.926 \times 10^{-4}$$

For the slab panel, at the edge.

$$I_s = \frac{1}{12} \times 4.5 \times 0.2^3$$

$$= 0.003 \text{ m}^4$$

$$\therefore K_s = \frac{I_s}{l_s} = \frac{0.003}{5.4}$$

$$= 5.556 \times 10^{-4} \text{ m}^3$$

$$\alpha_e = \frac{\Sigma K_c}{K_s}$$

$$= \frac{3.926 \times 10^{-4}}{5.536 \times 10^{-4}}$$

$$= 0.7066$$

Hence, the bending moments for the exterior panel, in a direction at right angles to the marginal beam according to 11.6.3, will be as follows:

(i) Negative design moment

Over interior support $= \left(0.75 - \dfrac{0.10}{1 + \dfrac{1}{0.7066}} \right) 118$

$$= 83.61 \text{ kN-m.}$$

Over exterior support $= \left(\dfrac{0.65}{1 + \dfrac{1}{0.7066}} \right) 118$

$$= 31.76 \text{ kN-m.}$$

(ii) Positive design moment

$$\text{At mid-span} = \left(0.63 - \frac{0.28}{1 + \dfrac{1}{0.7066}} \right) 118$$

$$= 66.66 \text{ kN-m.}$$

(d) Calculations are made as per 11.9.1 for the bending moment shared by the upper and lower columns together.

(i) Over exterior support

$$a_0 = 0.7066 \text{ already calculated in (c).}$$

Hence,

$$\text{moment M} = \frac{0.65}{1 + \dfrac{1}{0.7066}} \times \frac{8.125 \times 4.5 \times 5.081^2}{8}$$

$$= 31.76 \text{ kN-m.}$$

Over interior support

$$\alpha_0 = \frac{\Sigma K_e}{\Sigma K_s} = \frac{3.926}{2 \times 5.556} = 0.353$$

Hence, moment

$$M = \frac{0.08\{(5.625 + .5 \times 2.5)4.5 \times 5.081^2 - 5.625 \times 5.4 \times 4.181^2\}}{1 + \dfrac{1}{0.353}}$$

$$= 2.86 \text{ kN-m.}$$

12

Design of Footings

12.1. INTRODUCTION

Footings mean the structural foundation required to transmit the loads from a structure to the underlying soil without causing soil shear failure or excessive settlement. In the absence of properly designed footing a structure may tilt, or develop cracks and may even collapse.

Under excessive pressure, the soil may fail in shear on account of plastic flow or lateral expulsion from beneath the foundation. As such, a reasonable and sufficient factor of safety is assigned against such a failure. Hence, the definition:

$$\text{Safe bearing capacity} = \frac{\text{Bearing capacity of soil}}{\text{factor of safety}}$$

Soils behave like an elastic medium for small strains only. However, they are treated as elastic, isotropic, and homogeneous mediums for estimating settlements. A certain minimum depth of the foundation has to be provided keeping in view the effect of frost, erosion of soil, soil volume change due to moisture, organic nature of top soil, corrosion problem and other environmental considerations. As a rough guide the depth of footing in clay or sandy soil should not be less than 1.0. m

12.2. ISOLATED FOOTINGS OF UNIFORM THICKNESS

12.2.1. Pressure distribution beneath the footing.
A footing supporting a single column is termed as single or isolated footing. It may be of uniform thickness or sloped as shown in Fig.12.1 (a) and (b) respectively. It is also known as a spread footing.

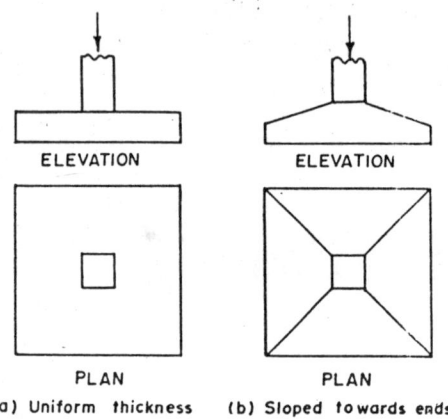

PLAN   PLAN

(a) Uniform thickness   (b) Sloped towards ends

Fig. 12.1 Isolated footing

In general, the pressure distribution beneath the symmetrically loaded footing is not uniform. Its variation depends upon the type of soil beneath and the elastic stiffness of the footing. On cohesionless material the grains tend to flow laterally under pressure at the edges, whereas, they remain confined at the centre. This results in a pressure distribution as shown in Fig. 12.2 (a). With covesive material high edge pressure is generated and the resulting pressure distrtbiution is shown in Fig. 12.2.(b). However,

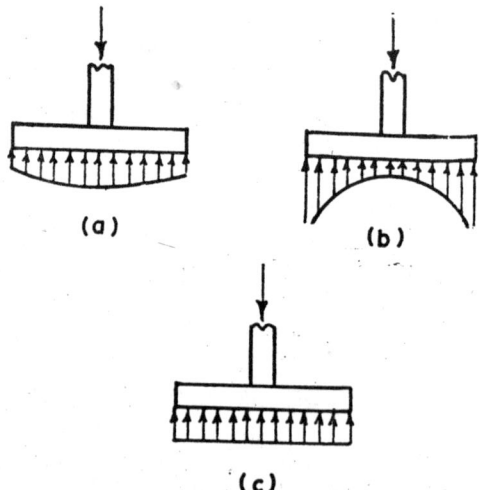

Fig. 12.2 Pressure distribution beneath the footing.

reinforced concrete footings are quite elastic with the result that the peak pressures are largely levelled off. Hence, it is a common practice to assume a linear pressure distribution beneath the footings, as shown in Fig.12.2. (c).

12.2. 2. Structural action and design. If the thickness of the footing is large compared with its lateral dimension, it may be designed as a plain concrete pedestal without the consideration of bending moment or shear force. Fig 12.3. The criterion is:

$$\text{Tan } \alpha \nless 0.9 \sqrt{\frac{100 \, q}{f_{ck}} + 1} \qquad (12.1)$$

Where

q = bearing pressure at the base of pedestal.

$$\text{Tan } \alpha = \frac{2D}{B\text{-}b}$$

No reinforcement is required in the pedestal. Only temperature and shrinkage reinforcement is needed.

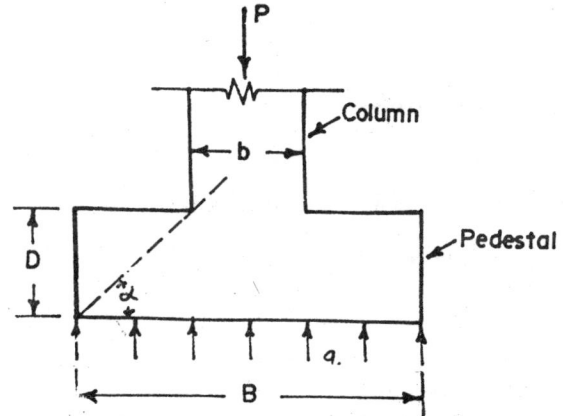

Fig. 12.3 Plain concrete pedestal footing

However, in footing in which the lateral dimension is large compared with the thickness, or more precisely, when angle α is smaller than that given by Eq. (12.1), the bending action is quite dominant and should be accounted for in the design along with the shear force. Hence, the footing will be of reinforced concrete. For its structural action please refer to the Fig. 12.4. The footing has a two way bending action. However, the bending moment in one direction is calculated at a section X-X which touches the face of column and extends for the full width of footing.

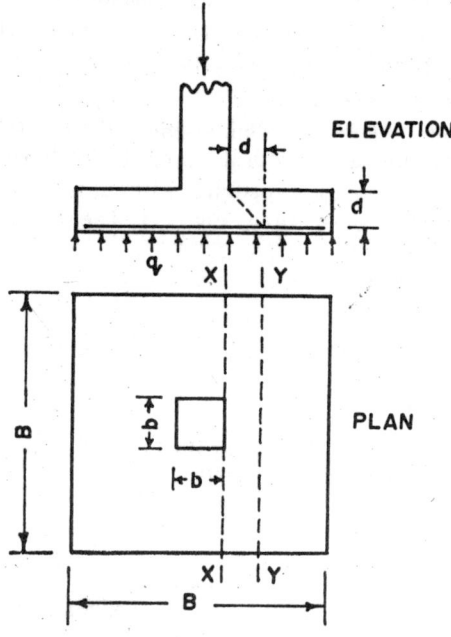

Fig. 12.4.

The bending moment at this section is

$$M = qB \frac{(B - b)}{2} \times \frac{(B - b)}{4}$$

$$= qB \frac{(B - b)^2}{8} \qquad (12.2)$$

Safety of the footing in shear is considered for two conditions:

(i) *One way action as a wide beam.*

The critical section for shear is taken along a section Y-Y distant an effective depth away from the face of the column and extending for the full width of footing. See Fig. 12.4 The shear force at this section is:

$$V = qB \frac{(B - b - 2d)}{2} \qquad (12.3)$$

Nominal shear stress is

$$\tau_v = \frac{V}{Bd}$$

$$= \frac{q(B - b - 2d)}{2d} \qquad (12.4)$$

Now,

if $\tau_v < \tau_c$ then the footing is safe in shear. Othewise, the depth of the slab may be increased to obtain increased shear strength.

(ii) *Two way shear action of footing slab.*

The potential diagonal cracking along the surface of a truncated pyramid below the concentrated load is illustrated in Fig 12.5 For the purpose of calculating the diagonal tension, the periphery of critical section is taken at a distance $\dfrac{d}{2}$ from the face of column.

Shear force is given by

$$V = q B^2 - q (b + d)^2$$
$$= q (B^2 - (b + d)^2)$$

Length of periphery for diagonal tension is

$$= 4 (b + d)$$

Hence, diagonal tension or shear stress is

$$= \frac{q (B^2 - (b + d)^2)}{4 (b + d)} \tag{12.5}$$

For safety in two way shear action, the calculated shear stress should be within the limits defined under para 12.10.3.

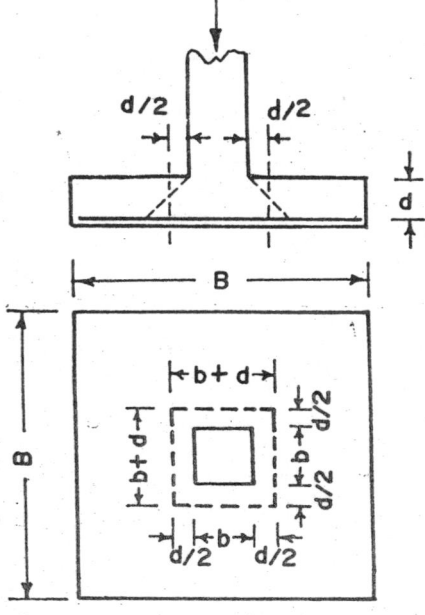

Fig. 12.5.

12.3. REQUIREMENT OF IS CODE: 456-1978

12.3.1. Lay-out of tension reinforcement. In a square footing the reinforcement in each direction shall be uniformly distributed aeross the full width. In a rectangular footing the reinforcement in the long direction shall be uniformly distributed. However, in the short direction a central band of width equal to the width of footing shall have a proportion of the total reinforcement uniformly distributed which is given as,

$$\frac{\text{Reinforcement in central band}}{\text{Total reinforcement in short direction}} = \frac{2}{\beta + 1}$$

where $\quad \beta = \dfrac{L}{B}$

The remainder reinforcement shall be uniformly distributed in the outer bands of the footing. See Fig. 12.6.

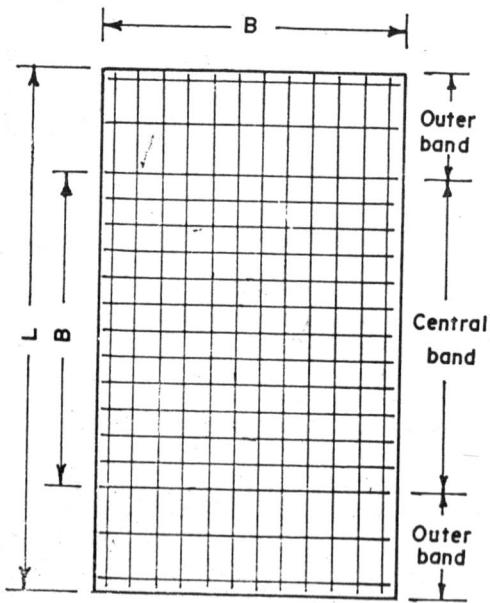

Fig. 12.6 Plan of reinforcement in a rectangular footing

12.3.2. Load transfer from the column to the footing. Transfer of the load from the base of column to the top of footing generates a bearing stress at the interface which should not exceed a value given by

$$0.25\ f_{ck}\ \sqrt{\frac{A_1}{A_2}}$$

where

$$\sqrt{\frac{A_1}{A_2}} \not> 2.0 \text{ in any case.}$$

A_1 = the area of the lower base of the largest frustum of a pyramid contained within the footing and having for its upper base the area loaded, and having the side slope of 1:2. See Fig. 12.7.

A_2 = loaded area at the column base.

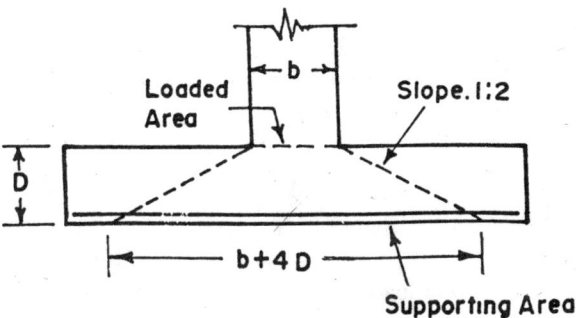

Fig. 12.7

Where the actual bearing stress at the interface exceeds the limit given above, reinforcement shall be provided for taking up the excess force. This may be achieved by extending the longitudinal bars into the footing, or by means of dowel bars. See Fig. 12.8.

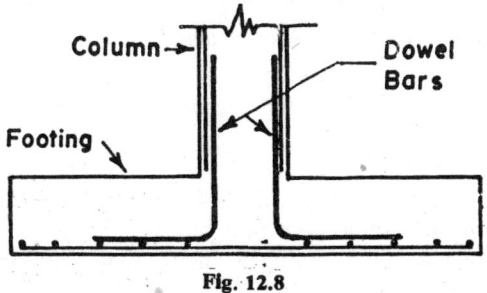

Fig. 12.8

The sectional area of extended longitudinal reinforcement or dowels shall not be less than 0.5% cf cross-sectional area of supported column. At least 4 bars shall be provided. Diameter of dowel bars shall not exceed that of column bars by more than 3 mm. The dowel bar shall extend into the column for a distance equal to the development length of the column bar and into the footing for a distance equal to the development length of the dowel bar.

12.3.3 Minimum thickness at the edge. The thickness at the edge shall not be less than 150 mm for footing resting on soil, and 300 mm for footing on piles

Example 12.1

Design an isolated square footing of uniform thickness for a R.C.C. column having a cross-section of size 400 × 400 mm. Load transmitted by column to the footing = 750 kN. Safe bearing capacity of soil = 180 kN/m². Depth of foundation below G.L. = 1 m. Concrete grade: M 20 for column and M 15 for footing, and steel grade: Fe 415 for both. Main reinforcement in column is 20 mm dia. bars — 4 nos.

Solution:

$$\begin{array}{ll} \text{Load from column} & = 750 \text{ kN} \\ \text{Estimated weight of foundation} & = \ \ 75 \text{ kN} \\ \hline \text{Total load} & = 825 \text{ kN} \end{array}$$

$$\text{Area of footing} = \frac{825}{180}$$

$$= 4.583 \text{ m}^2.$$

$$\text{Side of footing} = \sqrt{4.583}$$

$$= 2.141 \text{ m}.$$

Adopt 2.15 × 2.15 mm footing in plan.

$$\text{Net upward soil pressure} \quad q = \frac{750}{2.15 \times 2.15}$$

$$= 162.25 \text{ kN/m}^2.$$

Thickness of the footing will now be computed.
Bending moment consideration (Eq. 12.2):
Bending moment in the footing at the face of column is

$$M = \frac{qB\,(B-b)^2}{8}$$

$$= 162.25 \times 2.15 \times \frac{(2.15 - 0.4)^2}{8}$$

$$= 133.54 \text{ kN-m.}$$

$\sigma_{cbc} = 5 \text{ N/mm}^2.$

$\sigma_{st} = 230 \text{ N/mm}^2.$

$m = 18.67$

Design constants are:

$n = 0.289$

$j = 0.904$

$R = 0.653$

$$\text{Effective depth} = \sqrt{\frac{M}{Rb}} = \sqrt{\frac{133.54 \times 10^6}{0.653 \times 2150}}$$

$$= 308.41 \text{ mm.}$$

Two-way shear consideration (Eq. 12.5):

$$\text{Shear stress } \tau = \frac{q\,(B^2 - (b+d)^2)}{4\,(b+d)\,d}$$

$$= \frac{162.25(2.15^2 - (0.4+d)^2)}{4\,(0.4+d)\,d}$$

$$= 40.5625 \frac{(4.4625 - 0.8\,d - d^2)}{(0.4+d)\,d} \text{ kN/m}^2.$$

As per para 8.10.3 permissible shear stress is found as follows:

$$\beta_c = \frac{400}{400} = 1.0$$

$$k_s = 0.5 + \beta_c = 0.5 + 1.0 = 1.5 > 1.0$$

$$\therefore k_s = 1.0$$

$$\tau_c = 0.16\sqrt{f_{ck}} = 0.16\sqrt{15} = 0.62 \text{ N/mm}^2$$

Hence, permissible shear stress

$$= k_s\,\tau_c = 1.0 \times 0.62 = 0.62 \text{ N/mm}^2$$

$$\text{or } 620 \text{ kN/m}^2$$

For safety

$$\tau \leqslant 620 \text{ kN/m}^2$$

or $$\frac{40.5625\,(4.4625 - 0.8\,d - d^2)}{(0.4+d)\,d} \leqslant 620 \text{ kN/m}^2$$

or $d = 0.353$ m or 353 mm.

The effective depth calculated from two-way shear consideration is more than that calculated from bending consideration.

Adopting the higher value and rounding off

d $= 355$ mm.

Area of reinforcement is

$$A_{st} = \frac{133.54 \times 10^6}{230 \times .904 \times 355} = 1808.7 \text{ mm}^2.$$

12 mm dia. bars - 16 nos. may be provided in each of the two orthogonal directions. Then

$$\frac{100 \, A_{st}}{bd} = \frac{100 \times 16 \times 133.1}{2150 \times 355} = 0.2371$$

From Table 2.11

$$\tau_c = 0.22 \text{ N/mm}^2 \text{ or } 220 \text{ kN/m}^2.$$

One-way shear consideration (Eq. 12.4):

$$\tau_v = \frac{q \, (B - b - 2 \, d)}{2d}$$

$$= \frac{162.25 \, (2.15 - 0.4 - 2 \times 0.355)}{2 \times 0.355}$$

$$= 237.65 \text{ kN/m}^2 \text{ or } 0.2377 \text{ N/mm}^2.$$

Since $\tau_v > \tau_c$ the footing is unsafe in shear. Increased depth may be found by equating

$$\frac{q \, (B - b - 2 \, d)}{2 \, d} < \tau_c$$

or $\qquad \dfrac{162.25 \, (2.15 - 0.4 - 2 \, d)}{2 \, d} < 220$

or $\qquad d \geqslant 0.371$ m or 371 mm.

Finally, adopt d $= 375$ mm.

Tension reinforcement for this is:

$$A_{st} = \frac{133.5 \times 10^6}{230 \times .904 \times 375} = 1712.7 \text{ mm}^2.$$

Provide 16 mm dia. bars — 9 nos. is each of the two orthogonal directions. An overall depth of 425 mm will be appropriate.

Check for bearing stress:

From Fig. 12.7

$$A_1 = (b + 4 \, D)^2 = (0.4 + 4 \times 0.425)^2 = 2.1^2$$
$$= 4.41 \text{ m}^2$$
$$A_2 = 0.4^2.$$
$$= 0.16 \text{ m}^2$$

$$\sqrt{\frac{A_1}{A_2}} = \sqrt{\frac{4.41}{0.16}} = 5.25 > 2.0$$

Hence, take $\sqrt{\dfrac{A_1}{A_2}} = 2.0$

Permissible bearing stress

$$= 0.25 \, f_{ck} \sqrt{\dfrac{A_1}{A_2}}$$

$$= 0.25 \times 15 \times 2$$

$$= 7.5 \text{ N/mm}^2$$

Actual bearing stress

$$= \dfrac{750 \times 10^3}{400 \times 400}$$

$$= 4.69 \text{ N/mm}^2$$

This is less than the permissible value. Hence, safe.

Dowel bars:

Nominal dowel bars may be provided.

Sectional area $= \dfrac{0.5 \times 400 \times 400}{100} = 800 \text{ mm}^2$.

16 mm dia. dowel bars — 4 nos. may be provided.
Development lengths:

For main bar in column

$$l_d = \dfrac{\phi \, \sigma_s}{4 \, \tau_{bd}}$$

$$= \dfrac{20 \times 190}{4 \times 1.25 \times 1.4 \times 0.8}$$

$$= 678.57 \text{ mm}$$

For dowel bar in footing

$$l_d = \dfrac{16 \times 190}{4 \times 1.25 \times 1.4 \times 0.6}$$

$$= 724 \text{ mm}$$

The dowel bar may be extended into the column for 680 mm and into the footing for 725 mm. A 75 mm thick base of lime concrete (1:4:8) may be provided below the footing. Details of reinforcement are shown in Fig. 12.9.

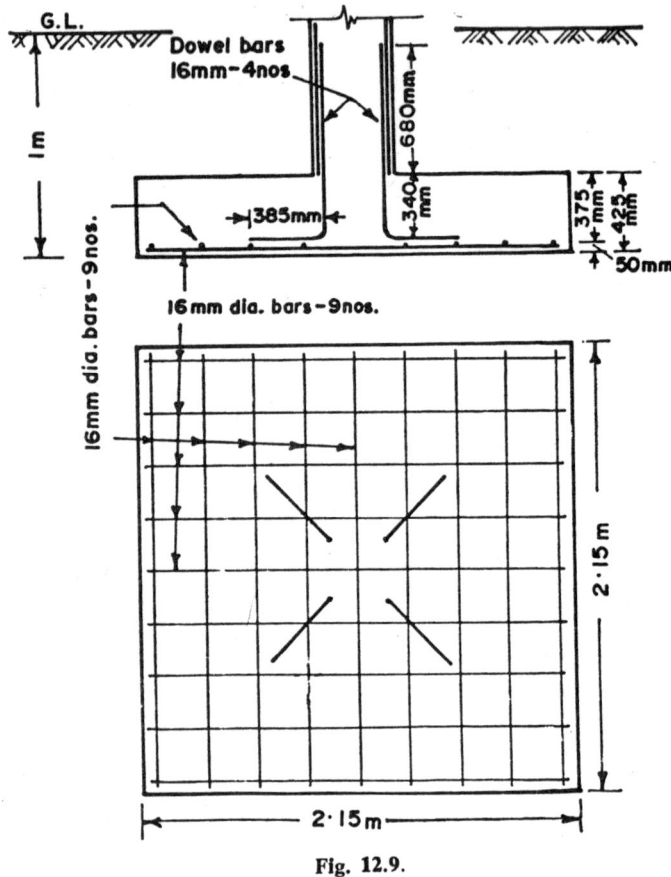

Fig. 12.9.

12.4 ISOLATED FOOTING WITH THICKNESS REDUCING TOWARDS EDGES

Towards the edges of the footing both bending moment and shear force sharply decrease. As such, the thickness can also be reduced. However, a minimum thickness of 150 mm has to be maintained at the edges. The section of the footing resisting bending moment near the face of column, has the compressive area of concrete reducing towards the top. See Fig. 12.10.

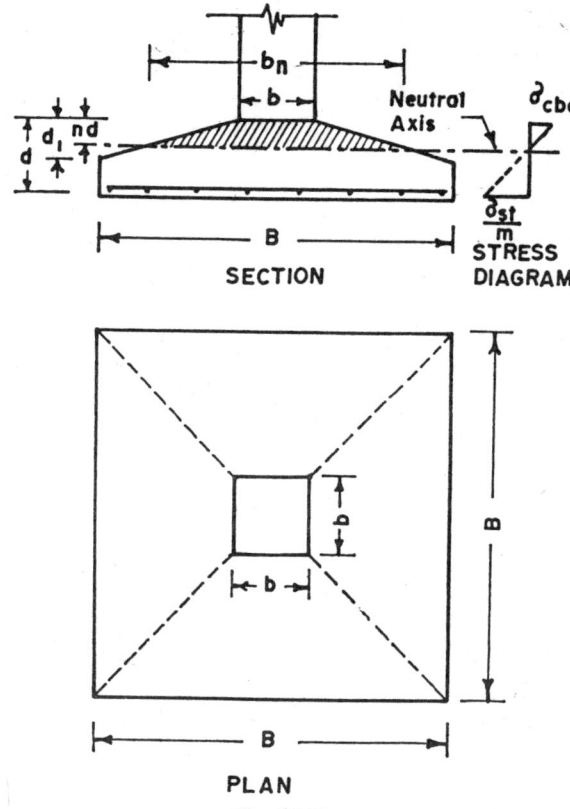

SECTION

STRESS DIAGRAM

PLAN

Fig. 12.10

12.4.1. Bending strength. Width of section at neutral axis is

$$b_n = b + (B - b) \ \frac{nd}{d_1} \tag{12.6}$$

Considereding the section to be balanced, total compressive force can be written as

$$C = bnd \ \frac{\sigma_{obc}}{2} + \frac{(B - b)}{d_1} \ n^2d^2 \ \frac{\sigma_{obc}}{6} \tag{12.7}$$

Tensile force is written as

$$T = A_{st} \ \sigma_{st} \tag{12.8}$$

For equilibrium $T = C$

$$\therefore A_{st} = \frac{C}{\sigma_{st}} \tag{12.9}$$

The moment of resistance can be found by taking the moment of the compresive force about the tension reinforcement

$$M = Rbd^2 + \frac{(B-b)}{d_1} n^2 (2-n) d^3 \frac{\sigma_{cbc}}{12} \qquad (12.10)$$

Steps for calculation:

Take a trial value for d.

Then $d_1 = d +$ concrete cover — edge thickness.

Calculate moment of resistance M from Eq. (12.10)

For safety, moment of resistance should be equal to or a little more than the bending moment, i.e. $M \geqslant$ B.M.

Compute C from Eq. (12.7)

Finally, compute A_{st} from Eq. (12.9).

12.4.2 Check for Shear. The depth of footing calculated for bending moment should be checked for one-way shear and two-way shear action as per the procedure described under para 12.2.2. However, reduced effective depth should be used for critical sections away from the column face.

Example 12.2

Redesign the isolated square footing of Example 12.1, with thickness reduced towards the edges to 250 mm. Concrete grade: M 20 and steel grade: Fe 415.

Solution:

Bending moment at column face = 133.54 kN-m as before.

Try an effective depth d   = 475 mm

Let the concrete cover   = 50 mm

Then

$$d_1 = 475 + 50 - 250$$
$$= 275 \text{ mm}$$
$$b = 400 \text{ mm, the width of column.}$$

Now

$$\sigma_{cbc} = 7 \text{N/mm}^2, \qquad \sigma_{st} = 230 \text{ N/mm}^2$$
$$n = 0.289, \; j = 0.904, \quad R = 0.914$$

Then, from Eq. (9.10), moment of resistance

$$M = 0.914 \times 400 \times 475^2 + \frac{(2150 - 400)}{275} \times 0.289^2$$
$$\times (2 - 0.289) \times 475^3 \times \frac{7}{12}$$

$= 82.49 \times 10^6 + 56.85 \times 10^6$ N-mm

$= 139.34$ kN-m > 133.54 kN-m, O.K.

From Eq. (12.7)

$$C = 400 \times 0.289 \times 475 \times \frac{7}{2} \times \frac{(2150 - 400)}{275}$$

$$\times 0.289^2 \times 475^2 \times \frac{7}{6}$$

$= 332.1 \times 10^3$ N

From Eq. (12.9)

$$A_{st} = \frac{332.1 \times 10^3}{230} = 1443.9 \text{ mm}^2$$

Provide 12 mm dia. bass — 13 nos. in each of the two orthogonal directions.

From Eq. (12.6) width of section at neutral axis

$$= 400 + (2150 - 400) \times \frac{0.289 \times 475}{275}$$

$= 1273.6$ mm.

Average width of compressive area

$= \frac{1}{2} (400 + 1273.6)$

$= 836.8$ mm.

$$\frac{100 \, A_{st}}{bd} = \frac{100 \times 13 \times 113.1}{836.8 \times 475} = 0.37$$

From Table 3.11

$\tau_o = 0.258$ N/mm^2

Check for one-way shear:

The critical section occurs an effective depth away from the column face.

Here, effective depth

$$d = 475 - \frac{(475 - 200)}{875} \times 475$$

$= 325.7$ mm.

Shear force at this section from Eq.(12.3) is

$$V = 162.25 \times 2.15 \frac{(2.15 - 0.4 - 2 \times 0.475)}{2}$$

$= 139.54$ kN.

∴ Nominal shear stress

$$\tau v = \frac{139.54 \times 10^3}{2150 \times 325.7} = 0.20 \text{ N/mm}^2$$

Evidently, $\tau_v < \tau_c$. Hence, the footing is safe.

Check for two-way shear:

Critical section occurs half effective depth away from column face all around. Magnitude of shear force

$$V = q (B^2 - (b + d)^2)$$
$$= 162.25 (2.15^2 - (0.4 + 0.475)^2)$$
$$= 625.8 \text{ kN.}$$

Periphery of critical section

$$b_o = 4 (b + d)$$
$$= 4 (0.4 + 0.475)$$
$$= 3.5 \text{ m}$$

Hence, shear stress is

$$\tau = \frac{V}{b_o d} = \frac{625.8 \times 10^3}{3.5 \times 10^3 \times 325.7}$$
$$= 0.549 \text{ N/mm}^2$$

Permissible shear stress, as already worked out in Example 12.1. is $= 0.62$ N/mm² which is more than the actual shear stress. Hence, the footing is safe.

Check for bearing stress has already been accomplished in Example 12.1. It will not be repeated here.

Development length of dowel bar in footing concrete

$$l_d = \frac{16 \times 190}{4 \times 1.25 \times 1.4 \times 0.8}$$
$$= 543 \text{ mm, say } 545 \text{ mm.}$$

Details of reinforcement are shown in Fig.12.11.

12.5 COMBINED FOOTING FOR TWO COLUMNS

Design of combined footings for two columns may be needed in the following instances:

(i) Column spacing may be close such that separate footings may not be possible, or a combined footing may be more economical than two isolated footings.

(ii) Column spacing may not be very close but there may be restrictions on the lengthwise projection of the footing beyond the face of one or both the columns on account of property line, or there may be a restriction on the width of footing considering the overall plan of building structure.

(iii) Practical considerations based on the problem of differential settlement as well as economy may necessitate a combined footing.

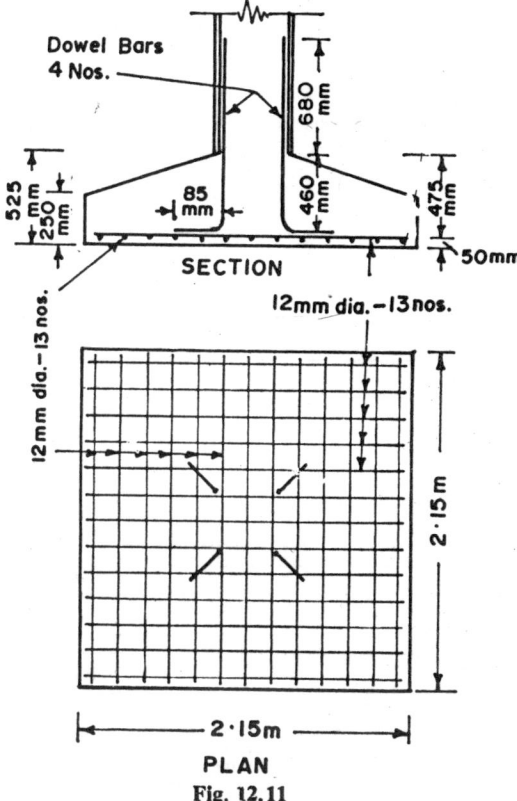

PLAN
Fig. 12.11

A rectangular combined footing may be provided for two equally loaded columns as well as for two unequally loaded columns. In the latter case, a uniform upward pressure can be achieved if the centroid of the footing area coincides with the centre of gravity of the two column loads. The footing may be designed as a single R.C.C. slab resting on soil and supporting the two columns, as shown in Fig.12.12(a). As an alternative, a longitudinal beam may be provided for the full length of the footing supporting the two columns, while the footing slab may cantilever out on the two sides over the soil, as shown in Fig.12.12 (b).

When one of the column loads is much heavier than the other, or if the former is an outer column such that little space is available within the property line, then a trapezoidal footing is called for. It

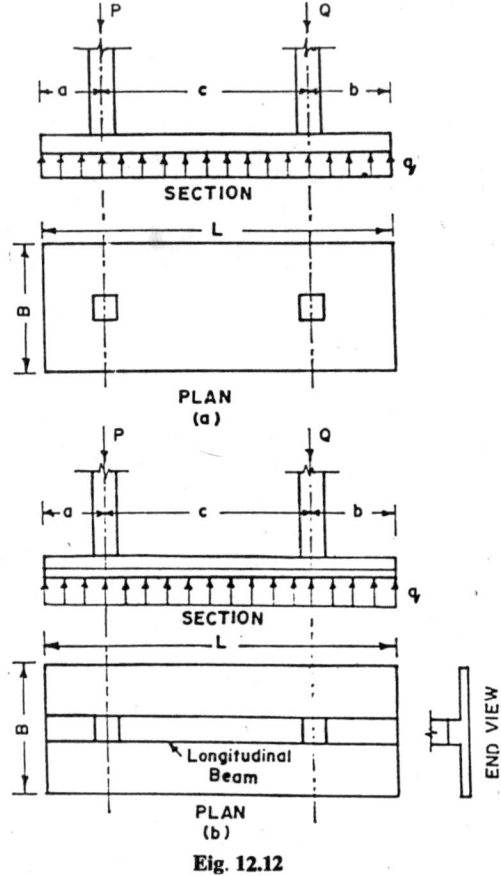

Fig. 12.12

may be possible in most of the cases to achieve coincidence of the centroid of footing area and the centre of gravity of the column loads, so that the upward soil pressure may be uniform. Footing having a single trapezoidal R.C.C. slab supporting two columns and resting on soil is shown in Fig. 12. 13(a). Footing having a longitudinal beam supporting the two columns from which the R.C.C. trapezoidal footing slab cantilevers out in the transverse direction, is shown in Fig.12.13(b).

Where trapezoidal footing is not practically possible due to large distance between two columns, each column may be provided with a separate square or rectangular footing which may be

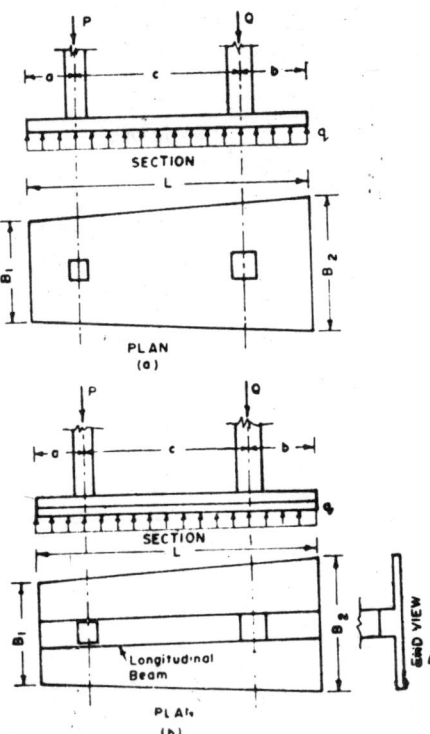

Fig. 12.13

connected together by a longitudinal strap beam. Fig. 12.14 illustrates such a footing in which the heavier column is located on the property line. As such, the footing under it is placed eccentrically due to space limitation, giving rise to a moment which is balanced by the load and reaction on the inner footing, while the strap beam acts as a lever. The footings under each of the two columns will have uniform upward soil pressure, though, the intensities of pressure may be different. The soil below the strap beam is kept loose so that there is no upward soil pressure acting on it. As such, the strap beam may be designed for the bending moment due to levering action only. The designs can be best illustrated by the following examples.

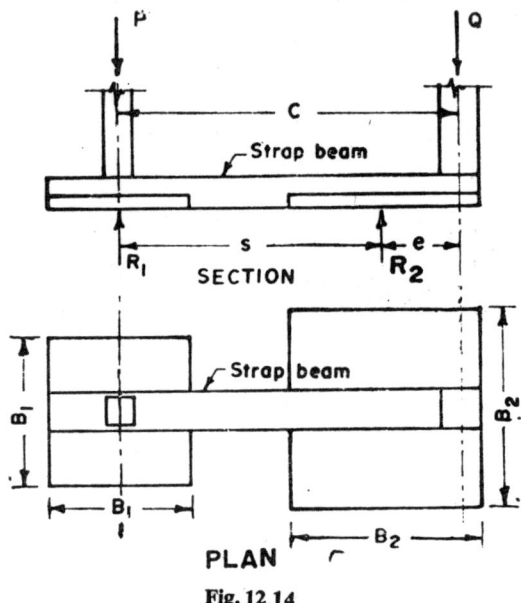

Fig. 12.14

Example 12.3

Two R.C.C. columns of a building have the same cross section i.e. 400 x 400 mm, and are spaced 2.8 m apart. Each column carries a design axial load of 720 kN. Main reinforcement in a column is 20 mm dia. bars — 4 nos. Safe bearing capacity of soil = 100 kN/m². Depth of foundation = 1 m below G.L. Design suitable foundation for the columns. Concrete grade: M 15 and Steel grade: Fe 415.

Solution:

First consider a single column with an independent square footing.

Load transmitted by a column   720 kN

Estimated self weight of foundation = 72 kN.

$$\text{Total load} \quad = 792 \text{ kN.}$$

$$\text{Area of foundation required} = \frac{792}{100}$$

$$= 7.92 \text{ m}^2.$$

$$\text{Length of side of foundation} = \sqrt{7.92}$$

$$= 2.81 \text{ m.}$$

Since the two columns are located 2.8 m centre to centre from each other, their independent foundations will overlap. As such, it would be more economical to provide a combined rectangular footing supporting the two columns together.

Footing area:

Load transmitted by two columns = 2 x 720

= 1440 kN

Estimated self weight of footing = 144 kN

Total load = 1584 kN

$$\text{Area of foundation required} = \frac{1584}{100}$$

$$= 15.84 \text{ m}^2$$

Provide a footing of size 5.66m x 2.83m.

$$\text{Net upward soil pressure} = \frac{2 \times 720}{5.66 \times 2.83}$$

$$= 89.9 \text{ kN/m}^2$$

The footing will be designed as a R.C.C. slab of uniform width and thickness resting on soil and supporting the two columns. See Fig.12.15. Consider a section X-X of the footing distant X from the shorter edge on left. The shear force and the bending moment at this section can be written as follows:

For $X < 1.43$ m.

$F_x = 89.9 \times 2.83 \text{ X}$

$= 254.42 \text{ X kN}.$

$$M_x = 89.9 \times 2.83 \frac{X^2}{2}$$

$$= 127.21 \text{ X}^2 \text{ kN-m}.$$

At $X = 1.43$ m, $F_x = 363.82$ kN and $M_x = 260.13$ kN-m

For $1.83 \text{ m} < X < 4.43$ m

$F_x = 89.9 \times 2.83 \text{ X} - 720$

$= 254.42 \text{ X} - 720 \text{ kN}.$

$$M_x = 89.9 \times 2.83 \frac{X^2}{2} - 720 (X - 1.63)$$

$$= 127.21 \text{ X}^2 - 720 (X - 1.63)$$

At $X = 1.83$ m, $F_x = 254.42$ kN and $M_x = 282.01$ kN-m.

At $X = 2.83$ m, $F_x = 0$, and $M_x = 154.81$ kN-m.

Diagrams for shear force and bending moment are also shown in Fig.12.15

Bending moment in transverse direction is found about the section Y-Y touching the two column faces. It is

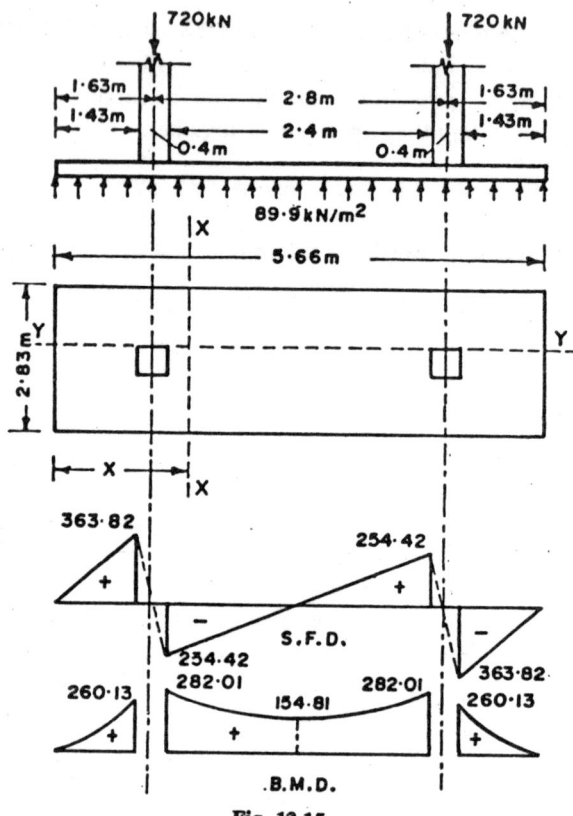

Fig. 12.15

$$= 89.9 \times 5.66 \times \frac{1.215^2}{2}$$

$$= 375.58 \text{ kN-m in a width of 5.66 m.}$$

Calculation for effective depth:

Consider bending moment per meter width.

$$\text{In longitudinal direction} = \frac{282.01}{2.83} = 99.65 \text{ kN-m/m.}$$

$$\text{In transverse direction} = \frac{375.58}{5.66} = 66.36 \text{ kN-m/m.}$$

Permissible stress and the design constants will be the same as in Example 12.1.

Taking the larger of the two values of bending moments

$$\text{Effective depth} = \sqrt{\frac{99.65 \times 10^6}{1000 \times 0.653}} = 390.64 \text{ mm, say 395 mm.}$$

Area of tension reinforcement in a width of 2.83 m

$$A_{st} = \frac{282.01 \times 10^6}{230 \times .904 \times 395} = 3433.77 \text{ mm}^2.$$

$$\frac{100 A_{st}}{bd} = \frac{100 \times 3433.77}{2830 \times 395} = 0.307$$

From Table 2.11

$$\tau_c = 0.236 \text{ N/mm}^2.$$

Check for one-way shear:

At the critical section located 0.395 m away from the column face towards the end

$$\begin{aligned} \text{S.F.} &= 363.82 - 89.9 \times 2.83 \times 0.395 \\ &= 263.33 \text{ kN.} \end{aligned}$$

Nominal shear stress due to this

$$\tau_v = \frac{263.33 \times 10^3}{2830 \times 395} = 0.236 \text{ N/mm}^2.$$

Since $\tau_v = \tau_c$, the footing slab Is just safe in shear.

Check for two-way shear:

Periphery of the critical section

$$\begin{aligned} b_0 &= 4 (b + d) = 4 (0.4 + 0.395) \\ &= 3.18 \text{ m.} \end{aligned}$$

Shear force, causing diagonal tension

$$\begin{aligned} V &= 720 - (b + d)^2 \times 89.9 \\ &= 720 - (0.4 + 0.395)^2 \times 89.9 \\ &= 663.2 \text{ kN.} \end{aligned}$$

Hence, diagonal tension, or shear stress

$$\tau = \frac{V}{b_0 d} = \frac{663.2 \times 10^3}{3180 \times 395} = 0.528 \text{ N/mm}^2$$

Permissible shear stress will be the same as in Example 12.1. i.e $= 0.62 \text{ N/mm}^2.$

Evidently, the footing is safe in two way shear.

An overall thickness of 430 mm may be adopted. The footing will be provided with a lime concrete base of thickness $= 75$ mm below it.

Calculation of tension reinforcement:

In longitudinal direction:

Below the column, heavier bending moment lies at the section on the right face. Its value $= 282.01$ kN-m.

$$A_{st} = \frac{282.01 \times 10^6}{230 \times .904 \times 395} = 3433.77 \text{ mm}^2.$$

Provide 16 mm dia. bars — 17 nos.

At mid-span also the nature of the bending moment is same, though the magnitude is lesser i.e. = 154.81 kN-m. However, it is proposed not to curtail the reinforcement towards the mid-span, and to continue all the 17 bars throughout the length of the footing.

In transverse direction:

Bending moment = 375.58 kN-m as calculated earlier.

Keeping transverse reinforcement above the longitudinal reinforcement

$$A_{st} = \frac{375.58 \times 10^6}{230 \times .904 \times 379} = 4766.4 \text{ mm}^2.$$

Provide 16 mm dia. bars — 24 nos.

Calculations for bearing stress will be exactly same as done in Example 12.1. Hence, the dowel bars will also be the same.

Details of reinforcement are shown in Fig.12.16.

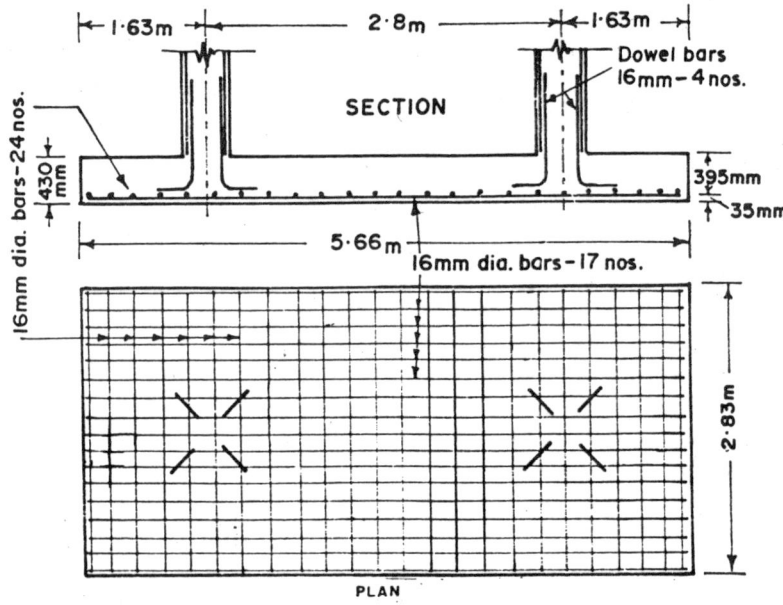

Fig. 12.16

Example 12.4

Two columns of a building structure are spaced 2.7 m on centres. The inner column is 400 × 400 mm in section and carries an axial load of 600 kN. The outer column is 500 × 500 mm in section and carries an axial load of 750 kN. The property line is at 1.25 m from the centre of the outer column. Safe bearing capacity of soil = 100 kN/m² Concrete grade: M 15 and steel grade: Fe 415. Design a suitable footing. Depth of footing below G.L. = 1 m.

Solution:

Even if there were no restriction of space due to the proximity of property line the two independent square footings for the two columns would have overlapped. Now that the property line is near the outer column an independent footing for it will be very much restricted in width. As such, it will be appropriate to provide a combined footing for the two columns. Since, the two columns have different loads, the footing area will be trapezoidal in shape. The line of action of the resultant of the two columon loads can be made to coincide with the centroid of the footing area, in order to give a uniform pressure distribution.

$$\text{Load from the two columns} = 600 + 750$$
$$= 1350 \text{ kN.}$$
$$\text{Estimated self weight of footing} = 135 \text{ kN.}$$
$$\text{Total load} = 1485 \text{ kN.}$$
$$\text{Area of footing required} = \frac{1485}{100}$$
$$= 14.85 \text{ m}^2.$$

Let the footing extend in the longitudinal direction by 1.25 m outwards from the centres of each of the two columns. Hence, the length of the footing is

$$= 1.25 + 2.7 + 1.25$$
$$= 5.2 \text{ m.}$$

The distance of the resultant load from the centre of the heavier column is

$$= 2.7 \times \frac{600}{1350}$$
$$= 1.2 \text{ m.}$$

Now

B_1 = length of the smaller end
B_2 = length of the larger end.

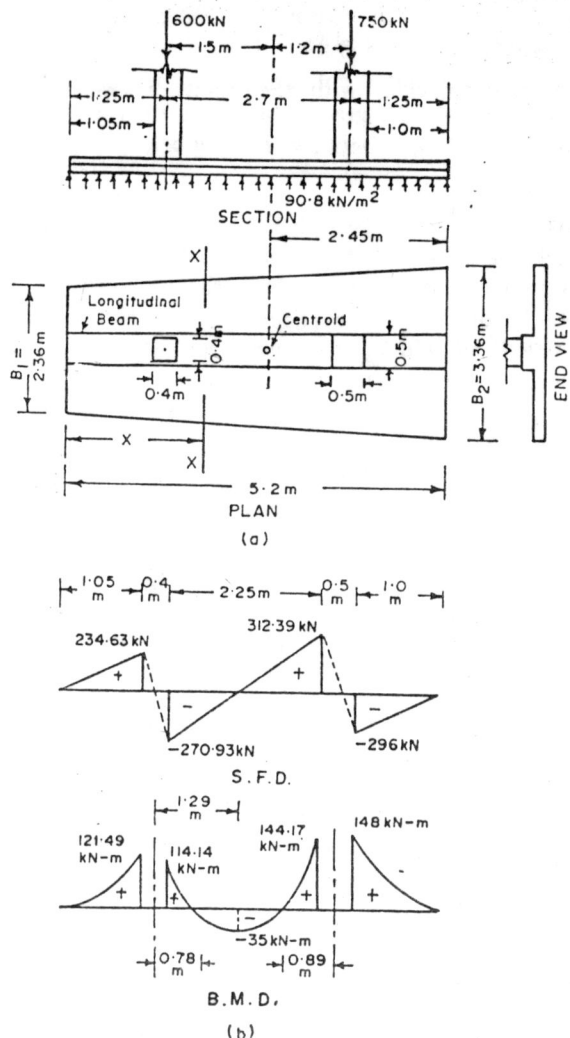

Fig..12.17

As distance of the resultant load from the longer end

$$= 1.2 + 1.25$$

$$= 2.45 \text{ m.}$$

The dimensions B_1 and B_2 are worked out as follows:

Equating the area

$$\frac{(B_1 + B_2)}{2} \times 5.2 = 14.85$$

As the centroid coincides with the point of application of the resultant load

$$\frac{(2B_1 + B_2)}{(B_1 + B_2)} \times \frac{5.2}{3} = 2.45$$

These give

$B_1 = 2.36$ m.

$B_2 = 3.36$ m.

Actual area of footing $= \dfrac{2.36 + 3.36}{2} \times 5.2$

$= 14.87$ m²

Net upward soil pressure $= \dfrac{600 + 750}{14.87}$

$= 90.8$ kN/m².

The schemetic diagram of the foundation is shown in Fig.12.17(a). A longitudinal beam has been proposed for the full length of the footing supporting the two columns. The footing slab cantilevers out on the two sides from the beam as shown in the figure. The bending moment and the shear force in the beam can now be worked out. Consider a section X-X distant X from the smaller end.

Width of this section

$$= 2.36 + \frac{(3.36 - 2.36)\, X}{5.2}$$

$$= 2.36 + 0.1923\, X$$

Area of footing upto the section

$$= \tfrac{1}{2} \times (2.36 + 2.36 + 0.1923\, X)\, X$$

$$= 2.36\, X + 0.09615\, X^2$$

Distance of centroid of this area from the section

$$\frac{(2 \times 2.36 + 2.36 + 0.1923\, X)}{(2.36 + 2.36 + 0.1923\, X)} \times \frac{X}{3}$$

$$= \frac{7.08 + 0.1923\, X^2}{3\,(4.72 + 0.1923\, X)}$$

Now

$F_x =$ shear force at the section

$M_x =$ bending moment at the section

Then

for $X < 1.05$ m.

$$F_x = (2.36\ X + 0.09615\ X^2) \times 90.8\ kN.$$

$$M_x = (2.36\ X + 0.09615\ X^2)\ \frac{(7.08\ X + 0.1923\ X^2)}{3\ (4.72 + 0.1923X)} \times 90.8$$

$$= (7.08\ X^2 + 0.1923\ X^3) \times 15.133\ kN\text{-}m.$$

At $X = 1.05$ m, $F_x = 234.63$ kN, $M_x = 121.49$ kN-m.

for $1.45 < X < 3.7$ m.

$$F_x = (2.36X + 0.09615\ X^2) \times 90.8 - 600\ kN.$$

$$M_x = (7.08\ X^2 + 0.1923\ X^3) \times 15.133 - 600\ (X - 1.25)$$
$$kN\text{-}m.$$

At $X = 1.45$, $F_x = 270.93$ kN, $M_x = 114.14$ kN-m.

At $X = 3.7$ m, $F_x = 312.39$ kN, $M_x = 144.47$ kN-m.

Maximum - ve value of M_x occurs at $X = 2.54$ m. Its value
$$= - 35\ kN\text{-}m$$

Points of zero bending moments occur at $X = 2.03$ m and $X = 3.06$ m.

for $4.2 < X < 5.2$ m.

$$F_x = (2.36\ X + 0.09615\ X^2) \times 90.8 - 600 - 750$$

$$M_x = (7.08\ X^2 + 0.1923\ X^3) \times 15.133 - 600\ (X - 1.25)$$
$$- 750\ (X - 3.95)$$

At $X = 4.2$ m, $F_x = - 296$ kN, $M_x = 148$ kN-m

The bending moments and shear force diagrams are shown in Fig.12.17(b).

Design of foundation slab:

Consider a strip of 1 m width of the slab projecting in transverse direction from the side of longitudinal beam.

Maximum length of strip at the longer end
$$= \tfrac{1}{2} (3.36 - 0.5)$$
$$= 1.43\ m.$$

Bending moment in the cantilever strip
$$= 90.8 \times \frac{1.43^2}{2}$$
$$= 92.84\ kN\text{-}m/m.$$

Permissible stress are
$$\sigma_{cbc} = 5\ N/mm^2,\ \sigma_{st} = 230\ N/mm^2\ and\ m = 18.67$$

Design constants are
$$n = 0.289,\ j = 0.904,\ R = 0.653$$

Effective depth $d = \sqrt{\dfrac{92.84 \times 10^6}{1000 \times 0.653}}$

$= 377$ mm, say 380 mm.

Tension reinforcement

$A_{st} = \dfrac{92.84 \times 10^6}{230 \times 0.904 \times 380} = 1175$ mm^2/m

12 mm dia. bars may be provided in transverse direction.

Spacing of bars

$= \dfrac{1000}{1175} \times 113.1 = 96.25$ mm, say 95 mm c/c

Actual $A_{st} = \dfrac{113.1 \times 1000}{95} = 1190.5$ mm^2/m.

Check for shear:

Consider a strip of footing slab of 1 m width projecting from the side of the longitudinal beam, near the larger end. Critical section occurs 0.38 m away from the side of the beam.

Shear force $V = 90.8 \times (1.43 - 0.38)$

$= 95.34$ kN.

Nominal shear stress

$\tau_v = \dfrac{95.34 \times 10^3}{1000 \times 380} = 0.25$ N/mm^2

$\dfrac{100\, A_{st}}{bd} = \dfrac{100 \times 1190.5}{1000 \times 380} = 0.313$

From Table 3.11

$\tau_c = 0.238$ N/mm^2

Since $\tau_v > \tau_c$, the slab is unsafe in shear.

The effective depth may be increased.

Increased effective depth is

$d = 380 \times \dfrac{0.25}{0.238} = 399.2$ mm, say 400 mm.

Then, overall depth = 435 mm.

Use balanced proportion of reinforcement

$p = 0.00314$ from Table 2.5.

Then, $A_{st} = 0.00314 \times 1000 \times 400$

$= 1256$ mm^2.

Provide 12 mm dia. bars — 90 mm c/c.

$\dfrac{100\, A_{st}}{bd} = 0.314$

From Table 3.11, $\tau_0 = 0.238$ N/mm².
Shear force at critical section
$$V = 90.8 \ (1.43 - 0.4)$$
$$= 93.52 \text{ kN.}$$
Nominal shear stress
$$\tau_v = \frac{93.52 \times 10^3}{1000 \times 4000} = 0.234 \text{ N/mm}^2$$
Since $\tau_v < \tau_c$ the section is safe.

Near the smaller end, projection of slab from the side face of longitudinal beam
$$= \tfrac{1}{2} \ (2.36 - 0.50)$$
$$= 0.93 \text{ m.}$$
Bending moment on a cantilever strip of 1 meter width
$$= 90.8 \times \frac{0.93^2}{2}$$
$$= 39.27 \text{ kN-m.}$$
Tension reinforcement
$$A_{st} = \frac{39.27 \times 10^6}{230 \times 0.904 \times 400}$$
$$= 472.2 \text{ mm}^2/\text{m.}$$
Temperature and shrinkage reinforcement
$$= \frac{0.12}{100} \times 435 \times 1000$$
$$= 522 \text{ mm}^2/\text{m} > 472.2 \text{ mm}^2/\text{m.} \quad \text{O.K.}$$
Provide 12 mm dia. bars — 215 mm c/c.
$$\text{Actual } A_{st} = \frac{113.1 \times 1000}{215}$$
$$= 526 \text{ mm}^2/\text{m.}$$
$$\frac{100 \ A_{st}}{bd} = \frac{100 \times 526}{1000 \times 400} = 0.132$$
From table 3.11, $\tau_0 = 0.22$ N/mm².
Shear force at critical section
$$V = 90.8 \ (0.93 - 0.4)$$
$$= 48.12 \text{ kN.}$$
Nominal shear stress
$$\tau_v = \frac{48.12 \times 10^3}{1000 \times 400}$$
$$= 0.12 \text{ N/mm}^2.$$
As $\tau_v < \tau_c$, the section is safe in shear.

Hence, the transverse reinforcement will be 12 mm dia. bars with a spacing = 90 mm c/c at the larger end, increasing to 215 mm c/c towards the smaller end..

Temperature and shrinkage reinforcement in longitudinal direction will be 12 mm dia. bars — 215 mm c/c.

Design of longitudinal beam:

For positive bending moments which occur under the columns, the beam will behave as a rectangular beam.

Maximum bending moment occurs under the heavier column; its value = 148 kN-m.

$$\text{Effective depth} = \sqrt{\frac{148 \times 10^6}{0.653 \times 500}} = 673.3 \text{ mm, say 675 mm.}$$

$$A_{st} = \frac{148 \times 10^6}{230 \times .904 \times 675} = 1054.5 \text{ mm}^2$$

Provide 12 mm dia. bars–10 nos.

These bars will be kept over the transverse bars.

Hence, cover on their centres

$$= 35 + 6 + 6$$
$$= 47 \text{ mm, say 50 mm.}$$

Over all depth of beam = 675 + 50 = 725 mm.

Under the lighter column the maximum bending moment = 121.49 k-Nm. For this

$$A_{st} = \frac{121.49 \times 10^6}{230 \times .904 \times 675} = 865.65 \text{ mm}^2.$$

Provide 12 mm dia. bars—8 nos.

In the position of beam in between the columns maximum negative bending moment = 35 kN-m. The beam will behave as a T — beam in this position. Taking j = 0.9, the tension reinforcement is

$$A_{st} = \frac{35 \times 10^6}{230 \times 0.9 \times 675} = 250.5 \text{ mm}^2$$

Provide 12 mm dia. bars — 3 nos.

Development length of 12 mm dia bars

$$l_d = \frac{12 \times 230}{4 \times 1.4 \times 0.6} = 821.43 \text{ mm.}$$

Check for shear:

Consider the critical section an effective depth away from the inner face of heavier column. Here, X = 3.025 m. S.F. = 128 kN.

Nominal shear stress

$$\tau_v = \frac{128 \times 10^3}{500 \times 675} = 0.379 \ N/mm^2.$$

$$\frac{100 \ A_{st}}{bd} = \frac{100 \times 113.1 \times 10}{500 \times 675} = 0.335$$

From Table 3.11

$$\tau_0 = 0.244 \ N/mm^2$$

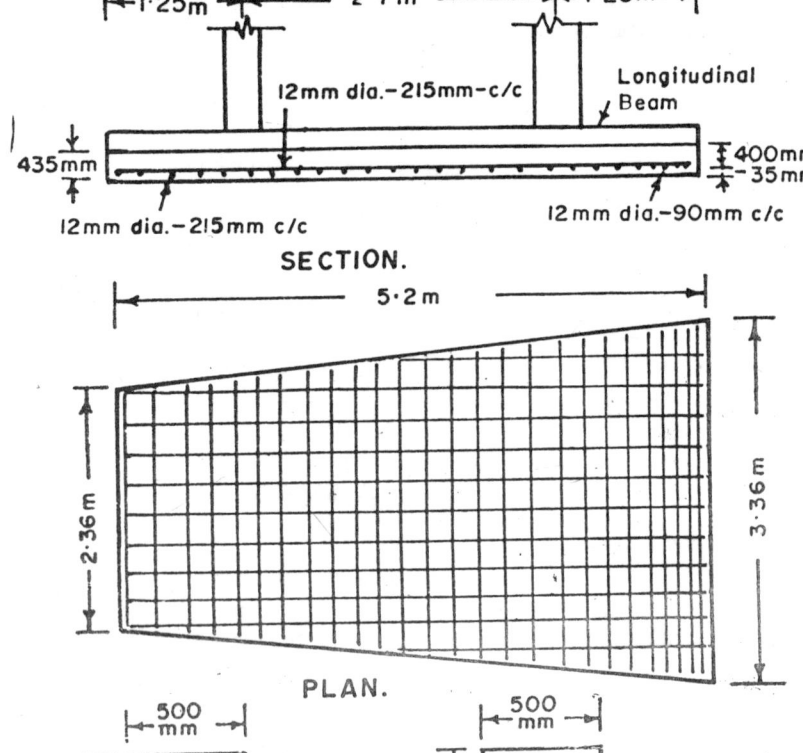

Fig. 12.18

As $\tau_v > \tau_c$ shear reinforcement is required.

$$V_c = \tau_c \, bd = 0.244 \times 500 \times 675 \times 10^{-3}$$
$$= 82.35 \text{ kN}$$
$$V_s = 128 - 82.35 = 45.65 \text{ kN}$$

Provide 8 mm dia. M.S. 2 -- legged stirrups. Spacing is given by

$$s_v < \frac{\sigma_{sv} \, A_{sv} \, d}{V_s}$$
$$< \frac{140 \times 2 \times 50.3 \times 675}{45.65 \times 10^3}$$
$$< 208.3 \text{ mm c/c, say 205 mm c/c.}$$

Nominal shear reinforcement:

Provide 8 mm dia M.S. 2 — legged stirrups. Spacing is given by

$$s_v \leqslant \frac{A_{sv} \, f_y}{0.4 \, b}$$
$$\leqslant \frac{2 \times 50.3 \times 250}{0.4 \times 500}$$
$$\leqslant 125.8 \text{ mm c/c.}$$

This is less than the spacing calculated above.

Hence, provide 8 mm dia. M.S. 2 — legged stirrups at a spacing of 125 mm c/c.

Check for bearing stress can be done in a similar manner as in Example 12.1.

Details of reinforcement are shown in Fig.12.18.

Example 12.5

Two columns of a building structure are spaced 5 m apart on centres. The cross-section of both the columns is the same i.e. 400 × 400 mm. Each column carries an axial load of 750 kN. One of the columns is situated with its face touching the boundary line of the property, and the other is situated inside the property. Safe bearing capacity of soil = 150 kN/m². Design a suitable foundation for the columns. Depth of foundation below G.L. = 1.0 m.

Solution:

An independent footing for the inner column poses no problem, but that for the outer column will be eccentric w.r.t. column axis and as such subjected to heavy moment. A combined rectangular footing or a trapezoidal footing are out of question since the distance between the two columns is large.

A strap beam footing offers a practical solution because the

moment on the outer column footing can be balanced by the
levering action of the strap beam. Consider Fig.12.19(a), showing
the section of the footing for two columns. The soil net upward
reactions on the footing pads under the inner and the outer
columns are R_1 and R_2. The weight of the strap beam is neglected.
R_2 is eccentric w.r.t the line of action of outer column load. A
trial value of this eccentricity is taken as $= 1.15$ m

$$\text{Then } s = 5 - 1.15$$
$$= 3.85 \text{ m}$$

Now, taking moment of the forces about the centre of the inner
column footing pad

$$R_2 \times 3.85 = 750 \times 5$$
$$R_2 = 974 \text{ kN}.$$

For equilibrium

$$R_1 + R_2 = 750 + 750$$
$$R_1 = 1500 - R_2$$
$$= 526 \text{ kN}.$$

Plan areas of footing pads:

(i) For inner column.

Net upward soil reaction	$= 526$ kN.
Estimated self weight of pad	$= 53$ kN.
Total	$= 579$ kN.

Required area of pad $= \dfrac{579}{150} = 3.86$ m².

Provide a square pad of size 2 m $\times$ 2 m.
Net upward soil pressure

$$q_1 = \frac{526}{4} = 131.5 \text{ kN/m}^2$$

(ii) For outer column.

Net upward soil reaction	$= 974$ kN.
Estimated self weight of footing	$= 98$ kN.
Total	$= 1072$ kN.

Required area of pad $= \dfrac{1072}{150} = 7.15$ m².

Provide a square pad of size 2.7 m $\times$ 2.7 m.
Net upward soil pressure

$$q_2 = \frac{974}{2.7 \times 2.7} = 133.6 \text{ kN/m}^2.$$

Bending moment and ahear force diagrams for the strap beam
are shown·in Fig.12.19 (b)

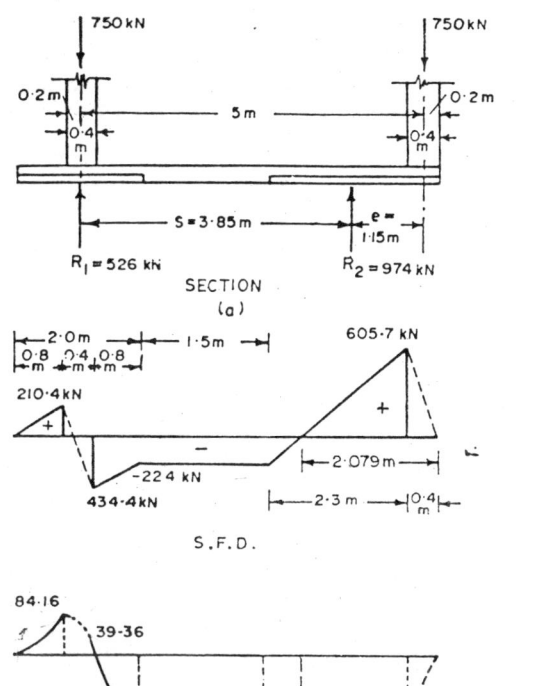

SECTION
(a)

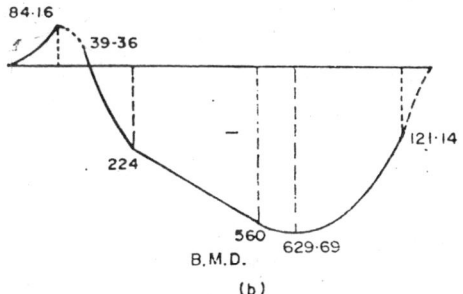

S.F.D.

B.M.D.
(b)

Fig.12.19

Design of strap beam.

It is proposed to keep the width of beam = 500 mm. and limit
its overall depth to 1000 mm. If the cover up to centre of reinforce-

ment be $= 35$ mm then effective depth $d = 965$ mm. Moment of resistance of a singly reinforced beam

$$= R b d^2$$
$$= 0.653 \times 500 \times 965^2 \times 10^{-6}$$
$$= 304 \text{ kN-m.}$$

Remainder of the bending moment will be taken up by additional tensile reinforcement balanced by compression reinforcement. Thus, the beam will be a doubly reinforced beam.

$$A_{st1} = \frac{304 \times 10^6}{230 \times .904 \times 965} = 1515 \text{ mm}^2.$$

$$M = M - M_b = 629.69 - 304 - 325.69 \text{ kN-m.}$$

$$A_{st2} = \frac{325.69 \times 10^6}{230 \times (965-35)} = 1522.6 \text{ mm}^2.$$

$$\therefore A_{st} = 1515 + 1522.6 = 3037.6 \text{ mm}^2.$$

Provide 20 mm dia. bars —10 nos. at top.

Compression reinforcement, from Eq. (4.6) is

$$A_{sc} = \frac{1522.6 \times 230}{(1.5 \times 18.67 - 1)\dfrac{(0.289 \times 965 - 35) \times 5}{0.289 \times 965}}$$

$$= 2966.4 \text{ mm}^2.$$

Provide 20 mm dia. bars —10 nos. at bottom.

Check for shear:

Consider a critical section distant an effective depth away from the inner face of outer column. Shear force

$$V = 605.7 - 133.6 \times 2.7 \times 0.965$$
$$= 257.6 \text{ kN.}$$

Nominal shear stress

$$\tau_v = \frac{257.6 \times 10^3}{500 \times 965}$$

$$= 0.534 \text{ N/mm}^2.$$

$$\frac{100 A_{st}}{bd} = \frac{100 \times 10 \times 314.16}{500 \times 965} = 0.651$$

From Table 3.11

$$\tau_c = 0.301 \text{ N/mm}^2.$$

As $\tau_v > \tau_c$ shear reinforcement is needed.

$$V_c = 0.301 \times 500 \times 965 \times 10^{-3}$$

$$= 149.6 \text{ kN.}$$

$$\therefore V_s = 257.6 = 149.6$$

$$= 108 \quad \text{kN.}$$

Try 8 mm dia. 2 legged stirrup.

$$\text{Spacing } s_v = \frac{\sigma_{st} A_{sv} d}{V_s}$$

$$= \frac{230 \times 2 \times 50.3 \times 965}{108 \times 10^3}$$

$$= 203.9 \text{ mm c/c.}$$

Spacing of nominal stirrups

$$s_v < \frac{A_{sv} f_y}{0.4 \, b}$$

$$< \frac{2 \times 50.3 \times 415}{0.4 \times 500}$$

$$< 208.75 \text{ mm c/c.}$$

Hence, 8 mm dia. 2 legged stirrups may be provided at 200 mm c/c throughout.

Design of footing pads:

(i) Under the inner column.

Projection of footing beyond the side face of strap beam

$$= \tfrac{1}{2} (2 - 0.5)$$

$$= 0.75 \text{ m}$$

Bending moment on projected part

$$= 131.5 \times 2 \times \frac{0.75^2}{2}$$

$$= 73.97 \text{ kN-m.}$$

Effective depth

$$d = \sqrt{\frac{73.97 \times 10^6}{0.653 \times 2000}}$$

$$= 237.9 \text{ mm, say 240 mm.}$$

$$A_{st} = \frac{73.97 \times 10^6}{230 \times 0.904 \times 240} = 1482.3 \text{ mm}^2.$$

Provide 12 mm dia bars — 14 nos. in transverse direction.

Shear force at critical section

$$V = (0.75 - 0.24) \times 2 \times 131.5$$

$$= 134.13 \text{ kN.}$$

. Nominal shear stress

$$\tau_v = \frac{134.13 \times 10^3}{2000 \times 240} = 0.28 \text{ N/mm}^2.$$

$$\frac{100 \, A_{st}}{b \, d} = \frac{100 \times 14 \times 113.1}{2000 \times 240} = 0.33$$

From Table 3.11

$$\tau_c = 0.24 \text{ N/mm}^2.$$

As $\tau_v > \tau_c$ the footing is unsafe in shear.
Hence, increase effective depth to

$$d = 240 \times \frac{0.28}{0.24}$$

$$= 280 \text{ mm}$$

Overall depth

$$= 280 + 50 = 330 \text{ mm}.$$

Temperature and shrinkage reinforcement

$$= \frac{0.12}{100} \times 2000 \times 330$$

$$= 792 \text{ mm}^2.$$

Provide 10 mm dia bars — 10 nos. in longitudinal direction.

(ii) Under the outer column.

Projection of footing beyond the side face of strap beam

$$= \tfrac{1}{2} (2.7 - 0.5)$$

$$= 1.1 \text{ m}.$$

Bending moment on projected part

$$= 133.6 \times 2.7 \times \frac{1.1^2}{2}$$

Effective depth

$$d = \sqrt{\frac{218.2 \times 10^6}{0.653 \times 2700}}$$

$$= 351.8 \text{ mm, say } 355 \text{ mm}.$$

$$A_{st} = \frac{218.2 \times 10^6}{230 \times .904 \times 355}$$

$$= 2956 \text{ mm}^2.$$

Provide 12 mm bars — 26 nos.

Shear force at critical section

$$V = (1.1 - 0.355) \times 2.7 \times 133.6$$

$$= 268.7 \text{ kN}.$$

Nominal shear stress

$$\tau_v = \frac{268.7 \times 10^3}{2700 \times 355}$$

$$= 0.28 \text{ N/m}^2.$$

$$\frac{100 A_{st}}{b \, d} = \frac{100 \times 26 \times 113.1}{2700 \times 355}$$

$$= 0.307 \text{ N/mm}^2.$$

From Table 3.11

$$\tau_c = 0.236 \text{ N/mm}^2.$$

As $\tau_v > \tau_c$ the section is unsafe in shear. Increase the effective depth to

$$d = 355 \times \frac{0.28}{0.236}$$

$$= 421 \text{ mm}, \text{ say } 425 \text{ mm}.$$

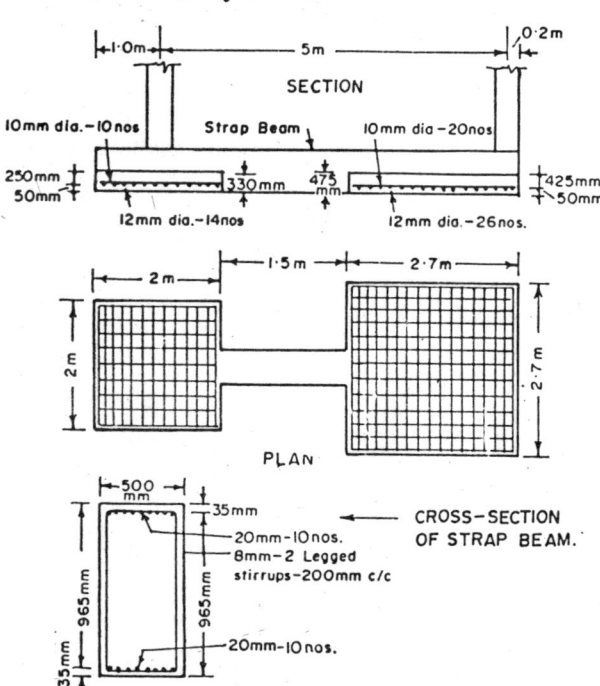

Fig. 12.20

Overall depth $= 425 + 50$

$\qquad = 475$ mm.

Temperature and shrinkage reinforcement

$$= \frac{0.12}{100} \times 2700 \times 475$$

$$= 1539 \text{ mm}^2.$$

Provide 10 mm dia bars — 20 nos. in longitudinal direction. Bearing stress may be checked in the same manner as done in Example 12.1. It is not repeated here to save space.

Details of reinforcement are shown in Fig. 12.20.

13

Design of Retaining Walls

13.1 INTRODUCTION

Retaining walls are the structures used to hold back earth, coal or ore and do not allow the mass to assume its natural slope. Naturally, the retained mass will exert lateral pressure on the retaining wall.

When built from stone, bricks or plain concrete, the retaining walls have to be so designed that the lateral earth pressure may not cause tensile stress at any point in it. Such structures become massive in size and depend upon their own weight for stability. They prove to be uneconomical as the height of retained earth increases. On the other hand, reinforced concrete retaining walls can be built with very slender structural components, and as such, prove to be very economical. Stability of such a wall is partially derived from the weight of the retained earth resting on the projected part of its base. Reinforced concrete retaintning walls are of two types:

(i) Cantilever type. This retaining wall has a vertical stem

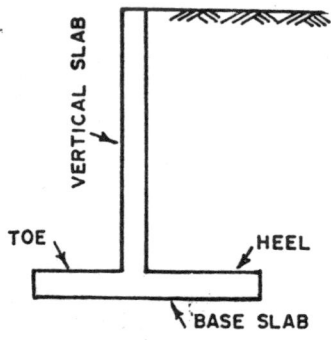

Fig. 13.1. Cross-section of cantilever type retaining wall.

which retains earth by cantilever action and is fixed to a horizontal base. See Fig. 13.1. The wall has a T — shaped cross-section in general and hence its name. However, L — shaped retaining walls have also been used but for smaller heights only. The stem as well as the base bend as cantilevers in vertical plane.

Such walls are economical up to a height of about 6 m only. For greater heights the bending moments in the vertical stem and in the horizontal base are rather high and require greater thicknesses which make the structure uneconomical.

(ii) Counterfort type. Counterfort type retaining wall may be used for heights of retained earth 6 m and more and for high magnitudes of earth pressure. The vertical structural component is tied to the horizontal base by means of counterforts which are built at certain intervals. See Fig. 13.2. Hence, the vertical structural component can not bend as a cantilever in the vertical plane. In fact its bending action is in horizontal plane due to earth pressure, while the counterforts act as supports. The horizontal base bends in vertical planes while the counterforts act as supports.

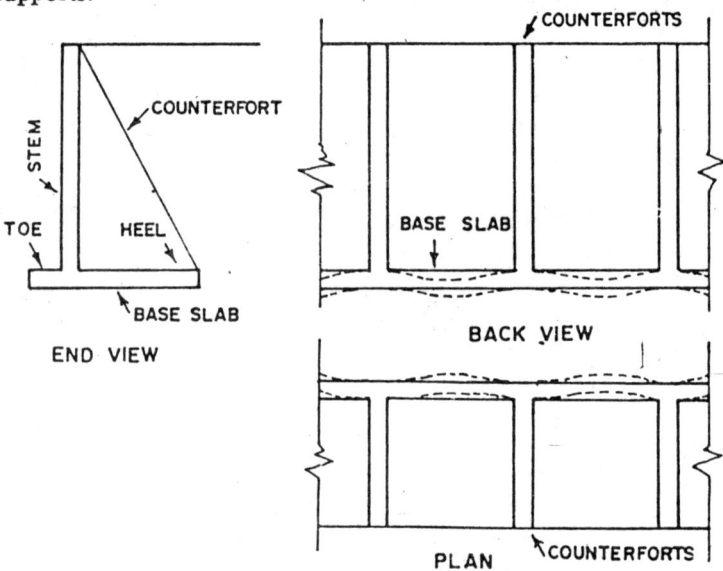

Fig. 13.2. Counterfort type retaining wall

13.2 EARTH PRESSURE

Rankine's theory is quite convenient as a rough guide for estimating the lateral earth pressure on retaining walls. It assumes that the earth is cohesionless and there is no adhesion and friction between the wall and the earth. This introduces approximation in the estimation of earth pressure. However, the Rankine Solution is popular because of its simplicity. Cohesive soils can be treated as equivalent cohesionless soils with a modified angle of internal friction. The lateral earth pressure on a wall is minimum when the wall moves away and rotates slightly. This minimum pressure is known as active earth pressure. Actual magnitudes of pressure against the wall are generally greater than the active pressure. However, due to slight movement of wall away from soil and due to rotation, the pressures tend to reduce.

Retaining walls do not retain water table on the back since spaced weep holes in the wall or back drainage pipes laid on the heel slab eventually drain off the water. At the most, saturated earth may be considered for computing the earth pressure. The latter can be found by considering the full pressure of water plus the pressure of earth of which the unit weight is reduced by the unit weight of water. The magnitude and distribution of lateral pressure of soil exerted on wall depends upon the type of soil, its moisture content, slope of backfill and surcharge due to any loads in proximity. Rankine's theory assumes a linear distribution of earth pressure on the back of the wall. See Fig. 13.3.

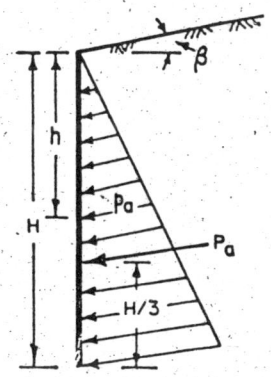

Fig. 13.3 Rankine's earth pressure.

Active earth pressure intensity at any point is given by

$$p_a = K_a \gamma \, h \qquad (13.1)$$

where h = depth of point from top

γ = unit weight of earth

K_a = Coefficient of active earth pressure

$$= \cos \beta \, \frac{\cos \beta - \sqrt{\cos^2 \beta - \cos^2 \phi}}{\cos \beta + \sqrt{\cos^2 \beta - \cos^2 \phi}} \qquad (13.2)$$

β = angle of surcharge or slope of the back fill

ϕ = angle of internal friction

The resultant total active force on the wall is given by

$$P_a = \tfrac{1}{2} K_a \, \gamma \, H^2 \qquad (13.3)$$

This force acts parallel to the slope of the back fill. If the top surface of the retained earth is horizontal, then $\beta = 0$.

Hence, the coefficient of active earth pressure becomes

$$K_a = \frac{1 - \sin \phi}{1 + \sin \phi} \qquad (13.4)$$

Passive earth pressure intensity at any point is given by

$$p_D = K_p \, \gamma \, h \qquad (13.5)$$

where

$$K_p = \cos \beta \, \frac{\cos \beta + \sqrt{\cos^2 \beta - \cos^2 \phi}}{\cos \beta - \sqrt{\cos^2 \beta - \cos^2 \phi}}$$

If the top surface of earth is horizontal, then $\beta = 0$. Hence, the coefficient of passive earth pressure becomes

$$K_\rho = \frac{1 + \sin \phi}{1 - \sin \phi} \qquad (13.6)$$

Forward movement of the wall due to earth pressure on its back can be partly resisted by the passive earth pressure offered by the soil in front. If this soil is excavated or eroded then the passive resistance may not be available. If there is certainty of no loss of the soil in front, then the passive resistance of the soil may be used while considering the stability of the wall against sliding.

13.3 STABILITY OF RETAINING WALL

13.3 1 Stability considerations. The stability of the retaining wall as a whole (See Fig. 13.4) has to be considered in respect of the following:

(i) *Overturning.* The earth pressure induces overturning moment on the retaining wall about its toe. Stabilizing moment is provided by the self weight of the retaining wall and the weight of the earth over the heel. The wall should be safe against overturning with adequate margin of safety.

(ii) *Sliding.* The earth pressure induces sliding force on the retaining wall and causes its horizontal movement. Resisting force is available from the friction existing between the base and the soil, and also the passive resistance of the soil in front if there is no possibility of its loss. Additional resisting force may be obtained by providing a key beneath the base. The wall should be safe against sliding with adequate margin of safety.

(iii) *Subsidence.* The pressure distribution beneath the base is nonuniform in general. It is maximum under the toe and minimum under the heel. The maximum pressure should be less than the safe bearing capacity of the soil. Otherwise, the wall will sink at the toe.

13.3.2 Provisions of IS Code: 456.
Consider the retaining wall shown in Fig. 13.4. The active earth pressure is considered for the full depth of the wall. The resultant active force is given by

$$P_a = \tfrac{1}{2} K_a \gamma H^2 \qquad (13.7)$$

This force acts at a point $\dfrac{H}{3}$ above the base.

There is a live load of intensity $= w$ per unit area on the back fill, which exerts a uniform lateral pressure on the wall of intensity $= K_a w$ acting over the full depth. The resultant lateral force due to this is given by

$$P_L = K_a w H \qquad (13.8)$$

This force acts at a point $\dfrac{H}{2}$ above the base.

Both the forces P_a and P_L induce an overturning moment about the point 0.

W_1, W_2 and W_3 are the self weights of the stem, the base and the earth over the heel respectively. All these loads contribute towards the stabilizing moment about the point 0. No contribution of live load towards stabilizing moment is considered since the arching effect in soil may not allow this load to be transmitted to the heel.

Both the forces P_a and P_L also induce sliding of the wall. Restoring force is available, partly from the friction between the base and the soil. The frictional force depends upon the vertical reaction.

The latter will be taken to be the sum of W_1, W_2 and W_3 only. Contribution of live load is neglected as explained earlier. Toe soil if present in front will provide passive resistance to sliding.

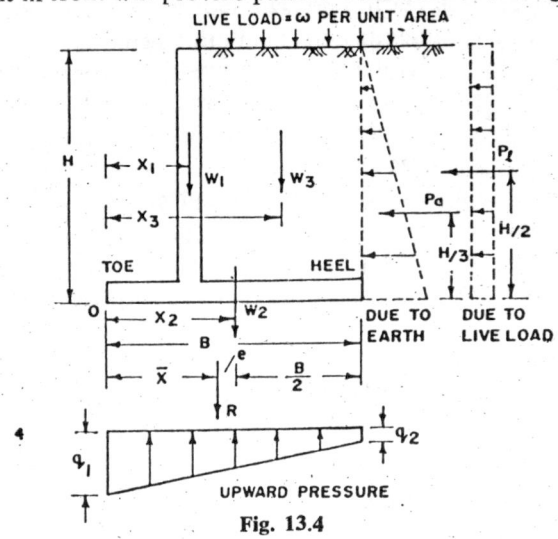

Fig. 13.4

(i) *Safety against overturning.* The retaining wall is considered to be safe against overturning if the following condition is satisfied:

$$M_R \nless 1.2 M_{OD} + 1.4 M_{OL} \qquad (13.9)$$

where M_R = Restoring moment

$$= 0.9 (W_1 X_1 + W_2 X_2 + W_3 X_3) \qquad (13.10)$$

M_{OD} = Overturning moment due to dead loads

$$= P_a \times \frac{H}{3} \qquad (13.11)$$

M_{OL} = Overturning moments due to live loads

$$= P_l \times \frac{H}{2} \qquad (13.12)$$

(ii) *Safety against sliding.* The retaining wall is considered to be safe against sliding if the following condition is satisfied:

$$\frac{F_R}{F_S} \nless 1.4 \qquad (13.13)$$

where F_R = Resisting force

$$= \mu \times 0.9 (W_1 + W_2 + W_3) + \text{Passive resistance of soil if available in front.}$$

$$F_S = P_a + P_L$$

μ = Coefficient of friction between the base and the soil. It depends upon the type of soil and its moisture content. Its value lies between 0.3 and 0.5.

(iii) *Safety against subsidence.* With reference to Fig. 13.4, total vertical load on the base is

$$R = W_1 + W_2 + W_3 + W_L$$

where W_L = resultant live load on earth over heel. Distance of the line of action of R from point 0 is

$$\overline{X} = \frac{W_1X_1 + W_2X_2 + W_3X_3 + W_LX_L - M_{OD} - M_{OL}}{R}$$

where X_L = distance of the W_L from O.

Then, eccentricity of the resultant from the centre of base

$$e = \frac{B}{2} - \overline{X}$$

Hence, pressure below the base

$$= \frac{R}{A} \pm \frac{R.e}{Z} \tag{13.14}$$

where $A = B \times 1$

$$Z = \frac{1}{6} \times B^2$$

Pressure under the toe is

$$q_1 = \frac{R}{A} + \frac{R.e}{Z} \tag{13.15}$$

Pressure under the heel is

$$q_2 = \frac{R}{A} - \frac{R.e}{Z} \tag{13.16}$$

For safety against subsidence,

$q_1 \ngtr$ safe bearing capacity of soil.

13.4 PRELIMINARY DESIGN PROPORTIONS

Tentative dimensions and proportions of the retaining wall are initially fixed up to carry out the stability analysis. This is a process of trial and the solutions satisfying the stability criteria are more than one. Once the stability is estabilished, the proper structural design can be taken up.

To faciliate the stability check, the preliminary design proportions of the cantilever and the counterfort type retaining wall are shown in Fig. 13.5. The spacing of the counterforts may be fixed up to be 0.3H to 0.5 H centre to centre.

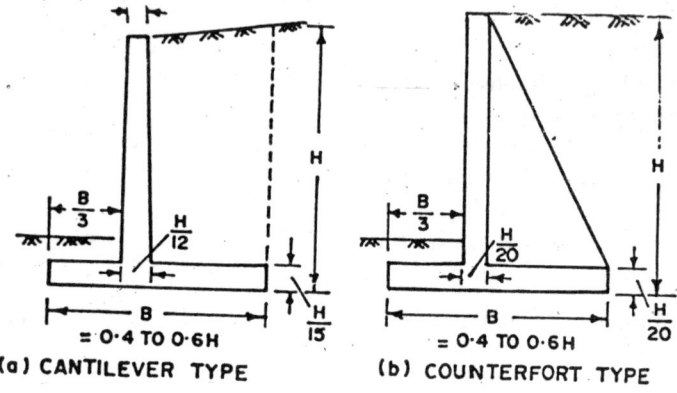

Fig. 13.5

Example 13.1

Design a retaining wall to retain earth for a height of 4 m. Depth of foundation = 1 m below G.L. Angle of surcharge = 7°. Unit weight of soil = 21 kN/m³. Angle of internal friction = 27°. Coefficient of friction between the concrete base and earth = 0.45. Safe bearing capacity of soil = 120 kN/m². Use concrete grade: M 15 and steel grade: Fe 415. There is certainty of the toe soil available in front for passive resistance.

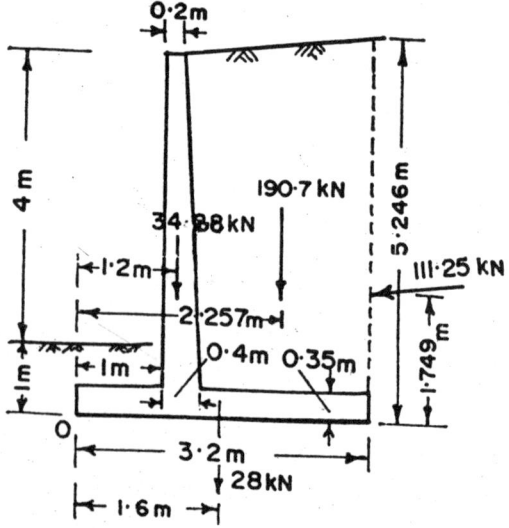

Fig. 13.6 Proportions for stability check

Solution

Total height of retaining wall = 5 m.

Cantilever type retaining wall will be designed. Preliminary dimensions and proportions of the wall are shown in Fig. 13.6

Now $\gamma = 21$ kN/m^3, $\beta = 7°$ and $\phi = 27°$

$$H = 5 + 2 \tan 7°$$
$$= 5.246 \text{ m}$$

Consider 1 m length of retaining wall.

(i) Weight of stem $W_1 = \frac{1}{2}(0.2 + 0.4) \times 4.65 \times 25$
$$= 34.88 \text{ kN.}$$

Distance of its line of action from O
$$X_1 = I + \frac{1}{3}(\frac{1}{3} \times 0.2 + \frac{2}{3} \times 0.4 \times 2)$$
$$= 1.2 \text{ m.}$$

(ii) Weight of base $W_2 = 0.35 \times 3.2 \times 25$
$$= 28 \text{ kN.}$$

Distance of its line of action from O
$$X_2 = 1.6 \text{ m.}$$

(iii) Weight of the earth over the heel
$$W_3 = (\frac{1}{2} \times 1.8 \times 4.65 + \frac{1}{2} \times 2 \times 4.896) \times 21$$
$$= 190.7 \text{ kN.}$$

Distance of its line of action from O
$$X_3 = 2.257 \text{ m}$$

Weight of the earth over the toe is not considered. Coefficient of active earth pressure is

$$K_a = \cos\beta \ \frac{\cos\beta - \sqrt{\cos^2\beta - \cos^2\phi}}{\cos\beta + \sqrt{\cos^2\beta - \cos^2\phi}}$$

$$= 0.993 \times \frac{0.993 - \sqrt{(0.993)^2 - (0.891)^2}}{0.993 + \sqrt{(0.993)^2 - (0.891)^2}}$$

$$= 0.385$$

Active earth pressure
$$P_a = \frac{1}{2} K_a \gamma H^2$$
$$= \frac{1}{2} \times 0.385 \times 21 \times 5.246^2$$
$$= 111.25 \text{ kN.}$$

This acts at a point 1.749 in above the base.

Horizontal component of P_a

$$P_h = P_a \cos \beta$$
$$= 111.25 \times 0.993$$
$$= 110.47 \text{ kN.}$$

This has an overturning effect about O.

Vertical component of P_a

$$P_v = P_a \sin \beta$$
$$= 111.25 \times 0.1219$$
$$= 13.56 \text{ kN.}$$

This has a stabilizing effect about O.

Overturning moment due to dead loads:

$$M_{OD} = P_h \times \frac{H}{3}$$
$$= 110.47 \times 1.749$$
$$= 193.21 \text{ kN-m.}$$

Overturning moment due to live loads

$$M_{OL} = \text{zero}$$

Restoring moment

$$M_R = 0.9 \, (W_1 X_1 + W_2 X_2 + W_3 X_3 + P_v \times B)$$
$$= 0.9 \, (34.88 \times 1.2 + 28 \times 1.6 + 190.7 \times 2.257 +$$
$$13.56 \times 3.2)$$
$$= 504.42 \text{ kN-m.}$$

Now

$$1.2 \, M_{OD} + 1.4 \, M_{OL} = 1.2 \times 193.21 + 1.4 \times \text{zero}$$
$$= 231.85 \text{ kN-m.}$$

Evidently

$$M_R > 1.2 \, M_{OD} + 1.4 \, M_{OL}$$

Hence, the wall is safe in overturning.

Sliding force $F_S = P_h + P_L$
$$= 110.47 + \text{zero}$$
$$= 110.47 \text{ kN}$$

Coefficient of passive earth pressure

$$K_p = \frac{1 + \sin \phi}{1 - \sin \phi}$$
$$= \frac{1 + 0.454}{1 - 0.454}$$
$$= 2.663$$

Resisting force,

$$F_R = \mu \times 0.9 \, (W_1 + W_2 + W_3 + P_v) + \frac{1}{2} K_p \, \gamma \, h^2$$
$$= 0.45 \times 0.9 \, (34.88 + 28 + 190.7 + 13.56) + \frac{2.663 \times 21 \times 1^2}{2}$$

$$= 108.19 + 27.96 \ \text{kN}$$
$$= 136.15 \ \text{kN}$$

Hence

$$\frac{F_R}{F_s} = \frac{136.15}{110.47} = 1.23 < 1.4$$

The retaining wall is unsafe in sliding.

A key may be provided below the base.

Then, resisting force

$$F_R = 108.19 + \tfrac{1}{2} \times 2.663 \times 21h^2$$
$$= 108.19 + 27.96h^2$$

Now, for safety

$$\frac{F_R}{F_s} < 1.4$$

i.e.

$$\frac{108.19 + 27.96h^2}{110.47} < 1.4$$

or $h \geqslant 1.29$ m.

Depth of key below the base

$$= 1.29 - 1$$
$$= 0.29 \ \text{m}.$$

Now, total vertical force is given by

$$R = W_1 + W_2 + W_3 + P_v$$
$$= 34.88 + 28 + 190.7 + 13.56 \ \text{kN}$$
$$= 267.14 \ \text{kN}.$$

The distance of the point from O where this resultant strikes the base is given by

$$\overline{X} = \frac{W_1 X_1 + W_2 X_2 + W_3 X_3 + P_v B - P_h \times \dfrac{H}{3}}{R}$$

$$= \frac{34.88 \times 1.2 + 28 \times 1.6 + 190.7 \times 2.257 + 13.56 \times 3.2 - 110.47 \times 1.749}{267.14}$$

$$= 1.375 \ \text{m from O}.$$

Eccentricity of the force from the centre of base

$$e = \tfrac{1}{2} \times 3.2 - 1.375$$
$$= 0.225 \ \text{mm towards O}.$$

Pressure distribution below the base is

$$= \frac{R}{A} \pm \frac{R.e}{Z}$$

$$= \frac{267.14}{3.2} \pm \frac{249.7 \times 0.225}{\frac{1}{6} \times 3.2^2}$$

$$= 83.48 \pm 35.22$$

Pressure below the toe

$$q_1 = 83.48 + 35.22$$
$$= 118.7 \text{ kN/m}^2 < 120 \text{ kN/m}^2, \text{ safe.}$$

Pressure below the heel

$$q_2 = 83.48 - 35.22$$
$$= 48.26 \text{ kN/m}^2 < 120 \text{ kN/m}^2, \text{ safe.}$$

The pressure distribution is shown in Fig. 13.7(a)

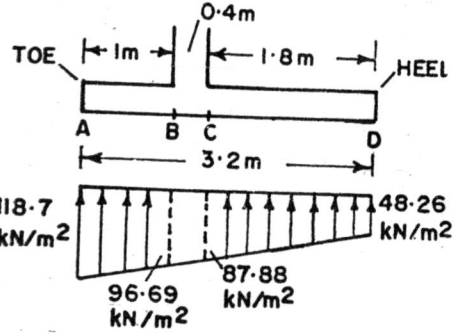

(a) PRESSURE DISTRIBUTION BELOW

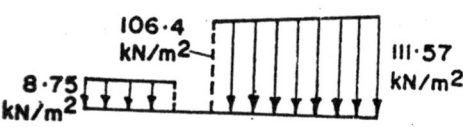

(b) LOAD INTENSITY FROM ABOVE.

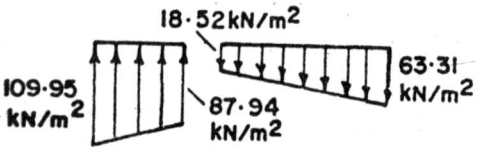

(c) RESULTANT LOAD INTENSITY
ON THE BASE SLAB.

Fig. 13.7

Load intensity on base due to self weight of base and supported earth:

At point A = $0.35 \times 25 = 8.75$ kN /m².

At point B = $0.35 \times 25 = 8.75$ kN/m².

At point C = 0.35 x $25 + 4.65 \times 21 = 106.4$ kN/m²

At point D = $0.35 \times 25 + 4.896 \times 21 = 111.57$ kN/m².

These are shown in Fig. 13.7(b).

The resultant load intensity on the base is shown in Fig. 13.7(c).

Design of heel slab.

The heel bends as a cantilever. Net load intensity on heel slab is shown in Fig. 13.7 (c).

Lines projected from the heel at 7° to horizontal enclose a part of the earth pressure diagram, which is shown shaded in Fig. 13.8.

Vertical distance covered by this diagram

$= 0.35 + 1.8 \tan 7°$

$= 0.571$ m.

Resultant force from the shaded diagram

$= \frac{1}{2} (37.80 + 42.41) \times 0.571$

$= 22.9$ kN, inclined to horizontal at 7°. It intersects the

mid-depth line of heel at a point, the distance of which from the back face of stem is

$= (0.571 - 0.291 - \frac{1}{2} \times 0.35) \cot 7°$

$= 0.945$

Vertical component of the resultant force

$= 22.9 \times \sin 7°$

$= 2.79$ kN

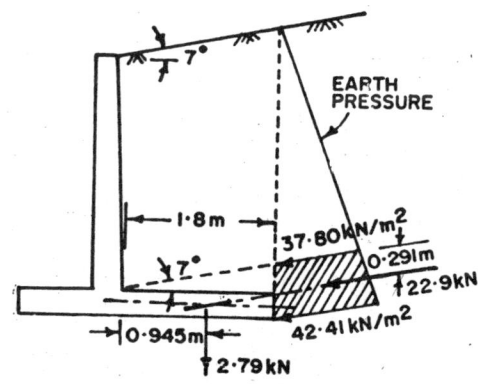

Fig. 13.8

This force will also induce bending moment in the heel.
Hence, bending moment in heel slab

$$M = (18.52 + 2 \times 63.31) \times \frac{1.8^2}{6} + 2.79 \times 0.945$$

$$= 81.01 \text{ kN-m.}$$

Shear force at the critical section taken at the face

$$V = (18.52 + 63.31) \frac{1.8}{2} + 2.79$$

$$= 76.44 \text{ kN.}$$

Permissible stresses are

$$\sigma_{obo} = 5 \text{ N/mm}^2 \text{ and } \sigma_{st} = 230 \text{ N/mm}^2$$
$$m = 18.67$$

Design constants are

$$n = 0.289, \ j = 0.904 \text{ and } R = 0.653$$

Effective depth of heel slab

$$= \sqrt{\frac{M}{R \ b}}$$

$$= \sqrt{\frac{81.01 \times 10^6}{0.653 \times 1000}}$$

$$= 352.2 \text{ mm, \quad say } 355 \text{ mm.}$$

Tension reinforcement

$$A_{st} = \frac{81.01 \times 10^6}{230 \times .904 \times 355}$$

$$= 1097.5 \text{ mm}^2.$$

Provide 12 mm dia. bars at 100 mm c/c.
Actual $A_{st} = 1131$ mm^2.
Development length of 12 mm bar in tension

$$l_d = 68.45 \times \text{dia. \quad from \ Table 3.9}$$

$$= 68.45 \times 12$$

$$= 821.4 \text{ mm.}$$

Nominal shear stress

$$\tau_v = \frac{76.44 \times 10^3}{1000 \times 355} = 0.215 \text{ N/mm}^2.$$

$$\frac{100 \ A_{st}}{bd} = \frac{100 \times 1131}{1000 \times 355} = 0.319$$

$$\therefore \quad \tau_c = 0.239 \text{ N/mm}^2 \text{ from Table 3.11.}$$

Since $\tau_v < \tau_c$, the slab is safe in shear. Overall thickness of slab=
355 + 45 = 400 mm.

Temperature and shrinkage reinforcement

$$= \frac{0.12}{100} \times 1000 \times 400$$

$$= 480 \text{ mm}^2/\text{m}.$$

Provide 10 mm dia. bars at 160 mm c/c at right angles to the main reinforcement.

The thickness of the heel slab may be reduced to 200 mm towards the free end. See details in Fig. 13.9.

Design of toe slab:

The toe slab bends as a cantilever due to the net upward load intensity shown in Fig. 13.7 (c).

Bending moment

$$M = (87.94 + 2 \times 109.95) \times \frac{1^2}{6}$$

$$= 51.31 \text{ kN-m/m}.$$

Effective depth required

$$d = \sqrt{\frac{51.31 \times 10^6}{0.653 \times 1000}}$$

$$= 280.0 \text{ mm}.$$

Tension reinforcement

$$A_{st} = \frac{51.31 \times 10^6}{230 \times .904 \times 280} = 881.35 \text{ mm}^2$$

12 mm dia. bars at a spacing of 125 mm c/c may be provided.

Shear force at the critical section taken 0.28 m from the face

$$V = \frac{1}{2} (109.95 + 94.10) \times 0.72$$

$$= 73.46 \text{ kN}.$$

$$\therefore \quad \tau_v = \frac{73.46 \times 10^3}{1000 \times 280} = 0.262 \text{ N/mm}^2$$

$$\frac{100 \, A_{st}}{bd} = \frac{100 \times 881.35}{1000 \times 280} = 0.315$$

$$\tau_c = 0.238 \text{ N/mm}^2 \text{ from Table 3.11}$$

As $\tau_v > \tau_c$, the toe is unsafe in shear.

Try an effective depth = 320 mm.

Shear force at the critical section

$$V = \frac{1}{2} (109.95 + 95.2) \times 0.68$$

$$= 69.68 \text{ kN}.$$

$$\therefore \quad \tau_v = \frac{69.68 \times 10^3}{1000 \times 330} = 0.218 \text{ N/mm}^2$$

This is less than the minimum value of τ_c which is 0.22 N/mm² from Table 3.11.

Hence, the toe is safe in shear.

Tension reinforcement recalculated is

$$A_{st} = \frac{51.31 \times 10^6}{230 \times .904 \times 320} = 771.18 \text{ mm}^2.$$

12 mm dia. bars at a spacing of 145 mm c/c may be provided.
Total thickness of toe = 320 + 35 = 355 mm.
Temperature and shrinkage reinforcement

$$= \frac{0.12}{100} \times 1000 \times 355$$

$$= 426 \text{ mm}^2/\text{m}.$$

Provide 10 mm dia. bars at 180 mm c/c at right angles to the main reinforcement bars.

The thickness of toe slab may be reduced to 200 mm towards the free end. See details in Fig..13.9.

Design of stem:

Clear height of stem above the base slab

$$= 5 - 0.4$$

$$= 4.6 \text{ m}.$$

Only the horizontal component of earth pressure will cause bending moment in the stem, Bending moment at a depth h below top

$$= \frac{1}{6} \times K_a \, \gamma \, h^3 \cos \beta \qquad\qquad (13.17)$$

Bending moment in stem where it meets the top of base

$$= \frac{1}{6} \times 0.385 \times 21 \times 4.6^3 \times 0.9925$$

$$= 130.18 \text{ kN-m/m}$$

Effective depth required,

$$d = \sqrt{\frac{130.18 \times 10^6}{0.653 \times 1000}}$$

$$= 446.5 \text{ mm., } \quad \text{say 450 mm.}$$

Tension reinforcement

$$A_{st} = \frac{130.18 \times 10^6}{230 \times .904 \times 450}$$

$$= 1391.3 \text{ mm}^2/\text{m}.$$

12 mm dia. bars at 80 mm c/c may be provided, in vertical direction on the back face Concrete cover of 35 mm upto the centre of the bars may be kept.. Hence, total thickness of

stem

$$= 450 + 35 = 485 \text{ mm.}$$

Critical shear force

$$V = \tfrac{1}{2} K_a \gamma \; h^2$$
$$= \tfrac{1}{2} \times 0.385 \times 21 \times 4.6^2$$
$$= 85.54 \text{ kN.}$$

Nominal shear stress

$$\tau_v = \frac{85.54 \times 10^3}{1000 \times 450} = 0.19 \text{ N/mm}^2.$$

This is less than the minimum value of τ_c given in Table 3.11. Hence, the stem is safe in shear.

The thickness may be reduced towards the top to 200 mm. Consider a point at a depth of 3.45 m below top:

Bending moment

$$= \frac{1}{6} \times 0.385 \times 21 \times 3.45^3 \times 0.9925$$
$$= 54.92 \text{ kN-m/m.}$$

Effective depth required

$$d = \sqrt{\frac{54.92 \times 10^6}{0.653 \times 1000}}$$
$$= 290 \text{ mm.}$$

Actual effective depth available at this point

$$= 200 + \frac{(485 - 200)}{4.6} \times 3.45 - 35$$
$$= 378.75 \text{ mm.} > 290 \text{ mm O.K.}$$

Tension reinforcement required

$$= \frac{54.92 \times 10^6}{230 \times .904 \times 378.75} = 697.4 \text{ mm}^2/\text{m}$$

12 mm dia. bars at 160 mm c/c may be enough.

Alternate bars of the main reinforcement may be cut off while the remaining may be allowed to continue to the top, beyond this point.

Actual point of cut of

$$= 3.45 - \text{effective depth}$$
$$= 3.45 - 0.37875$$
$$= 3.07 \text{ m, say } 3.0 \text{ m from top.}$$

Temperature and shrinkage reinforcement

$$= \frac{0.12}{100} \times \frac{(200 + 485)}{2} \times 4600$$
$$= 1890.6 \text{ mm}^2.$$

Portion of reinforcement on back face

$$= \tfrac{1}{3} \times 1890.6$$
$$= 630.2 \text{ mm}^2/\text{m}.$$

12 mm dia. bars spaced at 180 mm c/c may be provided horizontally. Portion of reinforcement on front face

$$= \tfrac{2}{3} \times 1890.6$$
$$= 1260.4 \text{ mm}^2/\text{m}.$$

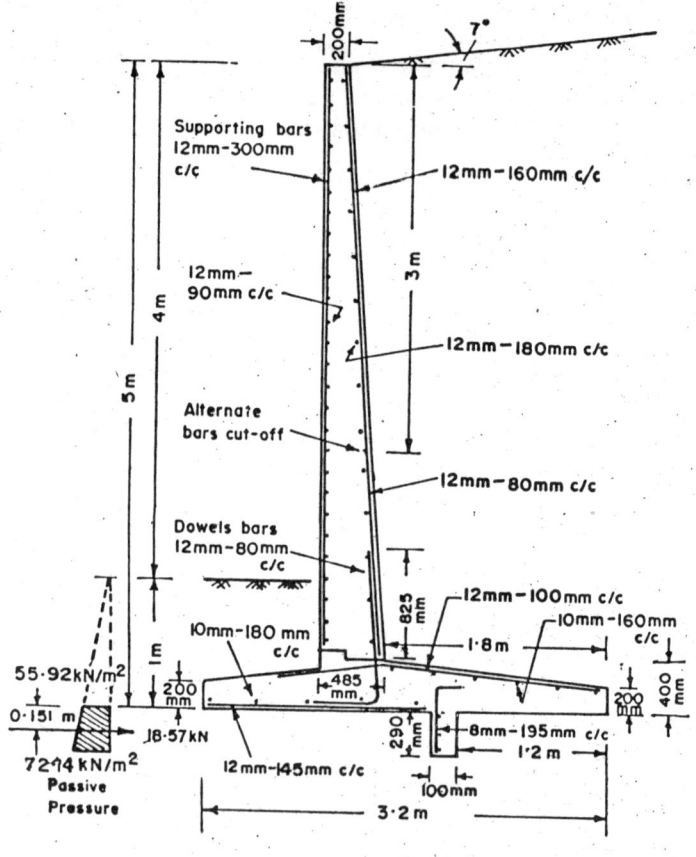

Fig. 13.9

12 mm dia. bars spaced at 90 mm c/c may be provided horizontally. Retaining wall is built in two stages, first the base and then the stem. Dowel bars of the same diameter and spacing as the main reinforcement in the stem are provided and extend both in the base

and the stem for a length equal to the development length. Main reinforcement terminates at the bottom of the stem and does not extend into the base. A key is provided to take shear at the joint between the stem and the base.

Details of reinforcement are shown in Fig. 13.9.

Design of key below the base

The lower portion of the passive earth pressure diagram is covered by the key. See Fig. 13.9.

The key moves against the earth, and as such, bears passive earth pressure.

Intensity of pressure is

$$= K_p \, \gamma \, h$$

Numerical value of intensity of pressure:
At the root of the key

$$= 2.663 \times 21 \times 1$$
$$= 55.92 \text{ kN/m}^2$$

At the lower tip of the key

$$= 2.663 \times 21 \times 1.29$$
$$= 72.14 \text{ kN/m}^2$$

Force on the key

$$= \tfrac{1}{2} \, (55.92 + 72.14) \times 0.29$$
$$= 18.57 \text{ kN}$$

Distance of its line of action below the base

$$= \frac{55.92 + 2 \times 72.14}{55.92 + 72.14} \times \frac{0.29}{3}$$
$$= 0.151 \text{ m.}$$

Bending moment on the key

$$= 18.57 \times 0.151$$
$$= 2.806 \text{ kN/m}$$

Effective depth

$$d = \sqrt{\frac{2.806 \times 10^6}{0.653 \times 1000}}$$
$$= 65 \text{ mm}$$

Tension reinforcement

$$A_{st} = \frac{2.806 \times 10^6}{230 \times .904 \times 65}$$
$$= 207.6 \text{ mm}^2$$

For 8 mm dia. bars the spacing works out to 242.3 mm.

The latter is more than 3 × d i.e. 195 mm.

Hence, adopt a spacing of 195 mm for 8 mm dia. bars

Example 13.2.

Design a retaining wall to retain earth for a height of 6 m. Depth of foundation = 1.5 m below G.L. Unit weight of soil = 18 kN/m³. Angle of internal friction = 28°. Coefficient of friction between earth and concrete = 0.50. Safe bearing capacity of soil = 155 kN/m². The back fill is horizontal. The soil above the toe is likely to be excavated at times. Concrete grade: M 15 and steel grade: Fe 415.

Solution:

Total height of retaining wall

$$= 6 + 1.5$$
$$= 7.5 \text{ m}.$$

A counterfort type retaining wall will be designed. Preliminary dimensions and proportions of the wall are shown in Fig. 13.10.

Now $\gamma = 18$ kN/m², $\phi = 28°$ and $\beta = $ zero.

∴ Coefficient of active earth pressure

$$K_a = \frac{1 - \sin \phi}{1 + \sin \phi} = \frac{1 - 0.4695}{1 + 0.4695} = 0.361$$

Total force due to earth pressure on the wall

$$P_a = \tfrac{1}{2} K_a \gamma H^2$$
$$= \tfrac{1}{2} \times 0.361 \times 18 \times 7.5^2$$
$$= 182.76 \text{ kN}.$$

This acts horizontally at a point 2.5 m above the foundation level. See Fig. 13.10.

Consider 1 m length of wall.

Weight of wall $W_1 = \tfrac{1}{2} (0.2 + 0.375) \times 7.125 \times 25$
$$= 51.211 \text{ kN}.$$

Distance of its line of action from O

$$X_1 = 1.648 \text{ m}.$$

Weight of the base

$$W_2 = 4.5 \times 0.375 \times 25$$
$$= 42.1875 \text{ kN}.$$

Distance of its line of action from O

$$X_2 = 2.25 \text{ m}$$

Weight of the earth over the heel slab

$$W_3 = \tfrac{1}{2} (2.625 + 2.8) \times 7.125 \times 18$$
$$= 347.9 \text{ kN}.$$

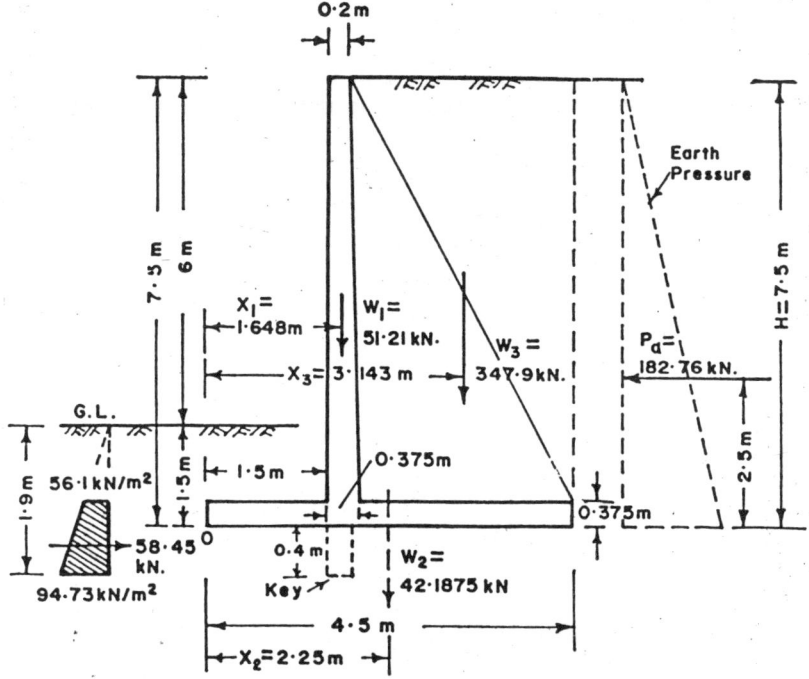

Fig. 13.10 Preliminary dimensions for stability check.

Distance of its line of action from O
$$X_3 = 3.143 \text{ m}.$$

There has been a little approximation here by not accounting for the self weight of the counterforts and substituting instead the weight of the earth. However, the error involved due to this is hardly appreciable.

Weight of earth over the toe is not considered in stability considerations.

Stability against overturning.

Overturning moment:

Due to dead load i.e. earth pressure
$$M_{OD} = 182.76 \times 2.5 = 456.9 \text{ kN-m}.$$

Due to live load
$$M_{OL} = \text{zero}.$$

$$1.2 \, M_{OD} + 1.4 \, M_{OL} = 1.2 \times 456.9 + \text{zero} = 548.28 \text{ kN-m.}$$

Restoring moment

$$M_R = 0.9 \, (W_1 \, X_1 + W_2 \, X_2 + W_3 \, X_3)$$
$$= 0.9 \, (51.21 \times 1.648 + 42.1875 \times 2.25 + 347.9 \times 3.143)$$
$$= 1145.4 \text{ kN-m}$$

Evidently

$$M_R > 1.2 \, M_{OD} + 1.4 \, M_{OL}$$

Hence, the structure is safe against overturning.

Stability against sliding.

Sliding force

$$F_S = P_a$$
$$= 182.76 \text{ kN.}$$

Resisting force available from friction between the base and the soil

$$F_R = \mu \times 0.9 \, (W_1 + W_2 + W_3)$$
$$= 0.5 \times 0.9 \, (51.211 + 42.1875 + 347.9)$$
$$= 198.59 \text{ kN.}$$

Then

$$\therefore \frac{F_R}{F_S} = \frac{198.59}{182.76} = 1.09 \quad \text{which is less than 1.4.}$$

Hence, the structure is unsafe in sliding.

A key may be provided below the base to obtain additional resisting force due to passive earth pressure in front. As the soil above the toe is likely to be excavated, it will be neglected from such a consideration. Remaining depth of soil will exert passive earth pressure. Try a key projecting 0.4 m below the base, which will reach up to 1.9 below the G.L. Available passive pressure force in front is indicated in Fig. 13.10. It is 58.45 kN.

Hence, total resisting force now will be

$$F_R = 198.59 + 58.45$$
$$= 257.04 \text{ kN.}$$

Then

$$\frac{F_R}{F_S} = \frac{257.04}{182.76} = 1.406 > 1.4$$

Hence, the structure is safe against sliding.

Pressure below the base.

Total vertical force on soil

$$R = W_1 + W_2 + W_3$$
$$= 51.211 + 42.1875 + 347.9$$
$$= 441.3 \text{ kN}$$

Distance of its line of action from O is

$$\overline{X} = \frac{W_1\, W_1 + W_2\, X_2 + W_3\, X_3 - P_a \times \dfrac{H}{3}}{R}$$

$$= \frac{51.211 \times 1.648 + 42.1875 \times 2.25 + 347.9 \times 3.143 - 182.76 \times 2.5}{441.3}$$

$$= 1.849 \text{ m}$$

Eccentricity of R from the centre of base
$$e = 2.25 - 1.849$$
$$= 0.401 \text{ m towards O.}$$

Hence, pressure below the base is

$$= \frac{441.3}{4.5} \pm \frac{441.3 \times 0.401}{\dfrac{1 \times 4.5^2}{6}}$$

$$= 98.07 \pm 52.43$$

Under the toe
$$q_1 = 98.07 + 52.43$$
$$= 150.5 \text{ kN/m}^2 < 155 \text{ kN/m}^2, \text{ safe.}$$

Under the heel
$$q_2 = 98.07 - 52.43$$
$$= 45.64 \text{ kN/m}^2 < 155 \text{ kN/m}^2, \quad \text{safe.}$$

The pressure distribution is shown in Fig. 13.11 (a).

Downward load intensity on the base due to its own weight and that of the supported earth:

At point A $= 0.375 \times 25$ $= 9.38$ kN/m^2

At point B $= 0.375 \times 25$ $= 9.38$ kN/m^2

At point C $= 0.375 \times 25 + 7.125 \times 18$ $= 137.63 \text{ kN/m}^2$

At point D $= 0.375 \times 25 + 7.125 \times 18$ $= 137.63 \text{ kN/m}^2$

These are shown in Fig. 13.11 (b).

The resultant load intensity on the base is shown in Fig. 13.11 (c).

Design of heel slab.

Let the spacing of counterforts be
$$= 0.4H$$
$$= 0.4 \times 7.5$$

= 3 m

Also, let thickness of counterforts be 250 mm.

The heel slab is a countinous horizontal slab spanning over counterforts bearing the net vertical load intensity shown in Fig. 13.11 (c).

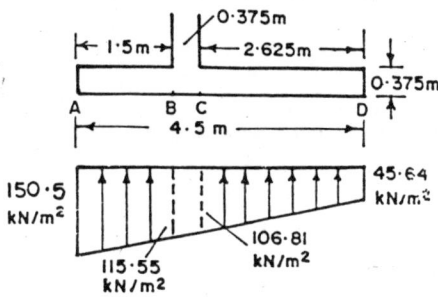

(a) Pressure Distribution below the Base.

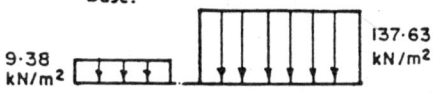

(b) Load Intensity from above.

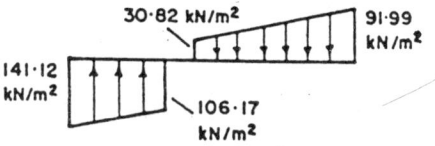

(c) Resultant Load Intensity on Base Slab.

Fig. 13.11

Consider 1 m wide strip of slab. See Fig. 13.2 also.

Negative bending moment near counterforts

$$= \frac{wl^2}{12}$$

$$= \frac{91.99 \times 3^2}{12}$$

$$= 69 \text{ kN-m/m}$$

Design stresses and constants are the same as in Example 13.1.

Effective depth of slab

$$d = \sqrt{\frac{69 \times 10^6}{0.653 \times 1000}} = 3\,25\text{mm}.$$

Negative reinforcement near top surface

$$A_{st} = \frac{69 \times 10^6}{230 \times .904 \times 325} = 1021.1 \text{ mm}^2/\text{m}.$$

Provide 12 mm dia. bars = 110 mm c/c.
Positive bending moment at mid-span

$$= \frac{wl^2}{16}$$

$$= \frac{91.99 \times 3^2}{16}$$

$$= 51.74 \text{ kN-m/m}.$$

Positive reinforcement near bottom surface

$$A_{st} = \frac{51.74 \times 16^6}{230 \times .904 \times 325} = 7\dot{0}5.7 \text{ mm}^2/\text{m}$$

Provide 12 mm dia. bars — 145 mm c/c.
Critical section for shear will be taken at the point where the slab joins with the counterfort. Shear force

$$V = \frac{1}{2}(3 - 0.25) \times 91.99$$

$$= 126.49 \text{ kN}.$$

Nominal shear stress

$$\tau_v = \frac{V}{bd}$$

$$= \frac{126.49 \times 10^3}{1000 \times 325}$$

$$= 0.389 \text{ N/mm}^2$$

$$\frac{100 A_{st}}{bd} = \frac{100 \times 1021.1}{1000 \times 325} = 0.314$$

From Table 3.11,

$$\tau_c = 0.238 \text{ N/mm}^2.$$

Since $\tau_v > \tau_c$, the slab is unsafe in shear.
Now

$$V_c = \tau_c \, bd = 0.238 \times 1000 \times 325 \times 10^{-3}$$

$$= 77.35 \text{ kN}.$$

Distance from the centre of the counterfort along the heel, where the slab is just safe in shear is found by equating

$$(3 - 2X) \times 91.99 = 77.35 \times 2$$

Hence $\qquad X = 0.66$ m.

Similarly, distance of the point from the edge of the heel, along the face of the counterfort, where the slab is just safe in shear is found by equating

$$91.99 - \frac{(91.99 - 30.82) Y}{2.625} = \frac{77.35 \times 2}{(3 - 0.25)}$$

Hence, $\qquad Y = 1.533$ m.

Triangular portions of the heel slab which are unsafe in shear are shown in Fig. 13.12(c). These portions can be strengthened by local thickening of the slab in these region. Increase the effective depth of slab at point 'a' to

$$= \frac{0.389}{0.238} \times 325$$

$$= 531.2 \text{ mm, say } 535 \text{ mm.}$$

This may be tapered off to 325 mm towards the points 'b' and 'c'. See Fig. 13.12 (a) and (b).

Elsewhere, the slab thickness may be kept uniform with an overall thickness $= 325 + 25 = 350$ mm.

Temperature and shrinkage reinforcement

$$= \frac{0.12}{100} \times 350 \times 1000$$

$$= 420 \text{ mm}^2 \text{ /m}$$

Provide 12 mm dia. bars — 270 mm c/c at right angles to the main reinforcement.

Load intensity on heel slab reduces towards the point C to 30.82 kN/m².

Corresponding negative reinforcement will also reduce to

$$= \frac{30.82}{91.99} \times 1021.1$$

$$= 342.1 \text{ mm}^2/\text{m. This is less than } 420 \text{ mm}^2/\text{m.}$$

Hence, provide at least 420 mm²/ m, i.e. 21 mm dia. bars—270 mm c/c. Point of inflexion may be taken at 1/5th of span from supports. Details of reinforcement are shown in Fig. 13.12

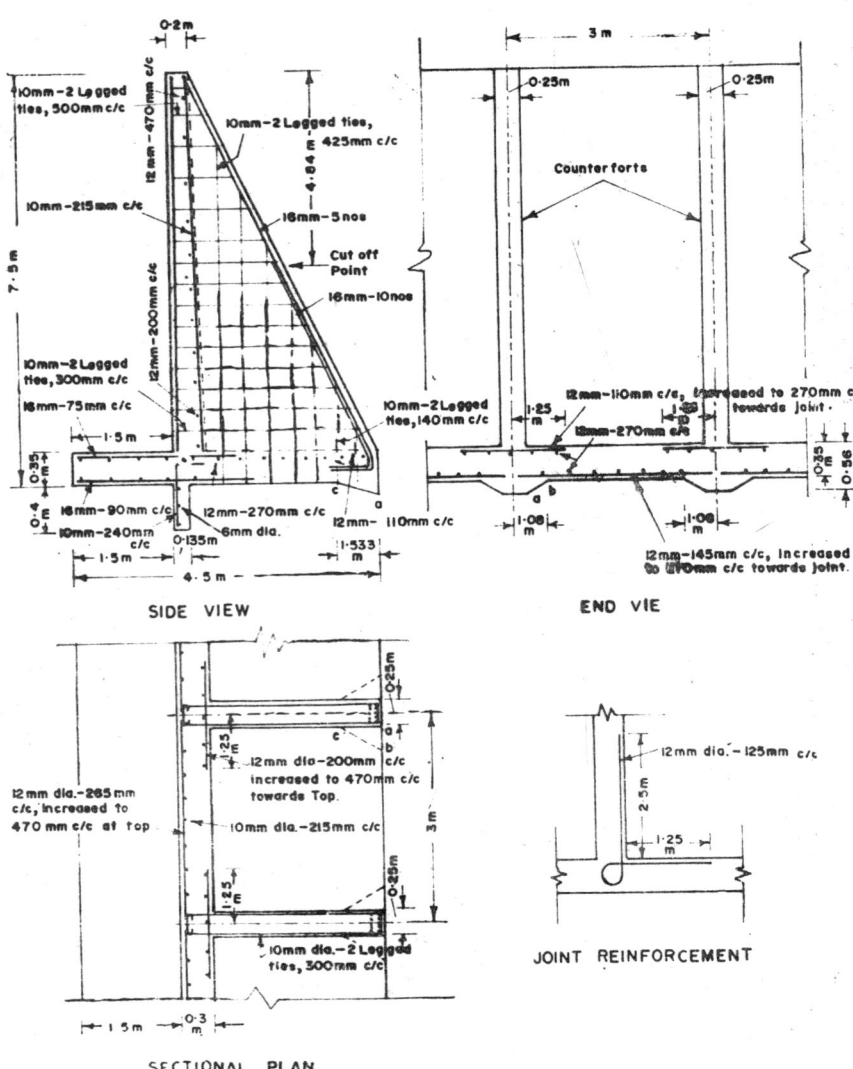

SIDE VIEW

END VIEW

SECTIONAL PLAN

JOINT REINFORCEMENT

Fig. 13.12

Design of vertical slab.

Clear height of vertical slab above the heel

$$= 7.5 - 0.35$$
$$= 7.15 \text{ m.}$$

Consider 1 m high strip running horizontally and continous over the counterforts. Effective span will be 3 m c/c, as was for the heel slab. Average intensity of horizontal pressure of earth on the strip

$$= K_a \, \gamma \, h$$

$$= 0.361 \times 18 \times \frac{(7.15 + 6.15)}{2}$$

$$= 43.21 \text{ kN/m}^2$$

Negative bending moment in the slab near counterforts

$$= 43.21 \times \frac{3^2}{12}$$

$$= 32.41 \text{ kN-m/m}$$

$$d = \sqrt{\frac{32.41 \times 10^6}{0.653 \times 1000}} = 222.8 \text{ mm, say 225 mm.}$$

$$A_{st} = \frac{32.41 \times 10^6}{230 \times .904 \times 225} = 692.8 \text{ mm}^2/\text{m.}$$

$$\frac{100 \, A_{st}}{bd} = \frac{100 \times 692.8}{225 \times 1000} = 0.308$$

From Table 3.11, $\tau_o = 0.236$ N/mm^2.
Critical shear force

$$V = \tfrac{1}{2} (3 - 0.25) \times 43.21$$
$$= 59.41 \text{ kN.}$$

$$\therefore \tau_v = \frac{59.4 \times 10^3}{1000 \times 225} = 0.264 \text{ N/mm}^2$$

As $\tau_v > \tau_o$, the wall is unsafe in shear.
Increase the effective depth to 275 mm.
Then

$$A_{st} = \frac{32.41 \times 10^6}{230 \times .904 \times 275} = 566.8 \text{ mm}^2/\text{m.}$$

Provide 12 mm dia. bars horizontally, at a spacing of 200 mm c/c.

$$\frac{100 \, A_{st}}{bd} = \frac{100 \times 566.8}{1000 \times 275} = 0.219$$

From Table 3.11, $\tau_c = 0.22$ N/mm^2.

$$\tau_v = \frac{59.41 \times 10^3}{1000 \times 275} = 0.216 \text{ N/mm}^2$$

As $\tau_v < \tau_c$, the wall is safe in shear.

Overall thickness $= 275 + 25 = 300$ mm.

Temperature and shrinkage reinforcement

$$= \frac{0.12}{100} \times 300 \times 1000$$

$$= 360 \text{ mm}^2/\text{m}.$$

Provide 10 mm dia. bars vertically, at a spacing of 215 mm c/c.

Positive bending moment at mid-span

$$= \frac{43.21 \times 3^2}{16}$$

$$= 24.3 \text{ kN} - \text{m/m}.$$

$$= \frac{24.3 \times 10^6}{230 \times .904 \times 275} = 425 \text{ mm}^2/\text{m}.$$

Provide 12 mm dia. bars horizontally, at a spacing of 265 mm c/c. The thickness of slab may be reduced towards the top to 200 mm. The spacing of the 12 mm dia. horizontal bars may be increased to 470 mm c/c towards top. This will give a reinforcement area

$$= \frac{113.1 \times 1000}{470} = 240.64 \text{ mm}^2/\text{m, where as, the temperature and}$$

shrinkage reinforcement requirement at top will be $= \dfrac{0.12}{100} \times 200 \times$ $1000 = 240$ mm^2/m only. Hence, it will be alright. Details of reinforcement are shown in Fig. 13.12.

Design of Toe.

The toe acts as a cantilever having upward load intensity varying as shown in Fig. 13.11 (c).

Bending moment is

$$M = (2 \times 141.12 + 106.17) \ \frac{1.5^2}{6}$$

$$= 145.65 \text{ kN-m/m}$$

As a singly reinforced section, the depth will come out to be excessive. Hence, it is proposed to design it as a doubly reinforced section. Keep the overall thickness same as that of the heel slab i.e. 350 mm. Then, effective depth $= 325$ mm.

Bending strength of a singly reinforced balanced section

$$M_b = 0.653 \times 1000 \times 325^2 \times 10^{-6}$$
$$= 68.97 \text{ kN-m/m}$$

Then, the bending moment taken up by the steel section is

$$M' = M - M_b = 145.65 - 68.97$$
$$= 76.68 \text{ kN-m/m}$$

Stress in concrete at the level of compression reinforcement

$$\sigma_{obc} = \frac{(nd - d_c)}{nd} \times \sigma_{cbc}$$

$$= \frac{(0.289 \times 325 - 25)}{0.289 \times 325} \times 5$$

$$= 3.67 \text{ N/mm}^2$$

Area of compression reinforcement

$$A_{sc} = \frac{M'}{(1.5 \, m - 1) \, \sigma'_{cbc} \, (d - d_c)}$$

$$= \frac{76.68 \times 10^6}{(1.5 \times 18.67 - 1) \times 3.67 \times (325 - 25)}$$

$$= 2579 \text{ mm}^2/\text{m}$$

Provide 16 mm dia. bars — 75 mm c/c at top.

Area of tension reinforcement

$$A_{st} = \frac{M_b}{\sigma_{st} \, jd} + \frac{M'}{\sigma_{st} \, (d - d_c)}$$

$$= \frac{68.97 \times 10^6}{230 \times .904 \times 325} + \frac{76.68 \times 10^6}{230 \times (325 - 25)}$$

$$= 1020.7 + 1111.30$$

$$= 2132 \text{ mm}^2/\text{m}$$

Provide 16 mm dia. bars — 90 mm c/c at bottom. Actual

$$A_{st} = 201.06 \times \frac{1000}{90} = 2334 \text{ mm}^2/\text{m}.$$

$$\frac{100 \, A_{st}}{bd} = \frac{100 \times 2334}{1000 \times 325} \quad 0.687$$

From Table 3.11, $\tau_c = 0.327 \text{ N/mm}^2$.

Critical section for shear occurs at 0.325 m from the fixed end.

Shear force

$$V = \tfrac{1}{2} \left(141.12 + 106.17 + \frac{(141.12 - 106.17)}{1.5} \times 0.325\right)$$
$$\times (1.5 - 0.325)$$

$$= 149.73 \text{ kN/m}$$

Then

$$\tau_v = \frac{149.73 \times 10^3}{1000 \times 325} = 0.461 \text{ N/mm}^2$$

As $\tau_v > \tau_0$ the slab is unsafe in shear.

Now

$$V_s = V - \tau_0 \, bd$$
$$= 149.73 - 0.327 \times 1000 \times 325 \times 10^{-3}$$
$$= 43.46 \text{ kN/m}$$

Inclind shear reinforcement bars may be provided which are given by

$$V_s = \sigma_s v \, A_s v \, \text{Sin } \alpha$$

Provide 10 mm dia. bars at 45°. The spacing is found by substituting the values in the above expression.

Hence

$$43.46 \times 10^3 = 230 \times 78.54 \times \frac{1000}{s_v} \times \sqrt{\frac{1}{2}}$$

$\therefore$ $s_v = 294$ mm, say 290 mm c/c.

Temperature and shrinkage reinforcement is same as for the heel slab i.e. 12 mm dia, bars — 270 mm c/c.

See details in Fig. 13.12.

Design for shear key

The key projects 0.4 m below the base, and its lowest point is 1.9 m below the G L. as shown in Fig. 13.10. Consider 1 m run in longitudinal direction.

Bending moment on key

$$= \int_{1.5}^{1.9} K_p \, \gamma \, h \, (h - 1.5) \, dh$$

$$= 2.77 \times 18 \int_{1.5}^{1.9} (h^2 - 1.5 \, h) \, dh$$

$$= 7.047 \text{ kN·m/m}$$

Hence

$$d = \sqrt{\frac{7.047 \times 10^6}{0.653 \times 1000}} = 103.9 \text{ mm, say 105 mm.}$$

$$A_{st} = \frac{7.047 \times 10^6}{230 \times .904 \times 105} = 322.8 \text{ mm}^2/\text{m}$$

Provide 10 mm dia. bars — 240 mm c/c.

Actual $A_{st} = 78.54 \times \dfrac{1000}{240} = 327.25$ mm^2.

$$\frac{100\, A_{st}}{bd} = \frac{100 \times 327.25}{1000 \times 105} = 0.312$$

From Table 3.11, $\tau_o = 0.238$ N/mm^2.

Critical section occurs 0.105 m below the base level.

Shear force at this section is

$$V = \int_{1.605}^{1.9} K_p\, \gamma\, h\, dh$$

$$= 2.77 \times 18 \int_{1.605}^{1.9} h\, dh$$

$$= 25.78 \text{ kN/m}$$

Hence $\tau_v = \dfrac{25.78 \times 10^3}{1000 \times 105} = 0.2455$ N/mm^2

As $\tau_v > \tau_o$ the section is unsafe in shear. Increase the effective depth to

$$= 105 \times \frac{0.2455}{0.238}$$

$$= 108.3 \text{ mm, say } 110 \text{ mm.}$$

Overall thickness $= 110 + 25 = 135$ mm.

Keep the reinforcement unchanged.

Temperature and shrinkage reinforcement

$$= \frac{0.12 \times 135 \times 1000}{100}$$

$$= 162 \text{ mm}^2/\text{m.}$$

Provide 6 mm dia. bars — 175 mm c/c. at right angles to main bars.

For details see Fig. 13.12.

Design of counterforts.

The counterfort acts as the web while the vertical slab acts as the flange of a T — beam. Due to earth pressure, the T — beam bends. Compression can be safely taken up by the vertical slab. To take up tension, steel reinforcement should be provided along the sloping edge of the counterfort. The angle which the sloping edge makes with the vertical is $\alpha = 21.45°$.

One counterfort supports earth pressure from 3 m width of vertical slab. Maximum bending moment at the base

$$= \frac{1}{6} K_a \gamma h^3 \times \text{spacing}$$

$$= \frac{1}{6} \times 0.361 \times 18 \times 7.125^3 \times 3$$

$$= 1175.2 \text{ kN-m.}$$

Even as a rectangular beam, the required effective depth will be

$$= \sqrt{\frac{1175.2 \times 10^6}{0.653 \times 250}} = 2683 \text{ mm.}$$

Overall depth of counterfort available at the base = 3000 mm and effective depth available will be at least, say = 3000 — 100 = 2900 mm, which is more than required.

Taking $j = 0.95$ for the T — beam, area of tension reinforcement

$$A_{st} = \frac{M}{\sigma_{st} \, j \, d \cos \alpha}$$

$$= \frac{1175.2 \times 10^6}{230 \times 0.95 \times 2900 \times \cos 21.45}$$

$$= 1992.7 \text{ mm}^2.$$

Provide 16 mm dia. bars — 10 nos., in two layers of 5 bars each.

Horizontal ties are required to bind the vertical slab to the counterfort. Consider a strip of unit width running horizontally. Force which is trying to separate the strip from counterfort

$$= K_a \gamma h \times (3 - 0.25)$$

$$= 0.361 \times 18 \times 2.75 \times h$$

$$= 17.87 \, h \text{ kN/m height.}$$

For unit strip at bottom, $h = 6.625$ m.

$$\therefore \quad \text{Force} = 17.87 \times 6.625$$

$$= 118.39 \text{ kN/m height}$$

Area of binder reinforcement required

$$= \frac{118.39 \times 10^3}{230}$$

$$= 514.74 \text{ mm}^2/\text{m}$$

Temperature and shrinkage reinforcement

$$= \frac{0.12}{100} \times 250 \times 1000$$

$$= 300 \text{ mm}^2/\text{m.}$$

At bottom, provide 10 mm dia. 2 — legged ties horizontally, with a spacing of 300 mm c/c in vertical direction.

The spacing may be increased towards top to 500 mm c/c. The counterfort may be checked in shear taking the critical section at the bottom. Total shear force

$$V = \tfrac{1}{2} K_a \, \gamma \, h^2 \times 3$$
$$= \tfrac{1}{2} \times 0.361 \times 18 \times 7.125^2 \times 3$$
$$= 494.81 \text{ kN.}$$

Now

$$V - \frac{M}{d} \tan \alpha = 494.81 - \frac{1175.2}{2.90} \tan 21.45°$$
$$= 335.6 \text{ kN.}$$

Then

$$\tau_v = \frac{V - \dfrac{M}{d} \tan \alpha}{bd}$$

$$= \frac{335.6 \times 10^6}{250 \times 2900}$$

$$= 0.463 \text{ N/mm}^2$$

$$\frac{100 \, A_{st}}{bd} = \frac{100 \times 10 \times 201.06 \times \cos 21.45°}{250 \times 2900}$$

$$= 0.258$$

From Table 3.11, $\tau_c = 0.222$ N/mm^2

$$V_o = \tau_c \, bd = 0.222 \times 250 \times 2900 \times 10^{-3}$$
$$= 160.95 \text{ kN.}$$

For the binders already provided

$$V_s = \frac{\sigma_{sv} \, A_{sv} \, d}{s_v} = \frac{230 \times 2 \times 78.54 \times 2900 \times 10^{-3}}{300}$$

$$= 349.24 \text{ kN.}$$

$$V_c + V_s = 160.95 + 349.24 = 510.19 \text{ kN.}$$

Evidently, $V_c + V_s > V - \dfrac{M}{d} \tan \alpha$

Hence, the section is safe in shear.

The bending moment in the counterfort decreases towards the top. The reinforcement can, therefore, be curtailed accordingly. At any point

$$A_{st} \, \alpha \, \frac{M}{d}$$

$$\alpha \, \frac{h^3}{h}$$

$$\alpha \, h^2$$

Hence, depth from top where 5 bars out of the total 10 are no longer required is given by

$$h = \frac{7.125}{\sqrt{2}} = 5.038 \text{ m.}$$

It is proposed to extend the 5 bars 12 times the diameter beyond this point and then curtail them. Hence, actual point of cut-off from top

$$= 5.038 - 12 \times 0.016$$
$$= 4.846 \text{ m, say } 4.48 \text{ m.}$$

Shear force at this point

$$V = \tfrac{1}{2} \times 0.361 \times 18 \times 4.84^2 \times 3$$
$$= 228.33 \text{ kN.}$$

Effective depth at this point

$$= 0.2 + 4.84 \tan 21.45 - 0.1$$
$$= 2.0 \text{ m.}$$
$$A_{st} = 5 \times 201.06 = 1005.3 \text{ mm}^2.$$

$$\therefore \frac{100 \, A_{st}}{bd} = \frac{100 \times 1005.3}{250 \times 2000} = 0.2$$

From Table 3.11, $\tau_c = 0.22$ N/mm^2

$$\therefore V_c = 0.22 \times 250 \times 2000 \times 10^{-3}$$
$$= 110 \text{ kN.}$$

For 10 mm dia. 2-legged binders spaced at 300 mm c/c

$$V_s = \frac{230 \times 2 \times 78.54 \times 2000}{300} \times 10^{-3}$$

$$= 240.86 \text{ kN.}$$
$$V_c + V_s = 110 + 240.86 = 350.86 \text{ kN.}$$

$$\tfrac{2}{3} (V_c + V_s) = \tfrac{2}{3} \times 350.86 = 232.91 \text{ kN.}$$

Evidently $V < \dfrac{2}{3} (V_c + V_s)$

Hence, the curtailment at the point is permissible.

Vertical ties are needed to connect the heel slab to the counterfort. Consider a 1 m wide strip of heel slab running horizontally and supported by the counterforts. Clear width of slab supported by a counterfort $= 3 - 0.25 = 2.75$ m. Resultant load intensity on heel slab is shown in Fig. 13.11 (c). Total downward force trying to separate the strip of unit width from the counterfort near point D is

$$= 91.99 \times 2.75$$
$$= 252.973 \text{ kN/m.}$$

Provide 10 mm dia. 2-legged stirrups in vertical direction and spaced at 140 mm c/c horizontally. Towards the point C the vertical load intensity is 30.82 kN/m only. The total downward force will be

$$= 30.82 \times 2.75$$
$$= 84.76 \text{ kN/m.}$$

For the 10 mm dia. 2-legged ties provided in vertical direction, the spacing horizontally will be 425 mm c/c. The spacing should be gradually increased from D to C. Details are given in Fig. 13.12.

Joint reinforcement.

At the junction of the upright slab and the heel, a rigid connection can be ensured by providing additional joint reinforcement. 0.3% of the concrete area may be provided in the vertical slab extending for a height = 2.5 m. These bars should be embedded in the heel slab also. Sectional area of joint reinforcement

$$= \frac{0.3}{100} \times 300 \times 1000$$
$$= 900 \text{ mm}^2/\text{m}$$

Provide 12 mm dia. bars — 125 mm c/c.

Details are shown in Fig. 13.12.

14

Beams Curved in Plan

14.1 ARCATE BEAMS WITH ONE END FIXED AND OTHER FREE

A beam curved in plan is known as an arcate beam. Consider the beam curved in plan as shown in Fig. 14.1 (a). It is free at the end A and fixed at the end B. The beam is curved at a constant radius = R and subtends an angle θ at the centre of curvature O. A point C on the beam is located at an angle ϕ anticlockwise with respect to the free end. A downward point load W is acting at the free end A. Due to this load three internal reactions will be generated in the beam at

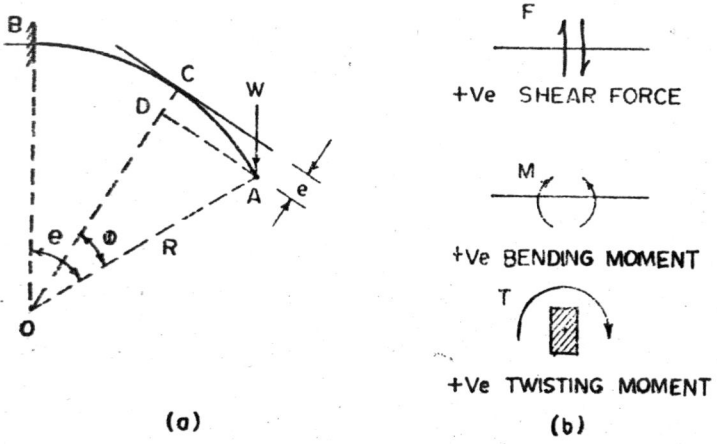

+Ve SHEAR FORCE

+Ve BENDING MOMENT

+Ve TWISTING MOMENT

(a) (b)

Fig. 14.1

point C. These are: shear force F_ϕ, bending moment M_ϕ, and twisting moment T_ϕ. The sign convention for the three internal reactions is shown in Fig. 14.1(b). Viewing at the beam section in the increasing direction of ϕ, a twisting moment with a clockwise sense about the beam axis is considered positive. Draw AD perpendicular to OC. Hence, AD is parallel to the tangent at C.

$$AD = R \sin \phi$$

and $\qquad e = R (1 - \cos \phi)$

the three internal reactions in the beam at point C are

$$F_\phi = W \tag{14.1}$$

$$M_\phi = - W \times AD$$
$$= - WR \sin \phi \tag{14.2}$$

$$T_\phi = - W \times e$$
$$= - WR (1 - \cos \phi) \tag{14.3}$$

Next, consider another curved beam, free at the end A, and fixed at the end B, as shown in Fig. 14.2(a). It carries a uniformly distributed load w per m over its portion AC only. Point C is located at an angle ϕ in anticlockwise sense with respect to the free end A. The centroid of the arc AC is at G which is also the centre of gravity of the load placed on it. The distance of the centroid of a circular arc with respect to the centre is illustrated in Fig. 14.2(b).

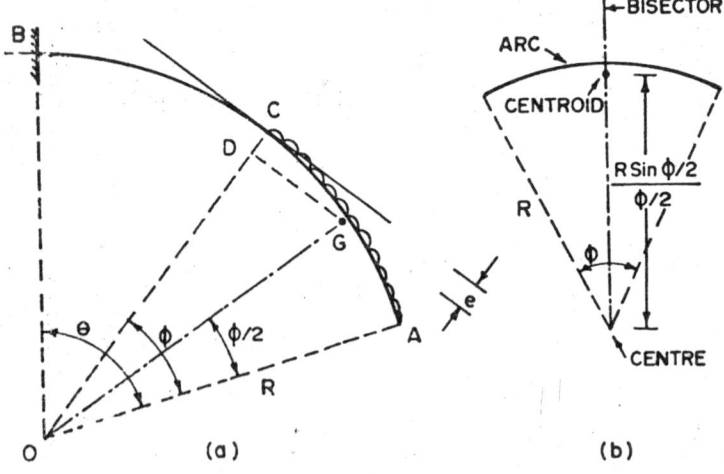

Fig. 14.2

Hence in Fig. 14.2(a)

$$OG = R \cdot \frac{\sin \phi/2}{\phi/2}$$

$$DG = R \cdot \frac{\sin \phi/2}{\phi/2} \times \sin \phi/2$$

$$= R \frac{(1 - \cos \phi)}{\phi}$$

$$e = R - R \frac{\sin \phi/2}{\phi/2} \times \cos \phi/2$$

$$= R \left(1 - \frac{\sin \phi}{\phi} \right)$$

$$= R \left(\frac{\phi - \sin \phi}{\phi} \right)$$

Load on AC $\quad = wR \phi$

Hence, the three internal reactions in the beam at point C are

$$F_\phi = wR \phi \qquad\qquad (14.4)$$

$$M_\phi = - wR \phi \times DG$$

$$= - wR \phi \times R \frac{(1 - \cos \phi)}{\phi}$$

$$= - wR^2 (1 - \cos \phi) \qquad\qquad (14.5)$$

$$T_\phi = - wR \phi \times e$$

$$= - wR \phi \times R \frac{(\phi - \sin \phi)}{\phi}$$

$$= - wR^2 (\phi - \sin \phi) \qquad\qquad (14.6)$$

Example 14.1

An acrate beam fixed at one end and free at the other, has a radius = 6 m and subtends an angle of 30° at the centre. It carries a uniformly distributed load of 30 kN/m. Draw the shear force, bending moment and twisting moment diagrams for the beam.

Solution

$$R = 6 \text{ m}$$

$$\theta = 30° \quad \text{or} \quad 0.5236 \text{ radians}$$

$$w = 30 \ kN/m$$
$$wR = 30 \times 6 = 180 \ kN$$
$$wR^2 = 30 \times 6^2 = 1080 \ kN\text{-}m.$$

Eqs. (14.4), (14.5) and (14.6) become

$$F_\phi = 180 \ \phi$$
$$M_\phi = -1080 \ (1 - \cos \phi)$$
$$T_\phi = -1080 \ (\phi - \sin \phi)$$

Values of the three internal reactions calculated for different values of ϕ are given below:

$\phi =$	30°	24°	18°	12°	6°	0° Degree
$=$	0.5236	0.4189	0.3142	0.2094	0.1047	0 Radians
$F_\phi =$	94.25	75.40	56.56	37.69	18.85	0 kN
$M_\phi =$	-144.69	-93.37	-52.86	-23.60	-5.92	0 kN-m
$T_\phi =$	-25.49	-13.14	-5.60	-1.61	-0.19	0 kN-m

Corresponding diagrams are shown in Fig. 14.3.

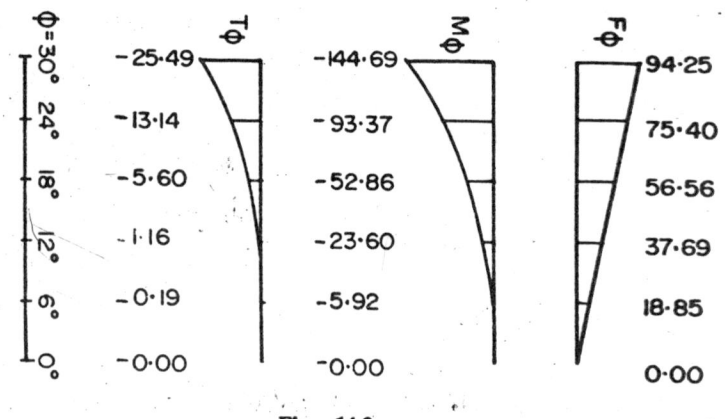

Fig. 14.3

14.2 UNIFORMLY LOADED ARCATE BEAM WITH BOTH ENDS FIXED

A beam curved in plan with both of its ends fixed is shown in Fig. 14.4(a). It is curved to a radius R subtending an angle 2θ at its centre of curvature O. The beam carries a uniformly distributed load w per m. In general, such a beam is statically indeterminate by three degrees. However, due to symmetry of the beam and loading, about the mid-span section C, both shear force and twisting moment will be zero at that section. The only internal reaction which exists at section C is the positive bending moment M_0, which is statically indeterminate. As such, the degree of indeterminacy of the beam is reduced to one only. M_0 can be determined from the principle of minimum strain energy.

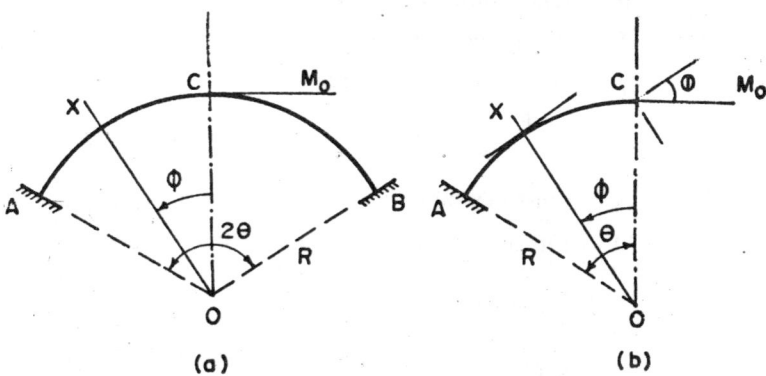

Fig. 14.4

Let X be a section of the beam, which is located at an angle ϕ in anticlockwise sense with respect to the mid-span section C. See Fig. 14.4(a). Anticlockwise angles are considered positive. Hence, for the complete beam $-\theta \leqslant \phi \leqslant \theta$. At section X of the beam, three internal reactions exist viz. shear force, bending moment and twisting moment. As the beam is symmetrical about the mid-span section C, only half portion AC of the beam may be considered for the purpose of analysis. Let the loaded beam be cut at the point C and a positive moment M_0 be applied in a vertical plane tangential to the beam at that point. See Fig. 14.4 (b). Thus, the loading conditions of the cut portion AC remain exactly the same as those in the original beam. The three internal reactions at section X can be written down in the light of Eqs. (14.4), (14.5) and (14.6) as follows:

$$F_\phi = wR\phi \qquad (14.7)$$

$$M_\phi = M_0 \cos \phi - wR^2 (1 - \cos \phi) \qquad (14.8)$$

$$T_\phi = M_0 \sin \phi - wR^2 (\phi - \sin \phi) \qquad (14.9)$$

M_0 is the unknown bending moment at mid-span section of beam. Strain energy of the beam due to bending and twisting is

$$u = 2 \int_0^\theta \frac{M_\phi^2 \, Rd\phi}{2 \, EI} + 2 \int_0^\theta \frac{T_\phi^2 \, Rd \, \phi}{2 \, GJ} \qquad (14.10)$$

The contribution of shear to strain energy is very small in comparison to the above, and hence, it is neglected. Here,

E = elastic modulus of concrete

I = second moment of area of section

G = shear modulus of concrete

$$= \frac{E}{2 \, (1 + \nu)}$$

ν = poisson's ratio

J = torsion constant of beam section

$$= \frac{1}{3} \, b^3 d \left(1 - 0.62741 \, \frac{b}{D} \, \sum_{n=1,3,5}^\infty \, \tanh \, \frac{n \pi D}{2b} \right)$$

$$= K_1 b^3 D \qquad (14.11)$$

Values of the coefficient K_1 calculated for different values of b/d are mentioned below:

D/b = 1.0	1.2	1.5	2.0	2.5	3.0
K_1 = 0.1406	0.166	0.196	0.229	0.249	0.263

For the strain energy to be a minimum,

$$\frac{\partial u}{\partial M_0} = 0$$

i.e.

$$2 \int_0^\theta \frac{M_\phi}{EI} \frac{\partial M_\phi}{\partial M_0} \, Rd\phi + 2 \int_0^\theta \frac{T_\phi}{GJ} \frac{\partial T_\phi}{\partial M_0} \, Rd\phi = 0$$

Substituting the values of M_ϕ and T_ϕ from Eqs. (14.8) and (14.9), dropping the common factors, and rewriting

$$\int_0^\theta M_0 \cos \phi - wR^2 (1 - \cos \phi) \cos \phi \, d \, \phi + \frac{EJ}{GJ} \, \times$$

$$\int_0^\theta M_0 \sin \phi - wR^2 (\phi - \sin \phi) \sin \phi \, d \phi = 0$$

On carrying out the integration

$$M_0 \left\{ \frac{\theta}{2} + \frac{\sin 2\theta}{4} + \lambda \left(\frac{\theta}{2} - \frac{\sin 2\theta}{4} \right) \right\} + wR^2$$

$$\left\{ \left(\frac{\theta}{2} - \sin \theta + \frac{\sin 2\theta}{4} \right) + \lambda \left(\cos \theta - \sin \theta + \frac{\theta}{2} \right. \right.$$

$$\left. \left. - \frac{\sin 2\theta}{4} \right) \right\} = 0 \qquad (14.12)$$

where, $\lambda = \dfrac{EI}{GJ}$

For a rectangular section

$$I = \frac{1}{12} bD^3$$

$$J = K_1 b^3 D$$

Hence, $\quad \lambda = E \times \dfrac{1}{12} \times bD^3 \times \dfrac{2(1 + v)}{E} + \dfrac{1}{K_1 b^3 D} \qquad (14.13)$

$$= \frac{(1 + v)}{6} \left(\frac{D}{b} \right)^2$$

where, $\quad D = $ overall depth of the section

$\quad b = $ width of section

From Eq. (14.12), the value of M_0 can be computed for the known values of θ, λ, R and w. The solution can be written as

$$M_0 = \alpha \, wR^2 \qquad (14.14)$$

where

$$\alpha = - \frac{\left(\dfrac{\theta}{2} - \sin \theta + \dfrac{\sin 2\theta}{4} \right) + \lambda \left(\theta \cos \theta - \sin \theta + \dfrac{\theta}{2} - \dfrac{\sin 2\theta}{4} \right)}{\dfrac{\theta}{2} + \dfrac{\sin 2\theta}{4} + \lambda \left(\dfrac{\theta}{2} - \dfrac{\sin 2\theta}{4} \right)}$$

$$(14.15)$$

Values of the coefficient α calculated from Eq. (14.15), taking $\nu = 0.15$, for the arcate beams of rectangular cross section, for different values of D/b ratio and half beam angle θ, are given in Table 14.1. Thus, knowing the value α for a given arcate beam, the mid-span

Table 14.1

CALCULATED VALUES OF THE COEFFICIENT α FOR UNIFORMLY
LOADED ARCATE BEAM FIXED AT THE TWO ENDS.

$(\nu = 0.15)$

Half beam angle	Ratio D/b					
	1.0	1.2	1.5	2.0	2.5	3.0
15	0.011194	0.011164	0.011113	0.011006	0.010878	0.010734
18	0.016203	0.016152	0.016064	0.015885	0.015674	0.015444
20	0.019792	0.019696	0.019531	0.019199	0.018814	0.017577
22½	0.024909	0.024771	0.024532	0.024061	0.023526	0.022967
25	0.030772	0.030600	0.030307	0.029738	0.029106	0.028462
30	0.043344	0.042963	0.042325	0.041129	0.039862	0.038633
35	0.057912	0.057268	0.056211	0.054299	0.052372	0.050589
40	0.074205	0.073238	0.071686	0.068983	0.066386	0.064090
45	0.091668	0.090296	0.088141	0.084528	0.081212	0.078401
60	0.150246	0.147660	0.143861	0.138125	0.133456	0.129875
67½	0.181417	0.178523	0.1744	0.168448	0.163831	0.160419
90	0.273237	0.273237	0.273237	0.273237	0.273237	0.273237

bending moment M_0 can be calculated from Eq. (14.14) and hence, the Eqs. (14.7), (14.8) and (14.9) are fully known.
Twisting moment will be maximum where

$$\frac{dT_\phi}{d\phi} = 0$$

i.e. $$M_0 \cos \phi - wR^2 (1 - \cos \phi) = 0$$

or $\qquad M_\phi = 0$

Or, where the points of contraflexure occur in the beam.

Example 14.2

An arcate beam fixed at its two ends is curved to a radius $= 6$ m and subtends an angle of $60°$ at its centre. For the beam section, ratio $D/b = 2.0$ Imposed load (Dead + Live) on the beam $= 27$ kN/m. Self weight of the beam $= 9$ kN/m. Analyse for the shear force bending moment and the twisting moment and draw diagrams showing their variation along the beam. Take $\nu = 0.15$.

Solution

Total distributed load on beam

$$w = 27 + 9$$
$$= 36 \text{ kN/m}$$

Radius $\qquad R = 6$ m

Angle $\qquad 2\theta = 60°$

$\therefore \qquad \theta = 30°$

Value of the coefficient α can now be calculated from Eq. (14.15). However, its calculated value is already given in Table 14.1, which, for $\theta = 30°$ and $D/b = 2.0$ is

$$\alpha = 0.041129$$

Hence, from Eq. (14.14)

$$M_0 = 0.041129 \times 36 \times 6^2$$
$$= 53.3 \text{ kN-m.}$$

The three internal reactions from Eqs. (14.7), (14.8) and (14.9) can now be written as

$$F_\phi = 36 \times 6\phi$$
$$= 216\phi \text{ kN}$$
$$M_\phi = 53.3 \cos \phi - 36 \times 6^2 (1 - \cos \phi)$$
$$= 53.3 \cos \phi - 1296 (1 - \cos \phi) \text{ kN-m}$$
$$T_\phi = 53.3 \sin \phi - 1296 (\phi - \sin \phi) \text{ kN-m}$$

Point of contraflexure is found by equating

$$M_\phi = 0$$

i.e. $53.3 \cos \phi - 1296 (1 - \cos \phi) = 0$

or $\cos \phi = 0.9605$

∴ $\phi = 16.158°$

At this point maximum twisting moment occurs. Its value is

$$T_{\phi max} = 53.3 \sin (16.158) - 1296 \left(\frac{16.158}{180} \times \pi - \sin (16.158) \right)$$

$$= 14.833 - 4.812$$

$$= 10.02 \quad \text{kN-m}$$

Values of F_ϕ, M_ϕ and T_ϕ calculated for the values of ϕ varying from one end of the beam to the other are given below in Table 14.2.

Table 14.2

Degrees	F_ϕ kN	M_ϕ kN-m	T_ϕ kN-m	Remarks
− 30	− 113.1	− 127.47	3.93	End B
− 25	− 92.25	− 73.12	− 4.75	
− 20	− 75.4	− 28.07	− 9.10	
− 16.158	− 60.92	0.00	− 10.02	Pt. of inflexion
− 15	− 56.55	7.32	− 9.93	
− 10	− 37.7	32.80	− 8.11	
− 5	− 18.85	48.17	− 4.50	
0	0.00	53.30	0.00	Mid-span
5	18.85	48.17	4.50	
10	37.7	32.80	8.11	
15	56.55	7.32	9.93	
16.158	60.92	0.00	10.02	Pt. of inflexion
20	75.4	− 28.07	9.10	
25	92.25	− 73.12	4.75	
30	113.1	− 127.47	− 3.93	End A.

The same values are diagrammatically shown in Fig. 14.5.

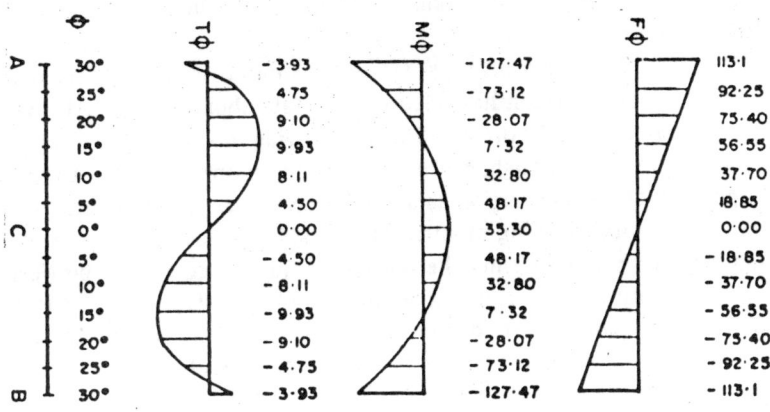

Fig. 14.5

14.3 UNIFORMLY LOADED RING BEAM HAVING EQUALLY SPACED SUPPORTS

The plan of a uniformly loaded ring beam having equally spaced supports is shown in Fig. 14.6(a). If the ring beam is rigidly connected

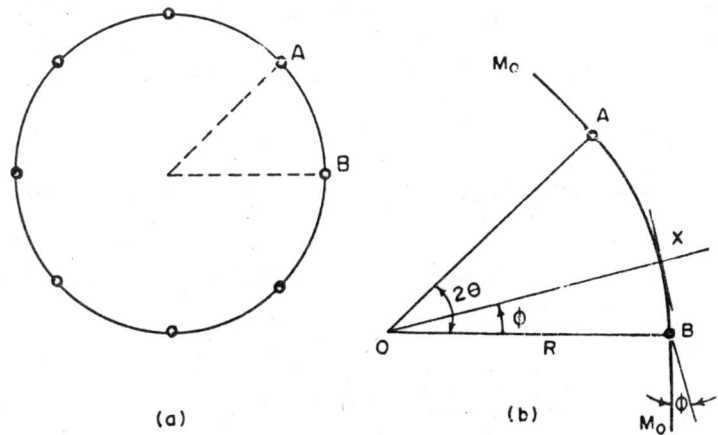

(a) (b)

Fig. 14.6

to the supports then a sector of the beam between two consecutive supports may be treated as a uniformly loaded arcate beam having the two ends fixed. The analysis for such a beam has already been dealt with under para 14.2.

However, if the ring beam is simply resting over equally spaced supports, then the beam has freedom to rotate about its own axis over the supports. As such, there will be no twisting moment in the beam over the supports, though, a negative bending moment will exist there. In general, at a beam section there will exist three internal reactions viz., shear force, bending moment and twisting moment. These can be analysed for by considering the portion of the loaded beam between two consecutive supports as shown in Fig. 14.6(b). The beam AB is symmetrical about its mid-span section. As such, the shear force and the twisting moment will be zero at mid-span section where only positive bending moment will exist. In the beam, over each of the supports A and B, negative bending moment M_0 exists in the vertical planes tangential to the beam at those points respectively. Twisting moment in beam over the supports A and B is zero. The beam AB is curved to radius R and subtends an angle 2θ at its centre of curvature. Consider a section X of the beam located at an angle ϕ in anticlockwise sense with respect to the support point B. The three internal reactions at section X can be written down as

$$F_\phi = - wR\,\theta + wR\,\phi$$

$$= - wR\,(\theta - \phi) \tag{14.16}$$

$$M_\phi = - M_0 \cos\phi + wR\,\theta \times R\sin\phi$$

$$- wR\,\phi \times R\,\frac{\sin\phi/2}{\phi/2} \times \sin\phi/2$$

$$= - M_0 \cos\phi + wR^2\,\theta\sin\phi - wR^2\,(1 - \cos\phi) \tag{14.17}$$

$$T_\phi = - M_0 \sin\phi + wR\,\theta \times R\,(1 - \cos\phi)$$

$$- wR\,\phi \times \left(R - R\,\frac{\sin\phi/2}{\phi/2}\cos\phi/2 \right)$$

$$= - M_0 \sin\phi + wR^2\,\theta\,(1 - \cos\phi)$$

$$- wR^2\,(\phi - \sin\phi) \tag{14.18}$$

In Eqs. (14.17) and (14.18) the only unknown quantity is M_0. Its value can be determined by considering that the twisting moment is zero at the mid-span section. On substituting $\phi = \theta$ in the R.H.S. of Eq. (14.18) and setting it to zero

$$M_0 \sin \theta + wR^2 \theta (1 - \cos \theta) - wR^2 (\theta - \sin \theta) = 0$$

Hence $\quad M_0 = wR^2 (1 - \theta \cot \theta)$ $\qquad\qquad$ (14.19)

The value of M_0 can now be determined from Eq. (14.19) above. Hence, the R.H.S. of Eqs. (14.17) and (14.18) become completely known and the values of F_ϕ, M_ϕ, and T_ϕ can be determined for any value of ϕ.

Example 14.3

A ring beam of radius = 5.7 m is simply resting over six equally spaced support points. Imposed load (Dead + Live) on the beam = 24 kN/m. Self weight of the beam = 6 kN/m. Analyse for the shear force, bending moment and twisting moment and show their variation along the beam.

Solution

Here

$$R = 5.7 \text{ m}$$
$$w = 24 + 6$$
$$= 30 \text{ kN/m}$$
$$\theta = 30° \quad \text{or} \quad 0.5236 \text{ radians}$$

From Eq. (14.19) support moment is

$$M_0 = 30 \times 5.7^2 (1 - 0.5236 \cot 30°)$$
$$= 90.74 \quad \text{kN-m}$$

From Eq. (14.17) bending moment at any section at an angular distance ϕ from the support is

$$M_0 = - 90.74 \cos \phi + 30 \times 5.7^2 \times 0.5236 \sin \phi$$
$$- 30 \times 5.7^2 (1 - \cos \phi)$$
$$= 883.96 \cos \phi + 510.35 \sin \phi - 974.70$$

From Eq. (14.18) twisting momei ⌐c that section

$$T_\phi = -90.74 \sin\phi + 30 \times 5.7^2 \times 0.5236 (1 - \cos\phi)$$
$$- 30 \times 5.7^2 (\phi - \sin\phi)$$
$$= 883.96 \sin\phi - 510.35 \cos\phi - 974.7\phi + 510.35$$

From Eq. (14.16) the shear force at that section is

$$= 30 \times 5.7 (0.5236 - \phi)$$
$$= 171 (0.5236 - \phi)$$

Values of M_ϕ, T_ϕ, and F_ϕ calculated from the above expressions are given below:

ϕ Degrees	M_ϕ kN-m	T_ϕ kN-m	F_ϕ kN
0	− 90.74	0	− 89.54
5	− 49.62	− 6.08	− 74.61
10	− 15.55	− 8.87	− 59.69
15	11.23	− 9.00	− 44.77
20	30.40	− 7.12	− 29.85
25	42.12	− 3.89	− 14.92
30	46.01	0.00	0.00
35	42.12	3.89	14.92
40	30.40	7.12	29.85
45	11.23	9.00	44.77
50	− 15.55	8.87	59.69
55	− 49.62	6.08	74.61
60	− 90.74	0	89.54

Graphical representation of these values is given in Fig. 14.7.

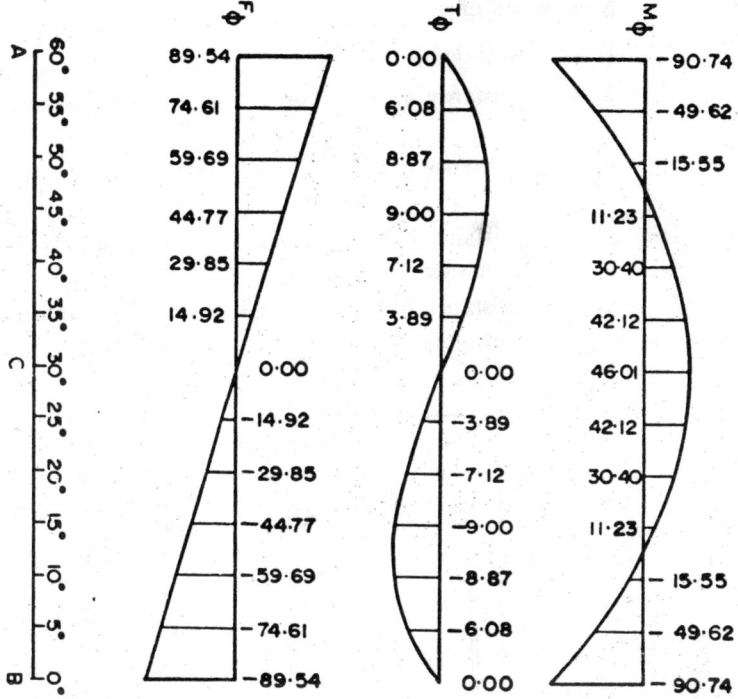

Fig. 14.7

14.4 DESIGN FOR COMBINED SHEAR AND TORSION

Torsion gives rise to shear stresses within the cross-section of a beam. In a rectangular section of a homogeneous elastic beam, the shear stresses linearly vary from zero at the centre to a maximum value at the mid point of the side. Maximum shear stress occurs at the middle of the longer side of the section. Applied shear stress along with the complimentary shear stress give rise to principal tensile stress (diagonal tension) inclind at 45° to the plane of the section. Inclind cracks develop on the sides as soon as the diagonal tensile stress exceeds the tensile strength of concrete. These cracks show up along a 45° helix around the beam surface. The most practical way to reinforce a concrete beam against twisting would be by means of longitudinal bars and stirrups, as shown in Fig. 14.8. The relevant dimensions marked in the diagram are defined below.

b = width of beam

D = overall depth of beam

b_1 = c/c distance between corner bars along the shorter side

d_1 = c/c distance between corner bars along the longer side

x_1 = c/c distance between the legs of stirrup along the shorter side

y_1 = c/c distance between the legs of stirrup along the longer side.

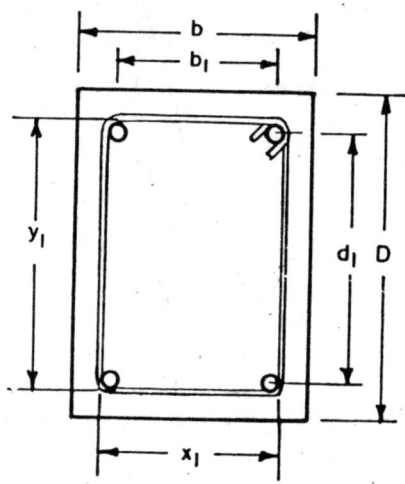

Fig. 14.8

14.4.1 Arrangement of reinforcement

(i) *Longitudinal reinforcement*

Longitudinal bars against torsion shall be placed as close as possible to the corners of the section. There shall be at least one longitudinal bar in each corner.

If the length of a side of the cross-section exceeds 450 mm, side face reinforcement shall also be provided in addition as per the rules discussed under Art. 3.5.6.

(ii) *Transverse reinforcement*

Rectangular closed stirrups shall be placed at right angles to the axis of the member enclosing the longitudinal bars. The spacing of stirrups is governed by the limits as follows:

Spacing $\not> x_1$

$$\not> \frac{x_1 + y_1}{4}$$

$\not> 300$ mm.

14.4.2 Equivalent shear force Shear force and Torsional moment both create shear stress within the cross-section of a beam. Their combined effect is considered in terms of an equivalent shear force V_e.

Thus,

$$V_e = V + 1.6 \frac{T}{b} \tag{14.20}$$

where $V =$ shear force

$T =$ torsional moment

Equivalent nominal shear stress is given by

$$\tau_{ve} = \frac{V_e}{bd} \tag{14.21}$$

14.4.3 Requirements of IS Code: 456 regarding reinforcement Calculated value of τ_{ve} is compared with the values of the permissible shear stress τ_c and the maximum shear stress τ_{cmax} for the concrete. As such, the following cases arise:

(i) The equivalent nominal shear stress does not exceed the permissible shear stress i.e. $\tau_{ve} < \tau_c$.

Provide minimum shear reinforcement according to the rule

$$\frac{A_{sv}}{bS_v} \geqslant \frac{0.4}{f_y} \tag{14.22}$$

(ii) The equivalent nominal shear stress exceeds the permissible shear stress i.e. $\tau_{ve} > \tau_c$.

Both longitudinal and transverse reinforcements should be provided. For this purpose the following procedure shall be adopted:

Equivalent bending moment M_{e1} is calculated by

$$M_{e1} = M + M_t \qquad (14.23)$$

where $\qquad$ M = bending moment at the section

$$M_t = \frac{(1 + D/b)\, T}{1.7} \qquad (14.24)$$

T = torsional moment

The beam section should be adequate to bear safely the bending moment M_{e1} for which longitudinal tension reinforcement is calculated by the usual procedure. The bars are placed along the tension face. If $M_t > M$ then the longitudinal reinforcement shall also be provided along the compression face of beam. This reinforcement can be calculated from another equivalent bending moment M_{e2} given by

$$M_{e2} = M_t - M \qquad (14.25)$$

This equivalent moment is taken to be acting in the opposite sense to the moment M_{e1}. The longitudinal reinforcement which is calculated for M_{e2} by the usual procedure is placed along the compression face. The corresponding transverse reinforcement shall be two legged closed stirrups enclosing the longitudinal bars as shown in Fig. 14.8. The spacing should be calculated from the three rules given below and the minimum adopted.

(a) $\qquad A_{sv} = \left(\dfrac{T}{b_1 d_1} + \dfrac{V}{2.5\, d_1} \right) \dfrac{S_v}{\sigma_{sv}} \qquad (14.26)$

(b) $\qquad A_{sv} = \dfrac{(\tau_{ve} - \tau_c) b S_v}{\sigma_{sv}} \qquad (14.27)$

(c) minimum shear reinforement consideration

$$\frac{A_{sv}}{b S_v} \geqslant \frac{0.4}{f_y}$$

where $\qquad A_{sv}$ = sectional area of the stirrup legs.

The minimum spacing adoped from the above three rules shall not exceed the limits given under para. 14.4.1 (ii).

(iii) The equivalent nominal shear stress exceeds the maximum permitted shear stress i.e. $\tau_{ve} > \tau_{cmax}$.

In this case the dimensions of the section should be increased.

Example 14.4

A rectangular beam of section 900 mm × 450 mm is required to resist a torsional moment = 50 kN-m along with a shear force = 100 kN. Design the reinforcement for the beam. Concrete grade: M20 and steel grade: Fe415.

Solution

Permissible stresses are,

$$\sigma_{cbc} = 7 \text{ N/mm}^2, \quad \sigma_{st} = 230 \text{ N/mm}^2 \text{ and } m = 13.33$$

Design constants are, $n = 0.289$, $j = 0.904$ and $R = 0.914$

Now $M = $ zero

$T = 61$ kN-m

$V = 100$ kN

$b = 450$ mm   or   0.45 m

$\therefore$

$$M_t = \frac{T\left(1 + \dfrac{D}{b}\right)}{1.7}$$

$$= 61\frac{(1 + 2.0)}{1.7}$$

$$= 107.65 \text{ kN-m}$$

Equivalent bending moments are

$$M_{e1} = M + M_t$$
$$= 0 + 107.65$$
$$= 107.65 \text{ kN-m}$$

$$M_{e2} = M_t - M$$
$$= 107.65 - 0$$
$$= 107.65 \text{ kN-m}$$

As $M_{e1} = M_{e2}$, the longitudinal reinforcements both at top and bottom will be the same. Taking effective depth

$$d = 900 - 35$$
$$= 865 \text{ mm}$$

the reinforcement area is given by

$$A_{st} = \frac{107.65 \times 10^6}{230 \times 0.904 \times 865}$$

$$= 598.55 \text{ mm}^2$$

Provide 16 mm – 3 nos. at top and the same at bottom. See Fig. 14.9.

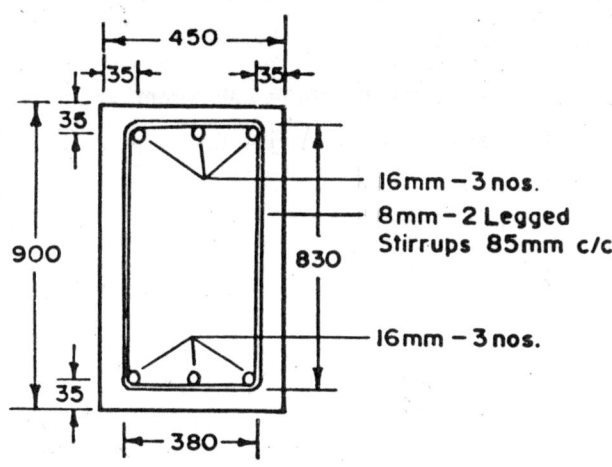

Fig. 14.9

Equivalent shear force is

$$V_e = V + 1.6 \frac{T}{b}$$

$$= 100 + 1.6 \times \frac{61}{3.45} = 316.9 \text{ kN}$$

$$\tau_{ve} = \frac{316.9 \times 10^3}{450 \times 865} = 0.814 \quad \text{N/mm}^2$$

$$\frac{100 \, A_{st}}{bd} = \frac{100 \times 3 \times 201.06}{450 \times 865}$$

$$= 0.155 \quad \text{N/mm}^2$$

From table 3.11

$$\tau_c = 0.22 \text{ N/mm}^2$$

As $\tau_{ve} > \tau_c$, the beam is unsafe in shear. Shear reinforcement should be provided.

It is proposed to provide 8 mm dia.−2 legged stirrups. Then

$$b_1 = 450 - 2 \times 35 = 380 \text{ mm}$$

$$d_1 = 900 - 2 \times 35 = 830 \text{ mm}$$

$$X_1 = 380 + 16 + 8 = 404 \text{ mm}$$

$$Y_2 = 830 + 16 + 8 = 854 \text{ mm}$$

The spacing of the stirrups is calculated as follows:

(i) $$A_{sv} = \left(\frac{T}{b_1 d_1} + \frac{V}{2.5 \, d_1} \right) \frac{S_v}{\sigma_{sv}}$$

or $$2 \times 50.26 = \left(\frac{61 \times 10^6}{380 \times 830} + \frac{100 \times 10^3}{2.5 \times 830} \right) \times \frac{S_v}{230}$$

∴ $$S_v = 95.7 \text{ mm}$$

(ii) $$A_{sv} = \frac{(\tau_{ve} - \tau_c)}{\sigma_{sv}} b \, S_v$$

or $$2 \times 50.26 = \frac{(0.814 - 0.22)}{230} \times 450 \times S_v$$

∴ $$S_v = 86.49 \text{ mm}$$

(iii) Minimum shear reinforcement.

$$\frac{A_{sv}}{b \, S_v} > \frac{0.4}{f_y}$$

or $$\frac{2 \times 50.26}{450 \times S_v} > \frac{0.4}{415}$$

∴ $$S \leqslant 231.74 \text{ mm}$$

Hence, a spacing of 85 mm c/c may be adopted. See Fig. 14.9.

Example 14.5

Design the arcate beam which has been analysed in Example 14.2. Concrete grade : M 15 and Steel grade : Fe.415.

Solution

Permissible stresses are:

$$\sigma_{cbc} = 5 \text{ N/mm}^2 \quad \text{and} \quad \sigma_{st} = 230 \text{ N/mm}^2$$

Modular ratio m = 18.67

Design constants are:

$$n = 0.289 \qquad j = 0.904 \qquad \text{and} \qquad R = 0.653$$

Shear force, bending moment and twisting moment diagrams for the beam are shown in Fig. 14.5.

Design of section at support

Maximum bending moment in the beam occurs at the support section. Its value is

$$M = 127.47 \quad \text{kN-m. (negative)}$$

Accompanying twisting moment at the same section is,

$$T = 3.93 \quad \text{kN-m.}$$

Hence, equivalent bending moment

$$
\begin{aligned}
M_{e1} &= M + M_t \\
&= M + T \frac{(1 + D/b)}{1.7} \\
&= 127.47 + 3.93 \frac{(1 + 2.0)}{1.7} \\
&= 134.4 \quad \text{kN-m.}
\end{aligned}
$$

Now, let $\qquad D = d + 35$ mm

Then

$$b = \tfrac{1}{2} D = \tfrac{1}{2} (d + 35)$$

Equating $\qquad Rbd^2 = M_{e1}$

i.e. $\qquad 0.653\, bd^2 = 134.4 \times 10^6$

or $\qquad 0.653 \dfrac{(d + 35)\, d^2}{2} = 134.4 \times 10^6$

or $\qquad (d + 35)\, d^2 = 411.64 \times 10^6$

or $\qquad d = 732.5$ mm. $\qquad$ and $\qquad b = \tfrac{1}{2} (732.5 + 35)$
$$= 383.75 \text{ mm.}$$

Adopt $\qquad b = 385$ mm

$\qquad\qquad d = 735$ mm

$\qquad\qquad D = 735 + 35 = 770$ mm.

Tension reinforcement

$$A_{st} = \frac{134.4 \times 10^6}{230 \times 0.904 \times 735}$$

$$= 879.5 \text{ mm}^2$$

Provide 12 mm — 8 nos. at the top face. Out of these, 6 bars may be cut off after extending them 735 mm (i.e. an effective depth) beyond the point of contraflexure and the remaining 2 bars continued throughout the length of the beam.

Since $M_t < M$ no reinforcement need be calculated for the compression face which is the lower face. However, 12 mm dia. bars 2 nos., one in each corner, may be provided along the lower face as per the minimum requirement.

Check for shear at the critical section

Critical section for both shear and tension occurs at a distance an effective depth away from the face of support. Angular distance of this section from mid-span of beam is given by

$$= 30° - \frac{0.735}{6} \times \frac{180}{\pi}$$

$$= 22.981°$$

or $\qquad = 0.4011$ radians.

Hence, at this section, shear force

$$V = wR \phi$$
$$= 36 \times 6 \times 0.4011$$
$$= 86.64 \text{ kN.}$$

Twisting moment

$$T = 53.3 \sin \phi - 1296 (\phi - \sin \phi)$$
$$= 53.3 \times 0.3904 - (0.4011 - 0.3904)$$
$$= 6.94 \text{ kN-m.}$$

Equivalent shear force,

$$v_e = V + 1.6 \frac{T}{b}$$

$$= 86.64 \times \frac{1.6 \times 6.94}{0.385}$$

$$= 115.48 \quad \text{kN-m}.$$

Nominal shear stress

$$\tau_{ve} = \frac{V_e}{bd}$$

$$= \frac{115.48 \times 10^3}{385 \times 735}$$

$$= 0.408 \quad \text{N/mm}^2$$

Now

$$\frac{100 \, A_{st}}{bd} = \frac{100 \times 8 \times 113.1}{385 \times 735}$$

$$= 0.32$$

From Table 3.11

$$\tau_c = 0.24 \quad \text{N/mm}^2$$

As $\tau_{ve} > \tau_c$, the section is unsafe in shear. Hence, transverse reinforcement should be designed and provided.

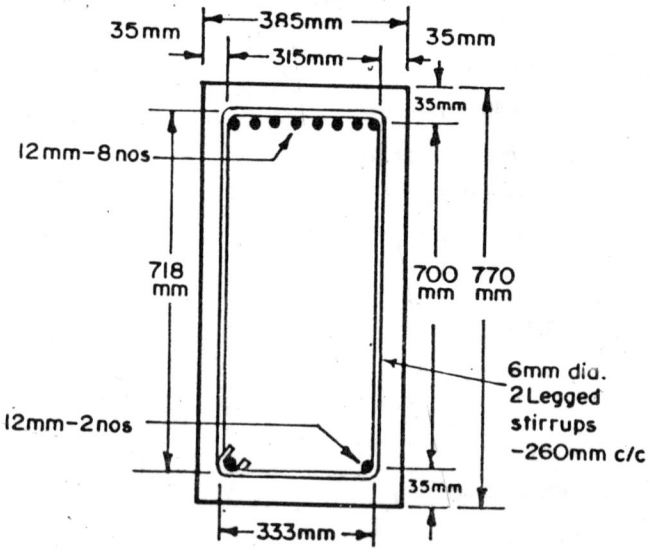

Fig. 14.10 Cross-section of beam near support.

Try 6 mm dia. 2 legged stirrups. For the arrangement of reinforcement shown in Fig. 14.10.

$$b = 385 \text{ mm}$$
$$d = 735 \text{ mm}$$
$$b_1 = 315 \text{ mm}$$
$$d_1 = 700 \text{ mm}$$
$$x_1 = 333 \text{ mm}$$
$$y_1 = 718 \text{ mm}$$

The spacing of stirrups can be calculated from the three considerations as follows:

(i) $\qquad A_{sv} = \left(\dfrac{T}{b_1 d_1 \, \sigma_{sv}} + \dfrac{V}{2.5 \, d_1 \, \sigma_{sv}} \right) S_v$

i.e. $\quad 2 \times 50.27 = \left(\dfrac{6.94 \times 10^6}{315 \times 700 \times 230} + \dfrac{86.64 \times 10^3}{2.5 \times 700 \times 230} \right) S_v$

or $\qquad S_v = 285.5 \text{ mm}$

(ii) $\qquad A_{sv} = \dfrac{(\tau_{ve} - \tau_c) \, b \, S_v}{\sigma_{sv}}$

i.e. $\quad 2 \times 50.27 = \dfrac{(0.408 - 0.24) \times 385}{230} \times S_v$

or $\qquad S_v = 357.52 \text{ mm}$

(iii) Minimum shear reinforcement

$$\frac{A_{sv}}{b S_v} > \frac{0.4}{f_y}$$

or $\qquad \dfrac{2 \times 50.27}{385 \, S_v} > \dfrac{0.4}{415}$

or $\qquad S_v \leqslant 270.94 \text{ mm}$

Of the three spacings calculated above, the least is 270.94 mm which corresponds to the minimum shear reinforcement.

The spacing of stirrups should be within the following limits:

$\qquad \not> x_1 \qquad$ i.e. $\qquad 333 \text{ mm}$

$\qquad \not> \dfrac{x_1 + y_1}{4} \qquad$ i.e. $\qquad \dfrac{333 + 718}{4} = 262.75 \text{ mm}$

$\qquad \not> 300 \text{ mm}$

Hence, stirrups may be spaced at 260 mm c/c throughout the length of the beam.

Check for shear at point of contraflexure

At this section

$$\phi = 16.158° \quad \text{or} \quad 0.282 \text{ radians}$$

Shear force

$$V = 60\ 92\ \text{kN}$$

Twisting moment

$$T = 10.02\ \text{kN-m}.$$

Bending moment

$$M = 0$$

$$M_t = T\frac{(1 + D/b)}{1.7}$$

$$= 10.02\frac{(1 + 2)}{1.7}$$

$$= 17.68\ \text{kN-m}.$$

Hence

$$M_{e1} = M + M_t$$
$$= 0 + 17.68$$
$$= 17.68\ \text{kN-m}.$$

$$M_{e2} = M_t - M$$
$$= 17.68 - 0$$
$$= 17.68\ \text{kN-m}.$$

Longitudinal reinforcement required near top as well as bottom face of beam is given by

$$A_{st} = \frac{M_t}{\sigma_{st}\ j\ d}$$

$$= \frac{17.68 \times 10^6}{230 \times 0.904 \times 735}$$

$$= 115.69\ \text{mm}^2$$

The beam has already 8 bars of 12 mm dia. near the top face. It will also have at least 2 bars of 12 mm dia. near the bottom face. Hence, the requirement of longitudinal reinforcement will be satisfied.

Now

$$V_e = V + 1.6 \frac{T}{b}$$

$$= 60.92 + 1.6 \times \frac{10.02}{0.385}$$

$$= 102.56 \text{ kN}$$

Nominal shear stress

$$\tau_{ve} = \frac{V_e}{bd}$$

$$= \frac{102.56 \times 10^3}{385 \times 735}$$

$$= 0.362 \text{ N/mm}^2$$

As found earlier,

$$\tau_c = 0.24 \text{ N/mm}^2$$

Since $\tau_v > \tau_c$ the section is unsafe in shear. Transverse reinforcement should be designed and provided. For 6 mm dia. 2 legged stirrups the spacing can be calculated from three considerations as follows:

(i) $$A_{sv} = \left(\frac{T}{b_1 d_1 \sigma_{sv}} = \frac{V}{2.5 d_1 \sigma_{sv}} \right) s_v$$

i.e. $$2 \times 50.27 = \left(\frac{10.02 \times 10^6}{315 \times 700 \times 230} + \frac{60.92 \times 10^3}{2.5 \times 700 \times 230} \right) s_v$$

or $$s_v = 288.14 \text{ mm}$$

(ii) $$A_{sv} = \frac{(\tau_{ve} - \tau_c) b s_v}{\sigma_{sv}}$$

i.e. $$2 \times 50.27 = \frac{(0.362 - 0.24) \, 385}{230} \times s_v$$

or $$s_v = 492.3 \text{ mm}$$

(iii) Minimum shear reinforcement

$$\frac{A_{sv}}{b s_v} > \frac{0.4}{f_y}$$

i.e.
$$\frac{2 \times 50.27}{385 \, s_v} > \frac{0.4}{415}$$

or
$$s_v \leqslant 270.94$$

The minimum value of the above three spacings works out to 270.94 mm. However, the maximum permitted value of spacing worked out earlier is 260 mm c/c. Hence, the latter may be adopted throughout the length of the beam.

Positive reinforcement at mid-span

At this section

> Bending moment = 53.3 kN-m.
> Shear force   = 0
> Twisting moment = 0

Tension reinforcement on the lower face

$$A_{st} = \frac{53.3 \times 10^6}{230 \times 0.904 \times 735}$$

$$= 348.8 \text{ mm}^2$$

Provided 12 mm − 2 nos. plus 10 mm − 2 nos. See Fig. 14.11.

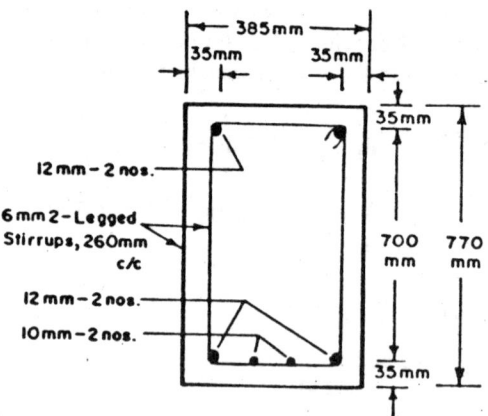

Fig. 14.11 Cross-section of beam at mid-span

While the 12 mm bars may continue throughout the length of beam, the 10 mm bars may be cut-off after extending them for 735 mm (i.e.

an effective depth) beyond the point of contraflexure. The details are shown in the Fig. 14.12.

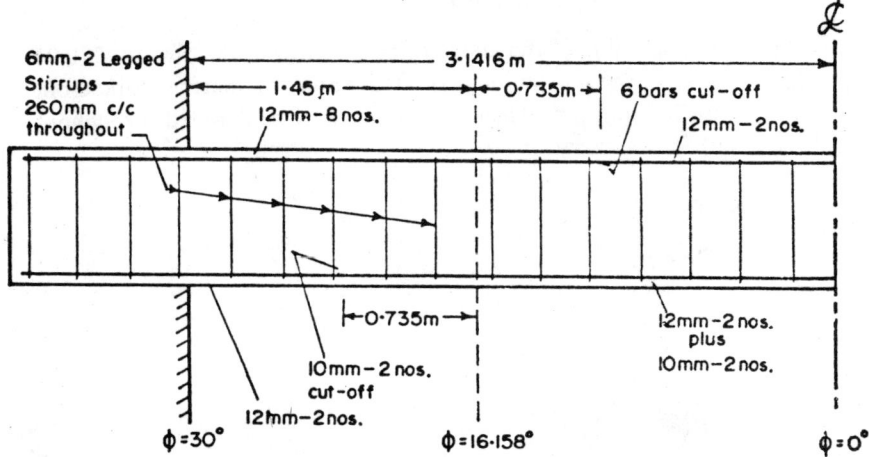

Fig. 14.12 Longitudinal section of arcate beam.

15

Spherical and Conical Domes

15.1 INTRODUCTION

A shell structure is a structure in space, made of solid material and bounded by two curved surfaces known as the faces. The thickness of the shell may be uniform, or it may vary from point to point. The geometry of the shell is completely known if the shape of the middle surface is defined and the thickness is known at all the points.

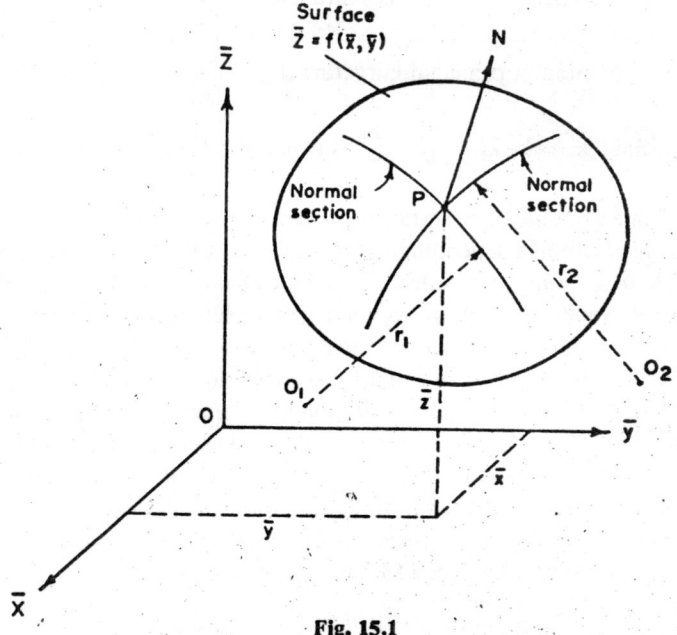

Fig. 15.1

A surface in space referred to rectangular cartesian coordinates is shown in Fig. 15.1. The coordinates $\bar{x}$, $\bar{y}$, and z refer to a point P on the surface. The surface can be represented by an equation of the form

$$\bar{z} = f(\bar{x}, \bar{y}) \qquad (15.1)$$

where, $\bar{x}$ and $\bar{y}$ are taken to be independent variables and $\bar{z}$ is taken to be dependent variable. Let NP be drawn a normal to the surface at point P. A plane containing the normal NP intersects with the surface along a curve which is called the normal section at the point P. There can be infinitely many such normal sections (or curves) passing through the point P, each having a different radius of curvature r. The radius of curvature of one of these normal sections (or curves) is minimum and that of another is maximum. These two normal sections (or curves) are orthogonal to each other. Let,

$$r_1 = \text{the minimum radius of curvature}$$

$$r_2 = \text{the maximum radius of curvature}$$

Then, the two principal curvatures at the point P are as follows:

$$\text{Maximum principal curvature } k_1 = \frac{1}{r_1} \qquad (15.2)$$

$$\text{Minimum principal curvature } k_2 = \frac{1}{r_1} \qquad (15.3)$$

The product $k = k_1 k_2 = \frac{1}{r_1} \cdot \frac{1}{r_2}$ is called the Gaussian curvature

of the surface at point P. A surface is called synclastic, anticlastic or developable at a point depending upon whether k is positive, negative or zero at that point. Consider an infinitesimally small element of shell with reference to a set of rectangular coordinates as shown in Fig. 15.2 (a). The orientation of the coordinate axes is so chosen at the point that the planes $x - z$ and $y - z$ coincide with the planes of the principal curvatures. The corresponding principal radii of curvature are denoted as r_x and r_y respectively. The shell element is contained between the pairs of adjacent planes parallel to the two principal directions respectively, and normal to the middle surface. On the sides of the element the stresses acting in the three orthogonal directions are shown in Fig. 15.2 (a) These are σ_x, σ_y, τ_{xy}, τ_{yx}, τ_{xz}

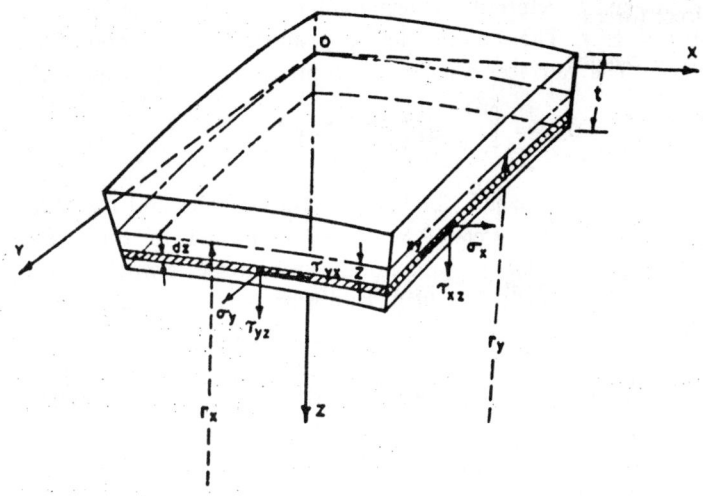

(a) Stresses in a shell element

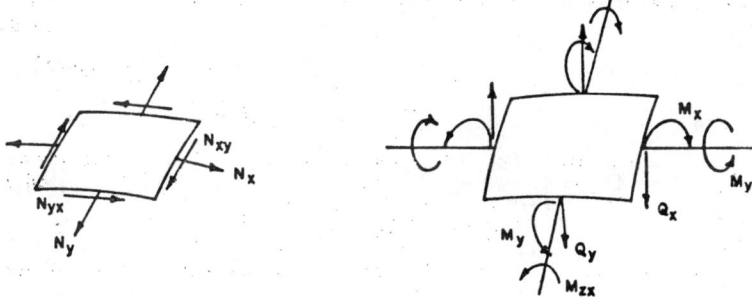

(b) Membrane forces on shell element

(c) Moments and transverse shears on shell element

Fig. 15.2

and τ_{yz}. It is well known that $\tau_{xy} = \tau_{yx}$. The stress resultants or the forces per unit width of the normal section can now be written down. The width of element at a distance z below middle surface is $\left(1 - \dfrac{z}{r_x}\right)$ and $\left(1 - \dfrac{z}{r_y}\right)$ in $x - z$ and $y - z$ planes respectively.

Hence, the stress resultants as indicated in Fig. 15.2 (b) and (c) can be written as follows:

(a) **Direct forces**

In x − z plane

$$N_x = \int_{-t/2}^{t/2} \sigma_x \left(1 - \frac{z}{r_y}\right) dz \qquad (15.4)$$

$$N_y = \int_{-t/2}^{t/2} \sigma_y \left(1 - \frac{z}{r_x}\right) dz \qquad (15.5)$$

(b) **Shear forces tangential to middle surface.**

In x − y plane

$$N_{xy} = \int_{-t/2}^{t/2} \tau_{xy} \left(1 - \frac{z}{r_y}\right) dz \qquad (15.6)$$

$$N_{yx} = \int_{-t/2}^{t/2} \tau_{yx} \left(1 - \frac{z}{r_x}\right) dz \qquad (15.7)$$

(c) **Shear forces transverse to middle surface.**

In y − z plane

$$Q_x = \int_{-t/2}^{t/2} \tau_{xz} \left(1 - \frac{z}{r_y}\right) dz \qquad (15.8)$$

In x − z plane

$$Q_y = \int_{-t/2}^{t/2} \tau_{yz} \left(1 - \frac{z}{r_x}\right) dz \qquad (15.9)$$

(d) **Bending moments.**

In x − z plane

$$M_x = \int_{-t/2}^{t/2} \sigma_x z \left[1 - \frac{z}{r_y}\right] dz \qquad (15.10)$$

In y − z plane

$$M_y = \int_{-t/2}^{t/2} \sigma_y z \left[1 - \frac{z}{r_x}\right] dz \qquad (15.11)$$

(e) **Twisting moments.**

In y − z plane

$$M_{xy} = \int_{-t/2}^{t/2} \tau_{xy} z \left[1 - \frac{z}{r_y}\right] dz \qquad (15,12)$$

In x − z plane

$$M_{yx} = \int_{-t/2}^{t/2} \tau_{yx} \, z \left[1 - \frac{z}{r_x} \right] dz \qquad (15.13)$$

For thin shells the thickness t is very small in comparison with the radii r_x and r_y. Hence, the terms z/r_x and z/r_y can be neglected in comparison with unity. Hence, from Eqs. (15.6) and (15.7)

$$N_{xy} = N_{yx}$$

15.2 MEMBRANE STRESSES IN THIN SHELLS

A thin shell may well resist the in-plane forces N_x, N_y and N_{xy} at any point on the middle surface. However the shell may have low bending and twisting stiffnesses, and therefore, its resistance against bending and twisting moments may be negligible. In all such cases the resultant transverse shearing forces Q_x and Q_y, the bending moments M_x and M_y, and the twisting moments M_{xy} and M_{yx} which are shown in Fig. 15.2(c) and given by Eqs. (15.8) through (15.13) need not be considered. Hence, the only stress resultants which remain to be considered are the in-plane forces N_x, N_y and N_{xy} which are shown in Fig. 15.2(b). These forces are known as the membrane forces and the analysis of shells neglecting the bending and twisting moments is called membrane analysis. The membrane forces in a thin shell can be written as:

$$N_x = \int_{-t/2}^{t/2} \sigma_x \, dz \qquad (15.14)$$

$$N_y = \int_{-t/2}^{t/2} \sigma_y \, dz \qquad (15.15)$$

$$N_{xy} = N_{yx} = \int_{-t/2}^{t/2} \tau_{xy} \, dz \qquad (15.16)$$

The three stress resultants N_x, N_y and N_{xy} can be obtained by establishing the equilibrium of a differential element such as shown in Fig. 15.2(a). The problem is statically determinate for any applied loading on the shell.

The basis for analysis of thin shells for membrane forces is that the deformations under load are so small that the changes in the geometry of the shell are negligible and the static equilibrium of the system is unaltered.

15.3 SHELLS OF REVOLUTION LOADED SYMMETRICALLY ABOUT THE AXIS

A plane curve revolving about an axis lying in its plane generates a surface of revolution. The curve is called meridian while the plane containing the curve and the axis of revolution is called a meridian plane. The position of a meridian is given by an angle ψ measured from a datum meridian plane. A point along the meridian is defined by the angle ϕ between the normal to the surface and the axis of revolution. See Fig. 15.3.

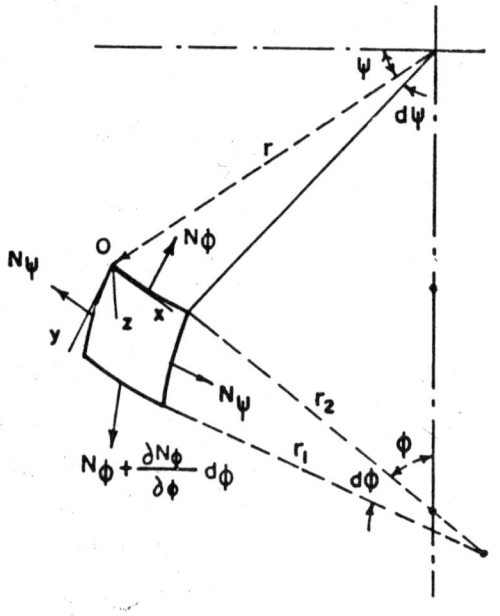

Fig. 15.3

A parallel circle (or latitude) is a locus of points on the surface so that the normals drawn at these points are equally inclined to the axis of revolution. A point on the surface is defined by the coordinates (ψ, ϕ). Consider an element of shell cut by two infinitesimally close meridians and two infinitesimally close parallel circles. At a point on the surface the meridian plane and the plane perpendicular to the meridian plane are the planes of principal curvature.

Let r_1 = the radius of curvature of the surface in the meridian plane

r_2 = the radius of curvature of the surface in a plane perpendicular to the meridian.

In general, the centre of curvature pertaining to r_1 does not lie on the axis of revolution, whereas, that pertaining to r_2 lies on the axis of revolution. Let, the radius of the parallel circle at the point 0 be = r. Then, $r = r_2 \sin \phi$. The lengths of the sides of the element are as follows :

Along meridian $= r_1 \, d\phi$ (15.17)

Along parallel circle $= r \, d\psi = r_2 \sin \phi \, d\psi$ (15.18)

The surface area of the element is

$$= r_1 r_2 \sin \phi \, d\phi \, d\psi$$
$$= r_1 r \, d\phi \, d\psi \qquad (15.19)$$

As both the surface and the loading is symmetrical about the axis, it is obvious that no shearing forces will exist along the sides of the element. The only membrane forces which exist are the meridonial force N_ϕ and the circumferencial force N_ψ. The local coordinate axes at the point O on the surface are Ox, Oy and Oz as shown in Fig. 15.3. Due to symmetry, the intensity of load which acts in the meridian plane, can be resolved in two components Y and Z, parallel to the y and z coordinate axes respectively. The component along the x-axis is zero.

Consider the equilibrium of the element. Resolving all the forces in y-direction

$$\left[N_\phi + \frac{dN_\phi}{d\phi} d\phi \right] \left[r + \frac{dr}{d\phi} d\phi \right] d_\psi - N_\phi r d_\psi - N_\psi r_1$$
$$\cos \phi \, d\phi d_\psi + Y r_1 r d_\phi d_\psi = 0$$

Neglecting small quantities of second order

$$N_\phi \frac{dr}{d_\phi} d_\phi \, d_\psi + \frac{d N_\phi}{d \phi} r \, d_\phi \, d_\psi - N_\psi r_1 \cos \phi \, d_\phi \, d_\psi$$
$$+ Y \, r_1 \, r \, d_\phi \, d_\psi = 0$$

or $N_\phi \dfrac{dr}{d_\phi} + \dfrac{d N_\phi}{d_\phi} r - N_\psi r_1 \cos \phi + y r_1 \, r = 0$

or $\quad\dfrac{d}{d\phi}\,(N_\phi\,r) - N_\psi\,r_1\cos\phi + Yr_1\,r = 0$ (15.20)

Resolving all the forces in z-direction

$$N_\phi\,r\,d\psi\cdot d\phi + N_\psi\,r_1\sin\phi\,d\phi d\psi + Z\,r_1\,r\,d\phi d\psi = 0$$

or $\qquad N_\phi r + N_\psi r_1\sin\phi + Zr_1 r = 0$

or $\qquad\dfrac{N_\phi}{r_1} + \dfrac{N_\varphi}{r}\sin\phi + Z = 0$

or $\qquad\dfrac{N_\phi}{r_1} + \dfrac{N_\psi}{r_2} + Z = 0$ (15.21)

From Eqs. (15.20) and (15.21) the forces N_ϕ and N_ψ can be evaluated at any point of the surface for the give values of the components Y and Z of the intensity of the external load.

Let the equilibrium of the portion of shell above a parallel circle defined by angle ϕ as shown in Fig. 15.4 be considered. If the result-

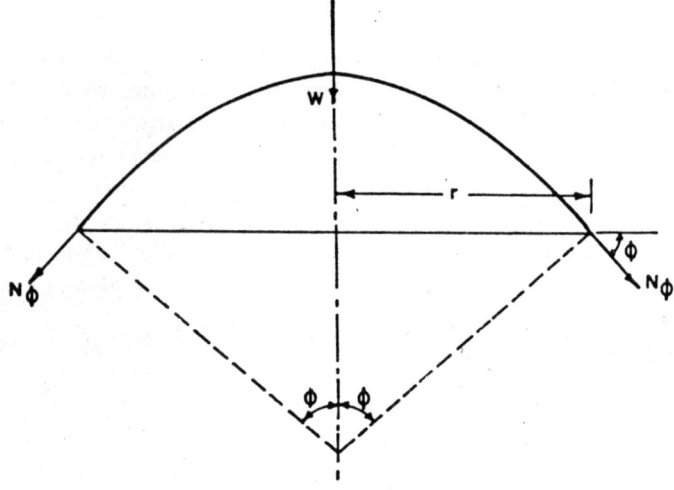

Fig. 15.4

ant of the total force on this portion of the shell acting parallel to the axis of revolution be W, then

$$2\pi\,r\,N_\phi\,\sin\phi + W = 0$$

or
$$N_\phi = \frac{-W}{2\pi r \sin \phi}$$
(15.22)

Equation (15.22) can also be derived from Eq. (15.20) by integration. It is comparatively simpler

15.3.1 Spherical dome

Consider a spherical shell shown in Fig. 15.5 (a).

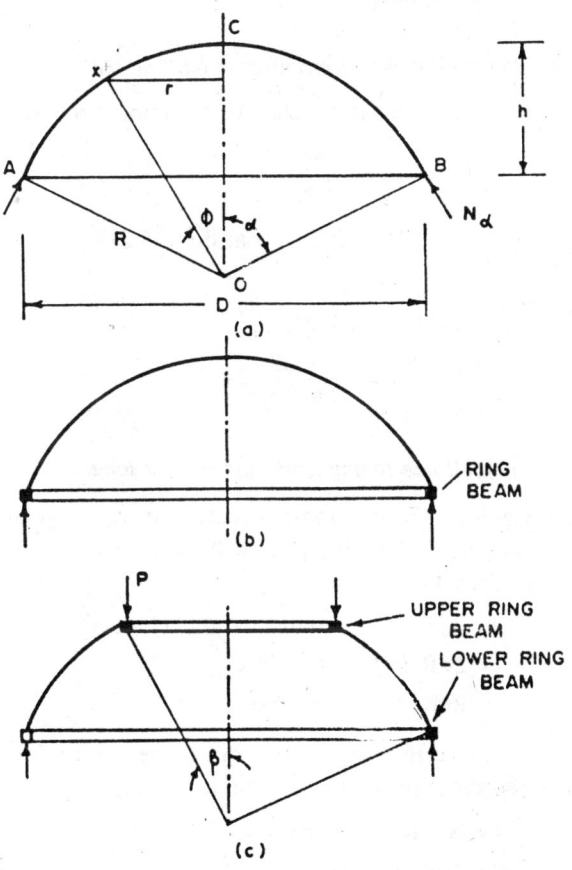

Fig. 15.5

Here,

R = the radius of the dome

D = the base diameter

h = rise of dome

= R (1 − cos α)

α = the semi vertical angle subtended at the centre of curvature.

r = R sin ϕ

(i) Stresses in shell due to concentrated load at top

Putting r = R sin ϕ in Eq. (15.22) the meridional thrust in the shell is

$$N_\phi = - \frac{W}{2\pi R \sin^2 \phi} \qquad (15.23)$$

Putting $r_1 = r_2 = R$ and Z = 0 in Eq. (15.21)

$$N_\phi + N_\psi = 0$$

∴

$$N_\psi = - N_\phi$$

$$= \frac{W}{2\pi R \sin^2 \phi} \qquad (15.24)$$

(ii) Stress in shell due to uniformly distributed load

Consider a point x on the dome situated at an angular distance ϕ from the top point C. The load on shell above the parallel circle passing through the point is

= surface area above parallel circle × q

= 2πR·R(1 − cos ϕ) q

= 2πR² (1 − cos ϕ) q

Resultant of the vertical component of the meridional force N_ϕ acting along the periphery of the parallel circle is

= N_ϕ sin ϕ · 2π Rsin ϕ

= 2πR N_ϕ sin² ϕ

Then, for equilibrium

2πR N_ϕ sin² ϕ + 2πR² (1 − cos ϕ) q = 0

or $\qquad N_\phi = \dfrac{- qR}{(1 + \cos \phi)}$ $\qquad\qquad$ (15.25)

Substituting the value of N_ϕ from Eq. (15.25) into Eq. (15.21) and putting $r_1 = r_2 = R$,

$$\dfrac{- q}{(1 + \cos \phi)} + \dfrac{N_\psi}{R} + q \cos \phi = 0$$

$\therefore \qquad N_\psi = qR \dfrac{(1 - \cos \phi - \cos^2 \phi)}{(1 + \cos \phi)} \qquad\qquad$ (15.26)

Meridional force N_ϕ is negative, that is, compressive for all values of ϕ and increases as ϕ increases. For $\phi = 0$, $N_\phi = - \dfrac{qR}{2}$ and for $\phi = \dfrac{\pi}{2}$, $N_\phi = - qR$. Circumferencial force N_ψ is negative, that is, compressive for only small values of ϕ. For $\phi = 0$, $N_\psi = \dfrac{- qR}{2}$.

The force N_ϕ becomes zero when

$$1 - \cos \phi - \cos^2 \phi = 0$$

i.e. for $\qquad\qquad\qquad \phi = 51° 50'$

For values of ϕ greater than $51° 50'$ N_ψ becomes positive that is, tensile.

Generally, at the base of the dome a ring beam is provided as shown in Fig. 15.5 (b). Supports provide only vertical reactions which neutralize the vertical component of N_ϕ there. The horizontal component of N_ϕ is taken up by the ring beam which comes under circumferencial tension. However, the expansion of the ring beam may be different from that of the adjacent parallel circle on the shell. As such, some bending of the shell occurs in the vicinity of the ring beam, which is localized in character for thin shells.

Often in the top portion of the dome, a circular opening is made for ventilation and lighting. An upper ring beam is provided to support the upper structure, along the periphery of the opening.

15.3.2 Conical shell Consider the shell element shown in Fig. 15.3. Let the length of the element along the meridian be denoted as ds. Then

$$ds = r_1 d\phi$$

$\therefore \qquad \dfrac{d}{d\phi} = r_1 \dfrac{d}{ds} \qquad\qquad$ (15.27)

As such, Eq. (15.20) can be rewritten as

$$r_1 \frac{d}{ds} (N_\phi r) - N_\psi r_1 \cos \phi + Y r_1 r = 0$$

or

$$\frac{d}{ds} (N_\phi r) - N_\psi \cos \phi + Y r = 0 \tag{15.28}$$

For a conical shell $r_1 = \infty$. Hence, from Eq. (15.21)

$$\frac{N_\psi}{r_2} + Z = 0$$

or

$$N_\psi + Z r_2 = 0 \tag{15.29}$$

Consider now a conical shell shown in Fig. 15.6 (a). From consideration of geometry

$$\phi = \frac{\pi}{2} + \alpha$$

$$\sin \phi = \cos \alpha \tag{15.30}$$
$$\cos \phi = - \sin \alpha \tag{15.31}$$
$$r = s \cdot \sin \alpha \tag{15.32}$$
$$r_2 = s \cdot \tan \alpha \tag{15.33}$$

Subsitituting into Eq. (15.28) the values of $\cos \phi$ and r as above and simplifying

$$\frac{d}{ds} (N_\phi s) + N_\psi + Y \cdot s = 0 \tag{15.34}$$

Substituting $r_2 = s \cdot \tan \alpha$ into Eq. (15.29)

$$N_\psi + Z s \cdot \tan \alpha = 0 \tag{15.35}$$

Circumferencial force can be first evaluated from Eq. (15.35) and then the meridlonal force can be found from Eq. (15.34) by integration and on applying the boundary condition.

(i) *Stresses in shell due to self weight*

Consider the element of the shell shown in Fig. 15.6 (b). Let the self weight of shell per unit area be w. It can be resolved normal to and parallel to the surface. Thus

$$Z = - w \sin \alpha$$
$$Y = w \cos \alpha$$

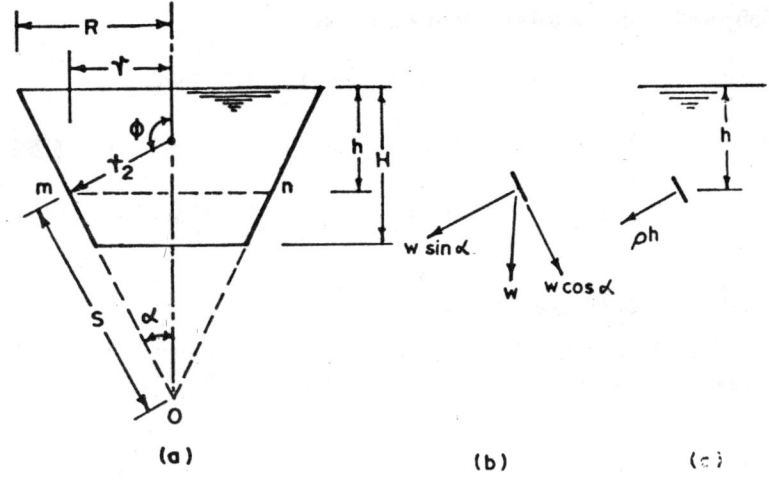

Fig. 15.6

Substituting the value of Z in Eq. (15.35)

$$N_\psi - w \sin \alpha \tan \alpha \cdot s = 0$$

$$\therefore \qquad N_\psi = w \sin \alpha \tan \alpha \cdot s \qquad (15.36)$$

Equation (15.36) gives the circumferencial force at any point on the shell. Substituting the values of N_ϕ and Y as above into Eq. (15.34)

$$\frac{d}{ds} (N_\phi s) + w \sin \alpha \tan \alpha \, s + w \cos \alpha \, s = 0$$

or $\qquad \dfrac{d}{ds} (N_\phi s) + \dfrac{ws}{\cos \alpha} = 0$

On integrating this

$$N_\phi s + \frac{ws^2}{2 \cos \alpha} + C = 0 \qquad (15.37)$$

At top of shell where $s = \dfrac{R}{\sin \alpha}$ the meridional force $N_\phi = 0$. Hence

$$\frac{wR^2}{2 \sin^2 \alpha \cos \alpha} + C = 0$$

$$\therefore \qquad C = \frac{- wR^2}{2 \sin^2 \alpha \cos \alpha}$$

Substituting the value of C into Eq. (15.37)

$$N_\phi s + \frac{w}{2 \cos \alpha} \left(s^2 - \frac{R^2}{\sin^2 \alpha} \right) = 0$$

$$\therefore \qquad N_\phi = \frac{w}{2s \cos \alpha} \left(\frac{R^2}{\sin^2 \alpha} - s^2 \right) \qquad (15.38)$$

Equation (15.38) gives the value of meridional force at any point on the shell.

(ii) *Stresses in the shell due to hydrostatic pressure*

Consider the element of the shell situated under water at a depth h below top as shown in Fig. 15.6 (c). The hydrostatic pressure on the elements is ρh. Hence

$$Y = 0$$
$$Z = - \rho h$$
$$= - \rho (R \cot \alpha - s \cos \alpha)$$

Substituting the values of Z in Eq. (15.35)

$$N_\psi - \rho (R \cot \alpha - s \cos \alpha) s \tan \alpha = 0$$

or $\qquad N_\psi - \rho (R - s \sin \alpha) s = 0$

$\therefore \qquad N_\psi = \rho (R - s \sin \alpha) s \qquad (15.39)$

Equation (15.39) gives the value of circumferencial force in the shell. Substituting the value of N_ψ and Y as above into Eq. (15.34)

$$\frac{d}{ds} (N_\phi s) + \rho (R - s \sin \alpha) s = 0$$

On integration of this

$$N_\phi s + \rho \left(\frac{Rs^2}{2} - \frac{s^3}{3} \sin \alpha \right) + C = 0 \qquad (15.40)$$

At top of the shell where $s = \dfrac{R}{\sin \alpha}$ the meridional force $N_\phi = 0$.

Hence

$$\rho \left(\frac{R^3}{2 \sin^2 \alpha} - \frac{R^3}{3 \sin^2 \alpha} \right) + C = 0$$

$$\therefore \qquad C = \frac{- \rho R^3}{6 \sin^2 \alpha}$$

Substituting the value of C in Eq. (15.40)

$$N_\phi s + \rho \left(\frac{Rs^2}{2} - \frac{s^3}{3} \sin \alpha \right) - \frac{\rho R^3}{6 \sin^2 \alpha} = 0$$

or

$$N_\phi s - \frac{\rho}{6} \left(2 s^3 \sin \alpha - 3Rs^2 + \frac{R^3}{\sin^2 \alpha} \right) = 0$$

$$\therefore \qquad N_\phi = \frac{\rho}{6} \left(2 s^2 \sin \alpha - 3Rs + \frac{R^3}{s \sin^2 \alpha} \right) \qquad (15.41)$$

Eq. (15.41) gives the value of meridional force at any point on the shell.

(iii) *Stresses in shell due to meridional force of uniform intensity at top*

Consider a conical shell as shown in Fig. 15.7. It is s subjected to

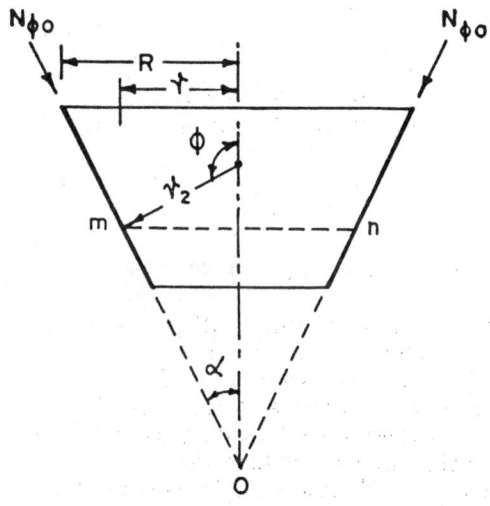

Fig. 15.7

meridional force $N_{\phi 0}$ of uniform intensity along the top periphery. There is no other loading. Hence

$$Y = 0$$
$$Z = 0$$

Substituting $Z = 0$ in Eq. (15.35)

$$N_\phi = 0 \qquad (15.42)$$

i.e. the circumferencial force in the shell is zero at all the points.

Substituting $N_\phi = 0$ and $Y = 0$ into Eq. (15.34) then

$$\frac{d}{ds} (N_\phi s) = 0$$

On integrating once

$$N_\phi s + C = 0$$

or $\qquad N_\phi \dfrac{r}{\sin \alpha} + C = 0$

or $\qquad N_\phi r + C_1 = 0$ $\hspace{4cm}$ (15.43)

where C and C_1 are constants. Now at top of shell $r = R$ and $N_\phi = N_{\phi 0}$. Substituting these values in Eq. (15.43)

$$N_{\phi 0} R + C_1 = 0$$
$$C_1 = - N_{\phi 0} R$$

Substituting the value of C_1 back in Eq. (15.43)

$$N_\phi r - N_{\phi 0} R = 0$$

or $\qquad N_\phi = N_{\phi 0} \cdot \dfrac{R}{r}$ $\hspace{4cm}$ (15.44)

Equation (15.44) gives the meridional force in the shell at any point.

15.4 DESIGN CONSIDERATIONS AND REQUIREMENTS

Wind force is taken to be acting normal to the surface of shell. As the latter has curvature in two directions, the exact analysis of stresses due to wind is cumbersome. In spherical domes the thickness of concrete and the area of reinforcement provided for practical reasons and for temperature and shrinkage is much more than that required for bearing stresses due to vertical loads. As such, it is not worthwhile to enter into complicated analysis to determine the stresses due to wind. It may be enough to consider a lead of 1 to 1.5 kN/m² of surface of dome to account for wind stresses.

Live load on spherical dome as per Table II of IS: 875 can be taken to be

$$= (0.75 - 3.45 \, r^2) \;\; kN/m^2$$

< 0.4 kN/m² in any case

where $\qquad r = \dfrac{h}{l}$

h = height of the highest point of dome above the springings

l = shorter plan dimension.

To safe guard against the possibility of failure by buckling of shell in a spherical dome the load per unit area of surface should not exceed 5.1 Et^2/R^2.

A minimum thickness of 75 mm should be provided to protect steel against corrosion. The minimum quantity of steel reinforcement that should be provided along both meridional and circumferencial directions is 0.12% of sectional area for HYSD bars and 0.15% for M.S. bars. This quantity of steel should be in addition to that required to bear tensile stresses. The reinforcement should be placed in the middle of the thickness of shell.

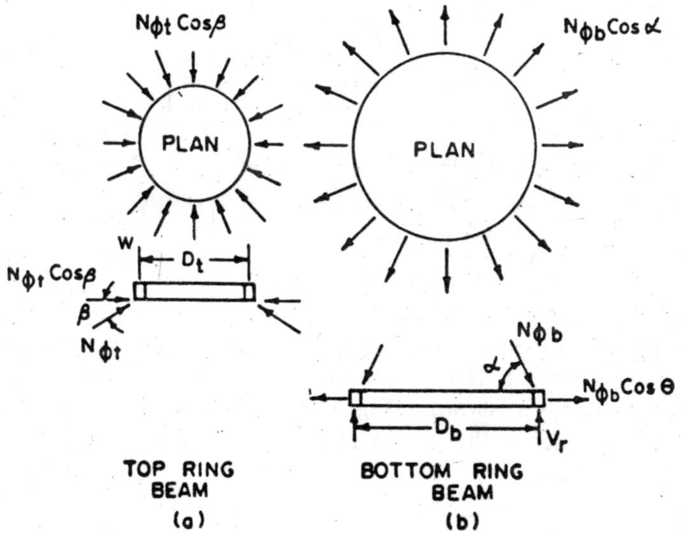

Fig. 15.8

The meridional thrust $N_{\phi b}$ at the base of the dome is inclind to the vertical. Its vertical component is balanced by the vertical reaction V_r

provided by the support. However, its horizontal component is $N_{\phi b}$ cos α which acts radially outwards as shown in Fig. 15.8 (b). To take up this force a ring beam is provided at the bottom edge of the dome. Tension in the bottom ring beam

$$= N_{\phi b} \cos \alpha \cdot D_b/z$$

The ring beam should be reinforced against the tension. As per IS: 456 para. 44.1.1 for a member in direct tension when the tension is entirely taken by the reinforcement, the tensile stress in concrete shall not exceed the values given below in Table 15.1.

Table 15.1

Concrete grade	M10	M15	M20	M25	M30	M35	M40
Tensile stress, N/mm²	1.2	2.0	2.8	3.2	3.6	4.0	4.4

Calculated tensile stress in concrete

$$= \frac{T}{A_c + m\, A_{st}}$$

$$= \frac{T}{A_g + (m-1)\, A_{st}} \tag{15.45}$$

where T = Tension

A_c = net area of concrete

A_{st} = sectional area of steel reinforcement

m = modular ratio

$A_g = A_c + A_{st}$ the gross area of section.

The meridional thrust $N_{\phi t}$ at the periphery of the opening at top is also inclind to the vertical. Its vertical component is balanced by the loading w from top. However, its horizontal component $N_{\phi t}$ cos β acts radially inwards as shown in Fig. 15.8(a). To take up this force a ring beam is provided at the top opening. Compression in the top ring beam is

$$= N_{\phi t} \cos \beta \cdot D_t/2$$

The size of the ring beam should be adequate to bear the compression.

Similar ring beams are provided in conical shells also to take up the horizontal component of the meridional force along top and bottom peripheries.

Example 15.1

Design a spherical dome having base diameter = 9.6 m, rise = 3.15 m. Diameter of the opening of lantern at top = 1.8 m. Self weight of lantern = 27 kN. Wind and occasional load etc. may be taken to be 1.5 kN/m². Use concrete grade: M15 and steel grade: Fe415.

Solution

First of all consider geometry with reference to Fig. 15.5. Here

$$D = 9.6 \text{ m}$$

$$h = 3.15 \text{ m}$$

Then $\quad h(2R - h) = \left(\dfrac{D}{2}\right)^2$

or $\quad 3.15(2R - 3.15) = \left(\dfrac{9.6}{2}\right)^2$

∴ $\qquad\qquad R = 5.232 \text{ m}$

$$\sin \alpha = \frac{D}{2R}$$

$$= \frac{9.6}{2 \times 5.232}$$

$$= 0.91741$$

∴ $\qquad\qquad \alpha = 66.55°$

$$\sin \beta = \frac{r}{R}$$

$$= \frac{1.8}{2 \times 5.232}$$

$$= 0.172$$

∴ $\qquad\qquad \beta = 9.905°$

Details are shown in Fig. 15.9

Surface area of the circular portion cut off in the opening

$$= 2\pi R^2 (1 - \cos \beta)$$

$$= 2 \times 5.232^2 (1 - \cos 9.905)$$

$$= 2.563 \text{ m}^2$$

To facilitate calculations it is assumed that there is no opening at top and the weight of this circular portion of dome may be deducted from

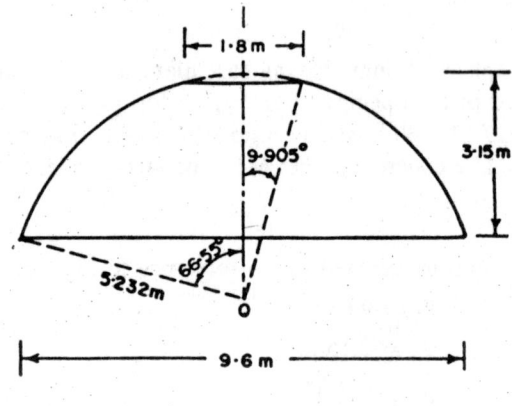

Fig. 15.9

the weight of the lantern.
Let the thickness of shell be = 100 mm.
Then, the self weight per unit area

$$= 0.1 \times 25$$
$$= 2.5 \text{ kN/m}^2$$

Two cases of analysis are considered as follows:

(i) *Occasional load is present on the dome*

In this case the intensity of load on the dome is

$$w = \text{self weight} + \text{occasional load}$$
$$= 2.5 + 1.5$$
$$= 4 \text{ kN/m}^2$$

Then, the effective weight of lantern is

$$W = 27 - 2.563 \times 4$$
$$= 16.748 \text{ kN.}$$

Total meridional force is obtained by combining Eq. (15.23) and (15.25). Thus

$$N\phi = -\frac{16.748}{2\pi \times 5.232 \; \sin^2\phi} - \frac{4 \times 5.232}{(1 + \cos\phi)}$$

$$= -\frac{0.5095}{\sin^2 \phi} - \frac{20.928}{(1 + \cos \phi)} \qquad (15.46)$$

Total circumferencial force is obtained by combining Eq. (15.24) and (15.26). Thus

$$N_\psi = \frac{16.748}{2\pi \times 5.232 \sin^2 \phi}$$

$$+ \frac{4 \times 5.232 (1 - \cos \phi - \cos^2 \phi)}{(1 + \cos \phi)}$$

$$= \frac{0.5095}{\sin^2 \phi} + \frac{20.928 (1 - \cos \phi - \cos^2 \phi)}{(1 + \cos \phi)} \qquad (15.47)$$

Calculated values of N_ϕ and N_ψ for different values of ϕ are given in Table 15.2.

Table 15.2

ANALYSIS WHEN OCCASIONAL LOAD IS PRESENT

Angle ϕ Degrees	Unit Force kN/m	
	Meridional	Circumferencial
9.905	−27.762	7.146
15	−18.251	−1.964
30	−13.249	−4.871
45	−13.279	−1.52
60	−14.632	4.167
66.55	−15.575	7.248

The angle for which circumferencial force is zero is found from Eq. (15.47) by putting $N_\psi = 0$. The angles are $13.2°$ and $49.5°$. Between these two points the circumferencial force is compressive.

(ii) *Occasional load not present on the dome*

In this case the intensity of load on the dome is

$$w = \text{self weight}$$
$$= 2.5 \text{ kN/m}^2$$

Then, the effective weight of lantern is

$$W = 27 - 2.563 \times 2.5$$

$$= 20.593 \quad \text{kN}$$

Total meridional force is obtained by combining Eq. (15.23) and (15.25). Thus

$$N_\phi = - \frac{20.593}{2\pi \times 5.232 \sin^2 \phi} - \frac{2.5 \times 5.232}{(1 + \cos \phi)}$$

$$= - \frac{0.6264}{\sin^2 \phi} - \frac{13.08}{(1 + \cos \phi)} \tag{15.48}$$

Total circumferencial force is obtained by combining Eq. (15.24) and (15.26). Thus

$$N_\psi = \frac{20.593}{2\pi \times 5.232 \sin^2 \phi}$$

$$+ \frac{2.5 \times 5.232 (1 - \cos \phi - \cos^2 \phi)}{(1 + \cos \phi)}$$

$$= \frac{0.6264}{\sin^2 \phi} + \frac{13.08 (1 - \cos \phi - \cos^2 \phi)}{(1 + \cos \phi)} \tag{15.49}$$

Calculated values of N_ϕ and N_ψ for different values of ϕ are given in Table 15.3.

Table 15.3

ANALYSIS WHEN OCCASIONAL LOAD IS NOT PRESENT

Angle Degrees	Unit Force kN/m Meridional	Circumferencial
9.905	− 27.759	14.874
15	− 16.004	3.370
30	− 9.515	− 1.813
45	− 8.915	− 0.334
60	− 9.555	3.015
66.55	− 10.101	4.896

The angle for which circumferencial force is zero is found from Eq. (15.49) by putting $N_\psi = 0$. The angles are 23.5° and 44.5°. Between these points the circumferencial force is compressive.

Maximum value of meridional force in shell is $= 27.762$ kN/m. Compressive stress due to this

$$= \frac{27.762 \times 10^3}{1000 \times 100}$$

$$= 0.278 \quad N/mm^2, \text{ safe.}$$

Maximum reinforcement against temperature and shrinkage

$$= \frac{0.12 \times 100 \times 1000}{100}$$

$$= 120 \quad mm^2/m$$

Provide 6 mm dia. bars—200 mm c/c, along meridians. Maximum circumfcrencial tension at top

$$= 14.874 \quad kN/m$$

Tension reinforcement against this

$$= \frac{14.874 \times 10^3}{230}$$

$$= 64.67 \quad mm^2/m.$$

Adding temperature and shrinkage reinforcement to this, the total reinforcement

$$= 64.67 + 120$$

$$= 184.67 \quad mm^2/m$$

Provide 6 mm dia. bars—150 mm c/c, circumferencially. Maximum circumferencial tension at bottom

$$= 7.248 \quad kN/m$$

Tension reinforcement against this

$$= \frac{7.248 \times 10^3}{230}$$

$$= 31.51 \quad mm^2/m$$

Adding temperature and shrinkage reinforcement to this, the total reinforcement is

$$= 31.51 + 120$$

$$= 151.51 \quad mm^2/m$$

Provide 6 mm dia. bars—185 mm c/c, circumferencially. From $\phi = 23.5°$ to $44.5°$ the circumferencial reinforcement will be 6 mm dia. bars—200 mm c/c.

At top, maximum value of $N_\phi = 27.762$ kN/m. Its inclination to horizontal is $\phi = 9.905°$. Therefore, circumferencial compression in top ring beam

$$= 27.762 \cos 9.905° \times \frac{1.8}{2}$$

$$= 24.613 \text{ kN.}$$

A section 100×150 mm for the top ring beam may be adopted. Compressive stress in concrete will be

$$= \frac{24.613 \times 1000}{100 \times 150}$$

$$= 1.641 \text{ N/mm}^2, \text{ safe.}$$

Two rings of 8 mm dia. bars may be provided.

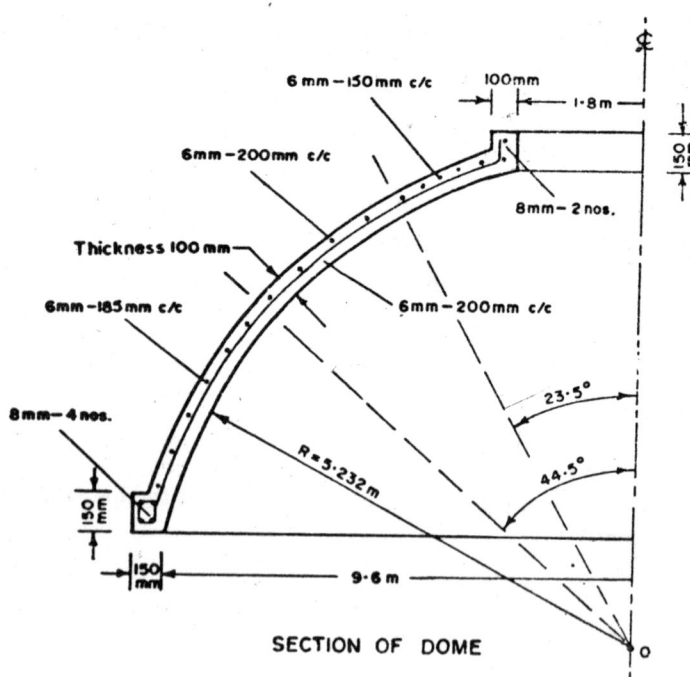

SECTION OF DOME

Fig. 15.10

At bottom maximum value of $N_\phi = 15.575$ kN/m. Its inclination to horizontal is $\phi = 66.55°$. Therefore, circumferencial tension in bottom ring beam

$$= 15.575 \cos 66.55° \times \frac{9.6}{2}$$

$$= 29.751 \text{ kN.}$$

Tensile reinforcement

$$= \frac{29.751 \times 10^3}{230}$$

$$= 129.35 \text{ mm}^2$$

Provide 8 mm dia. bars—4 nos.

Then, tensile stress in concrete

$$= \frac{29.751 \times 10^3}{A_g + (18.67 - 1) \times 4 \times 50.3}$$

$$= \frac{29.751 \times 10^3}{A_g + 3555.2}$$

The permissible tensile stress in concrete is 2 N/mm². Equating

$$\frac{29.751 \times 10^3}{A_g + 3555.2} = 2.0$$

$$A_g = 11320.5 \text{ mm}^2$$

Provide a ring beam of size 150 × 150 mm at bottom. Details of reinforcement are shown in Fig. 15.10.

INDEX